The New Quantum Universe

Tony Hey
Southampton University

Patrick Walters
University of Wales, Swansea

CAMBRIDGE
UNIVERSITY PRESS

PUBLISHED BY THE PRESS SYNDICATE OF THE UNIVERSITY OF CAMBRIDGE
The Pitt Building, Trumpington Street, Cambridge, United Kingdom

CAMBRIDGE UNIVERSITY PRESS
The Edinburgh Building, Cambridge CB2 2RU, UK
40 West 20th Street, New York, NY 10011-4211, USA
477 Williamstown Road, Port Melbourne, VIC 3207, Australia
Ruiz de Alarcón 13, 28014 Madrid, Spain
Dock House, The Waterfront, Cape Town 8001, South Africa

http://www.cambridge.org

First published 2003
Reprinted 2004

Printed in the United Kingdom at the University Press, Cambridge

Typefaces Swift 9.25/12.5 pt and Gill Sans   *System* LaTeX 2_ε [TB]

A catalogue record for this book is available from the British Library

Library of Congress Cataloguing in Publication data

Hey, Anthony J. G.
    The new quantum universe / Tony Hey, Patrick Walters.
        p.   cm.
    Includes bibliographical references and indexes.
    ISBN   0 521 56418 2 (hardback) – ISBN 0 521 56457 3 (paperback)
    1. Quantum Theory. I. Walters, Patrick, 1949–   II. Title.
QC174.12 .H478 2003
530.12–dc21   2002074047

ISBN 0 521 56418 2 hardback
ISBN 0 521 56457 3 paperback

Contents

Preface

The popularization of science is now an established business for booksellers and publishers. The traditional formula for a 'trade' book on popular science is a text of about 100 000 words or so (around 200 pages) with relatively few diagrams or pictures. The target audience is the educated reader with a general interest in science. At the other end of the scale we have popular science reference books such as encyclopedias and atlases. Our target audience lies in between these two extremes. We wish to write a book that will not only interest the 'educated reader' as above but, more importantly, capture the interest and imagination of young people. We believe that it is vitally important to give young people a glimpse of the excitement of physics so that they may be motivated to take up the challenge themselves. Nowadays, there are many more alternatives for young people and there is a general perception that science and mathematics are 'hard' subjects. It is certainly true that understanding in these subjects does not come without effort and real mastery may indeed take years. So we cannot promise instant gratification. What we can promise is that the study of science and mathematics will provide a gateway to a deeper understanding of a fascinating universe – our universe, a quantum universe. And paradoxically, as our world becomes more and more dependent on science and technology, it has also become increasingly technologically fragile in that fewer people understand the technology on which we all depend. Civilization requires us to inspire and motivate young people to take up the challenge of science. This is the true target audience for this book. But we hope that our text and our extensive use of diagrams, colour photographs and biographies of great scientists will also be interesting and entertaining to the 'educated reader'!

Our first book on quantum mechanics, *The Quantum Universe*, was published in 1987. At the time, it seemed to us that there was a clear need to communicate the strange ideas of quantum mechanics to a wider audience, since this is the theory that underpins the operation of many 'high tech' objects in daily use. So, after a look at the fundamentals, we concentrated on explaining how quantum mechanics gives us an understanding of not only atoms and nuclei, but also all the elements and even the stars. Quantum mechanics makes possible the incredible silicon chip as well as all the myriad of applications of lasers that we see today. It explains not only the structure of Jupiter but also provides us with an understanding of the mechanism of energy generation in our Sun and other stars. Because of the bizarreness of quantum theory at a fundamental level, we deliberately avoided all philosophical issues and followed Richard Feynman's advice in adopting a very pragmatic stance. We therefore concerned ourselves with demonstrating that the theory, no matter how strange it may seem, clearly works in practice. Since quantum theory was developed in the 1920s by Niels Bohr, Erwin Schroedinger, Werner Heisenberg, Paul Dirac and others, it seemed that, apart from more applications, there was little new to be discovered.

To our surprise, the last fifteen years have been years of great advances in quantum technology. Although no new results have arisen to challenge the supremacy of the underlying quantum theory, there have been many exciting new discoveries. In the main, these developments all demonstrate our increasing control of quantum systems. So much so that we believe that we are seeing the emergence of a new field of scientific endeavour – 'quantum engineering'. This term signifies our belief that this new century will see our increasing mastery over manipulating matter at the quantum level leading to new and spectacular applications of such 'nanotechnology'. There will certainly be significant implications for the semiconductor industry. We will see the end of 'Moore's Law' – the prediction that the number of transistors on computer chips, and hence their computational speed and memory capacity, doubles every eighteen months. In ten years or so, the dimensions of features on a silicon chip will have shrunk to such a size that the properties of individual atoms and electrons will play a determining role. Such quantum objects do not behave in a classically describable way. Unless quantum engineers are able to come up with some competitive new technology, Moore's Law will come to an end – along with the necessity to upgrade PCs every 18 months! One possible new technology on the horizon is 'quantum computing'. Instead of bits of information restricted to be either '1' or '0', as in present-day 'classical' computers, a quantum computer would allow the possibility of algorithms using quantum bits – 'qubits' – that are somehow simultaneously '1' and '0'! This observation has led to the development of a whole new area of research – 'quantum information theory' – and there are already possibilities of its practical application in cryptography. Although we retain our original approach to quantum mechanics in this book, the chapters on quantum applications have been extensively re-written and updated. In addition, a new chapter on 'quantum engineering' introduces the ideas and technologies of nanotechnology and quantum information.

As we have said, in our earlier book on quantum mechanics, we followed Feynman and avoided asking the question 'But how can it be like that?'. However, the last fifteen years have seen an upsurge of interest in understanding what quantum mechanics implies about the physical reality of the world in which we live. We have therefore included a chapter on 'quantum paradoxes' in which we introduce the reader to the unfinished debate between Niels Bohr and Albert Einstein. It was Bohr who formulated the orthodox 'Copenhagen' view of quantum mechanics and who was its most robust defender. According to Bohr's interpretation, uncertainty and unpredictability are intrinsic features of quantum theory, and the actual physical reality of quantum objects is debateable. Against such orthodoxy, Albert Einstein, Bohr's long-term friend and colleague, fought for the remainder of his life. He summed up his opposition to the Copenhagen interpretation in the memorable phrase 'God does not play dice!' After a lengthy but ultimately inconclusive debate, Einstein died still a non-believer in quantum theory. Soon after his death, the Irish physicist John Bell came up with a way to distinguish between the orthodox quantum mechanics of Bohr and the deterministic approach favoured by Einstein. Experiments to test 'Bell's Inequality' have now come down in favour of quantum mechanics and Einstein would have to think again! Bell's result is of such importance for quantum mechanics that we include an intuitive explanation of the Bell Inequality. Our presentation closely follows one given by John Bell himself in a meeting in Geneva. The other essential creature in any discussion of the interpretation of quantum mechanics is Schrödinger's Cat. The paradox of the cat graphically illustrates the so-called 'measurement problem' in quantum mechanics. We discuss how this problem is resolved – to a greater or lesser degree – by the ever popular 'Many Worlds' interpretation of quantum mechanics of Hugh Everett or by the 'Decoherence' mechanism favoured by Wojtek Zurek and others.

Finally, as a light-hearted 'afterword', we look at the treatment of quantum mechanics in Science Fiction. H.G. Wells led the way with his account of an atomic-bomb-induced Armageddon in his book *The World Set Free*. In the early years of quantum mechanics, SF writers struggled to incorporate the new understanding of the atom into a fictional context. Modern SF has now moved on to include multiple universes and nanotechnology as part of its standard technology base. Finally, in Michael Crichton's recent book, *Timeline*, quantum computers, teleportation and time travel are woven together to create yet another new dimension for Science Fiction to explore.

The distinguished theoretical physicist and author Paul Davis has made the following prediction:

> The nineteenth century was known as the machine age, the
> twentieth century will go down in history as the information age.
> I believe the twenty-first century will be the quantum age.

In the course of the next decades we will see how far this vision will be realized. Certainly, we believe that the influence on our society of this coming nanotechnology revolution, underpinned by quantum mechanics, will be at least as substantial as the fall-out from the present bio-informatics explosion. We hope that this book will assist in stimulating the imagination of a new generation of quantum engineers.

Some acknowledgements are in order. Once again we wish to thank our families for their invaluable support and forbearance – Marie Walters, and Jessie, Nancy, Jonathan and Christopher Hey. We are also grateful to colleagues who have read and commented on draft chapters, especially Phil Charles, Malcolm Coe, Jeff Mandula and Steve King. In Southampton, we thank Maggie Bond and Juri Papay for their invaluable help in getting the new photographs and permissions together. At Cambridge University Press we are grateful to Rufus Neale and Simon Mitton who initiated the project, and to Simon Capelin and Jacqueline Garget and the rest of the team, who cheerfully assisted us in seeing a complex project through to completion. Finally, Tony Hey wishes especially to thank Ray Browne, of the UK Department of Trade and Industry, and Juri Papay, at the University of Southampton, both for their boundless enthusiasm for science and for their energy and support in assisting me to bring this project to a successful conclusion.

Prologue

Poets say science takes away from the beauty of the stars – mere globs of gas atoms. Nothing is 'mere'. I too can see the stars on a desert night, and feel them. But do I see less or more? The vastness of the heavens stretches my imagination – stuck on this carousel, my little eye can catch one-million-year- old light … Or see them [the stars] with the greater eye of Palomar, rushing all apart from some common starting point when they were perhaps all together. What is the pattern, or the meaning, or the why? It does not do harm to the mystery to know a little about it. For far more marvellous is the truth than any artists of the past imagined! Why do the poets of the present not speak of it?

Finally, may I add that the main purpose of my teaching has not been to prepare you for some examination – it was not even to prepare you to serve industry or the military. I wanted most to give you some appreciation of the wonderful world and the physicist's way of looking at it, which, I believe, is a major part of the true culture of modern times. (There are probably professors of other subjects who would object, but I believe they are completely wrong.) Perhaps you will not only have some appreciation of this culture; it is even possible that you may want to join in the greatest adventure that the human mind has ever begun.

Richard Feynman

Route map

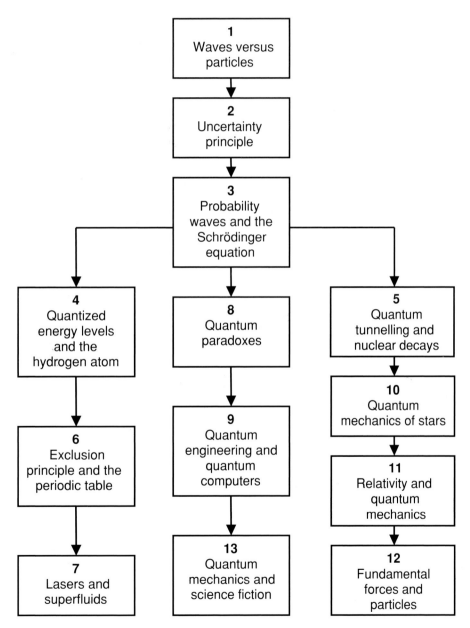

The three major strands of interconnected topics through the book. Broadly speaking, the left-hand strand is concerned with the quantum mechanics of the solid state; the right-hand strand focuses on the quantum mechanics of stars and elementary particles; and the middle strand explores quantum paradoxes and quantum engineering, before looking at the fictional realization of these ideas.

Waves versus particles

| 1 |

... I think I can safely say that nobody understands quantum mechanics.

Richard Feynman

Science and experiment

Science is a special kind of explanation of the things we see around us. It starts with a problem and curiosity. Something strikes the scientist as odd. It doesn't fit in with the usual explanation. Maybe harder thinking or more careful observation will resolve the problem. If it remains a puzzle, it stimulates the scientist's imagination. Perhaps a completely new way of looking at things is needed? Scientists are perpetually trying to find better explanations – better in the sense that any new explanation must not only explain the new puzzle, but also be consistent with all of the previous explanations that still work well. The hallmark of any scientific explanation or 'theory' is that it must be able to make successful predictions. In other words, any decent theory must be able to say what will happen in any given set of circumstances. Thus, any new theory will only become generally accepted by the scientific community if it is able not only to explain the observations that scientists have already made, but also to foretell the results of new, as yet unperformed, experiments. This rigorous testing of new scientific ideas is the key feature that distinguishes science from other fields of intellectual endeavour – such as history or even economics – or from a pseudoscience such as astrology.

In the seventeenth century Isaac Newton and several other great scientists developed a wonderfully successful explanation of the way things move. This whole theoretical framework is called 'classical mechanics', and its scope encompasses the motion of everything from billiard balls to planets. Newton's explanation of motion in terms of forces, momentum and acceleration is encapsulated in his 'laws of motion'. These principles are incorporated into so many of our machines and toys that classical mechanics is familiar from our everyday experience. We all know what to expect in

Isaac Newton (1642–1727) published his book *Optics* in 1704 that explained the rainbow and put forward the 'corpuscular' theory of light. In his 1687 book *Mathematical Principles of Natural Philosophy* Newton set down the principles of mechanics and gravity that guided science until the mid nineteenth century.

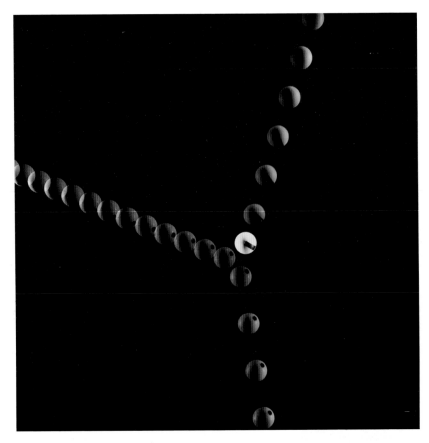

Fig. 1.1 A multi-flash photograph of a billiard ball collision. The motions of the balls can be calculated using Newton's laws but we have a good feel for what will happen from watching snooker on television or playing ourselves.

the collision of two billiard balls. Perhaps the most spectacular application of classical mechanics is in the exploration of space. Nowadays, it surprises no one that the astronaut and the space shuttle float side by side and neither falls dramatically to Earth. A hundred years ago it was not so 'obvious', and in Jules Verne's famous story *A Trip Around The Moon* the passengers of the spacecraft were amazed to find the body of a dog that died on takeoff, and which they had jettisoned outside the craft, floating side by side with them all the way to the Moon. Today, you may not know how Newton's theory works in detail but you can see that it works. It is part of our daily experience.

All this brings us to the problem most of us have in coming to terms with 'quantum mechanics'. It is just this. At the very small distances involved in the study of atoms and molecules, things do *not* behave in a familiar way. Classical mechanics is inadequate and an entirely new explanation is needed. Quantum mechanics is that new explanation, and it is cunningly constructed so that it not only works in the quantum realm of very-short length scales, but also so that, for larger distances, its predictions are identical with those of Newton. An atom is a typical quantum thing – it cannot be understood from the standpoint of classical physics. One popular

Fig. 1.2 Astronaut Bruce McCandless floats in space during the first untethered space walk on February 7th, 1984. The astronaut is essentially an independent spacecraft in orbit near the shuttle. McCandless commented 'Well that may have been one small step for Neil [Armstrong] but it's a heck of a big leap for me!'

Fig. 1.3 In the story by Jules Verne *A Trip Around The Moon*, published in 1865, the dog 'Satellite' died on take-off and was jettisoned from the space-ship. Much to the surprise of the occupants the dog's body floated along with them all the way to the Moon!

visualization of an atom imagines electrons orbiting the nucleus of the atom much in the way planets orbit the Sun in the solar system. In fact, for negatively charged electrons in orbit round a positively charged nucleus, this simple model is unstable! According to classical physics the electrons would spiral into the centre and the atom would collapse. This nice and comforting model of the atom cannot account for even the existence of real atoms, let alone predict their expected behaviour. It is important to be aware at the outset that there is *no* simple picture that can accurately describe the behaviour of electrons in atoms. This is the first hurdle faced by the newcomer to the quantum domain: the inescapable and unpalatable fact that the behaviour of quantum objects is totally unlike anything you have ever seen.

How can we convince you that quantum mechanics is both necessary and useful? Well, a physicist, just like a good detective, sifts through the evidence and remembers the old maxim of Sherlock Holmes that 'when you have excluded the impossible, whatever remains, however improbable, must be the truth'. Nonetheless, it was only with much reluctance that twentieth-century physicists became convinced that the whole magnificent edifice of classical physics was not 'almost right' for describing the behaviour of atoms, but had, instead, to be radically rebuilt. Nowhere, was the confusion generated by this painful realization more evident than in their attempts to understand the nature of light.

Fig. 1.4 The interference pattern produced by two vibrating sources in water.

Fig. 1.5 George Gamow's rather whimsical view of the planetary model of the atom in *Mr Tompkins Explores The Atom.*

Light and quantum mechanics

Thomas Young (1773–1829) was an infant prodigy who could read at the age of two. During his youth he learnt to speak a dozen languages. He is best remembered for his work on vision and for establishing the wave theory of light. However, he was also the first to make progress on deciphering the hieroglyphic language of the ancient Egyptians.

In the seventeenth century, Isaac Newton suggested that light should be regarded as a stream of particles, rather like bullets from a machine gun. Such was Newton's reputation that this view persisted, apart from some isolated pockets of opposition, until the nineteenth century. It was then that Thomas Young and others conclusively showed that the particle picture of light must be wrong. Instead, they favoured the idea that light was a kind of wave motion. One property of waves that is familiar to us is that of 'interference', to use the physicists' term for what happens when two waves collide. For example, in Fig. 1.4 we show the 'interference' patterns produced by two sources of water waves on the surface of the water. Using his famous 'double-slit' apparatus to make two sources of light, Thomas Young had observed similar interference patterns using light.

Alas, physicists were not able to congratulate themselves for long. Experiments at the end of the nineteenth century revealed effects that were inexplicable by a wave theory of light. The most famous of such experiments concerns the so-called 'photo-electric' effect. Ultraviolet light shone onto a negatively charged metal caused it to lose its charge, while shining visible light on the metal had no effect. This puzzle was first explained by Albert Einstein in the same year that he invented the 'theory of relativity' for which he later became famous. His explanation of the photo-electric effect resurrected the particle view of light. The discharging of the metal was caused by electrons being knocked out of the metal by light energy concentrated into individual little 'bundles' of energy, which we now call 'photons'. According to Einstein's theory, ultraviolet photons have more energy than visible-light ones, and so no matter how much visible light you shine on the metal, none of the photons has enough energy to kick out an electron.

After several decades of confusion in physics, a way out of this dilemma was found in the 1920s with the emergence of quantum mechanics, pioneered by physicists such as Heisenberg, Schrödinger and Dirac. This theory is able to provide a successful explanation of the paradoxical nature of light, atoms and much else besides. But there is a price to pay for this success. We must abandon all hope of being able to describe the motion of things at atomic scales in terms of everyday concepts like waves or particles. A 'photon' does not behave like anything anyone has ever seen. This does not, however, mean that quantum mechanics is full of vague ideas and lacks predictive power. On the contrary, quantum mechanics is the only theory capable of making definite and successful predictions for systems of atomic sizes or smaller, in much the same way that classical mechanics makes predictions for the behaviour of billiard balls, rockets and planets. The difficulty with quantum things such as the photon is that, unlike billiard balls, their motion cannot be visualized in any accurate pictorial way. All we can do is summarize our lack of a picture by saying that a photon behaves in an essentially quantum mechanical way.

J. J. Thomson (1856–1940) measured the charge-to-mass ratio of the electron thus establishing it as a new elementary particle. He was awarded the Nobel Prize in 1906.

There is one sense in which Nature has been kind to us. Viewed from the perspective of classical physics, photons and electrons are very different kinds of objects. Remarkably, in the quantum domain both photons and electrons, and indeed all quantum objects, behave in the same strange quantum mechanical way. This is at least some compensation for our inability to picture quantum things! There is a curious little irony in the history of our attempts to understand the nature of electrons. In 1897 J. J. Thomson measured the charge-to-mass ratio of the electron and established the electron as a new elementary particle of Nature. Thirty years later, his son, G. P. Thomson, and also Davisson and Germer in the USA, performed a beautiful series of experiments that conclusively revealed that electrons behave like waves. The historian Max Jammer wrote: 'One may feel inclined to say that Thomson, the father, was awarded the Nobel Prize for having shown that the electron is a particle, and Thomson, the son, for having shown that the electron is a wave'.

Our intention in this book is to impress even the most skeptical reader with the enormous range and diversity of the successful predictions of quantum mechanics. The apparently absurd ideas of de Broglie, Schrödinger and Heisenberg have now led to whole new technologies whose very existence depends on the discoveries of these pioneers of quantum mechanics. The modern electronics industry, with its silicon chip technology, is all based on the quantum theory of materials called semiconductors. Likewise, all the multitude of applications of lasers are possible only because of our understanding, at the fundamental quantum level, of a mechanism for radiation of light from atoms first identified by Einstein in 1916. Moreover, understanding how large numbers of quantum objects behave when packed tightly together leads to an understanding of all the different types of matter ranging from 'superconductors' to 'neutron stars'. In addition, although originally invented to solve fundamental problems concerned with the existence of atoms, quantum mechanics was found to apply with equal success to the tiny nucleus at the heart of the atom, and this has led to an understanding of radioactivity and nuclear reactions. As everyone knows, this has been a mixed blessing. Not only do we now know what makes the stars shine, but we also know how to destroy all of civilization with the awesome power of nuclear weapons.

Before we can explain how quantum mechanics made all these things possible, we must first attempt to describe the strange quantum mechanical behaviour of objects at atomic distance scales. This task is clearly difficult given the absence of any accurate analogy for the mathematical description of quantum behaviour. However, we can make progress if we use a mixture of analogy and contrast. Young's original 'double-slit' experiment used a screen with two slits in it to make two sources of light which could interfere and produce his famous 'interference fringes' – alternating light and dark lines (Fig. 1.6). We shall describe the results of similar 'double-slit' experiments carried out using bullets, water waves and electrons. By

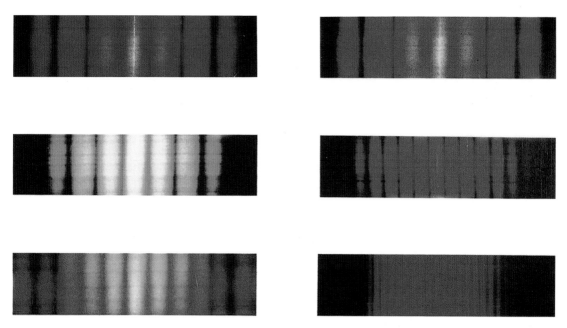

Fig. 1.6 Double-slit interference patterns for light, usually taken as demonstrating that light is a wave motion. In the left-hand pictures, as the wavelength of the light is decreased and the colour changes from red to blue, the interference fringes become closer together. On the right, for red light, the decrease in the fringe separation is caused by increasing the separation of the slits.

comparing and contrasting the results obtained with the three different materials we shall be able to give you some idea of the essential features of quantum mechanical behaviour. Quantum mechanics textbooks contain detailed discussion of many types of experiment, but this double-slit experiment is sufficient to reveal all the mystery of quantum mechanics. All of the problems and paradoxes of quantum physics can be demonstrated in this single experiment.

A word of warning before we begin. To avoid running into a frustrating psychological cul-de-sac, try to be content with mere acceptance of the observed experimental facts. Try not to ask the question 'but how can it be like that?' As Richard Feynman says 'nobody understands quantum mechanics'. All we can give you is an account of the way Nature appears to work. Nobody knows more than that. Only **after** we have convinced you that quantum mechanics really works will we examine what quantum mechanics has to say about the very nature of reality, with a discussion about Schrödinger's cat, Einstein and dice.

The double-slit experiment

This section may be rather hard going first time through. If so, just glance at the pictures and pass on quickly to the next chapter!

With bullets

Source: a wobbly machine gun that, as it fires, spreads the bullets out into a cone, all with the same speed but with random directions.

Screen: armour plate with two parallel slits in it.

Detector: small boxes of sand to collect the bullets.

Results: the gun fires at a fixed rate and we can count the number of bullets that arrive in any given box in a given period of time. The bullets that go through the slits can either go straight through or else bounce off one of the edges, but must always end up in one of the boxes. The bullets we are using are made of a tough enough metal so that they never break up – we can never have half a bullet in a box. Moreover, no two bullets ever arrive at the same time – we have only one gun, and each bullet is a single identifiable 'lump'.

If we let the experiment run for an hour and then count the bullets in each of the boxes, we can see how the 'probability of arrival' of a bullet varies with the position of the detector box. The total number of bullets arriving at any given position is clearly the sum of the number of bullets going through slit 1 plus the number going through slit 2. How this 'probability of arrival' varies with position of the sand boxes is shown in Fig. 1.7. We shall label this result P_{12} – the probability of arrival of bullets when both slits are open. We also show in Fig. 1.7 the results obtained with slit 2 closed, which we call P_1, and those obtained with slit 1 closed, which we call P_2. Looking at the figures, it is evident that the curve labelled P_{12} is obtained by adding curves P_1 and P_2. We can write this mathematically as the equation:

$$P_{12} = P_1 + P_2$$

For reasons that will become apparent in a moment, we call this result the case of *no interference*.

With water waves

Source: a stone dropped into a large pool of water.

Screen: a jetty with two gaps in it.

Detector: a line of small floating buoys whose jiggling up and down gives a measure of the amount of energy of the wave at that position.

Results: Ripples spread out from the source and reach the jetty. On the far side of the jetty ripples spread out from each of the gaps. At the detector, the resulting disturbance of the water is given by the sum of the disturbances of the ripples coming from both gaps. As we look along the line of buoys, there will be some places where the crest of a wave from slit 1 coincides with the arrival of a crest from slit 2, resulting in

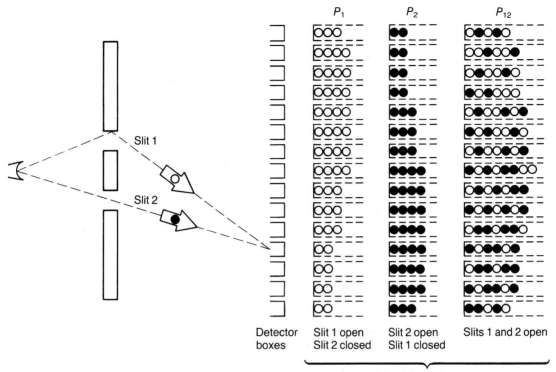

Fig. 1.7 A double-slit experiment with bullets. The experimental set-up is shown on the left of the figure and the results of three different experiments indicated on the right. We have shown bullets that pass through slit 1 as open circles and bullets through slit 2 as black circles. The column labelled P_1 shows the distribution of bullets arriving at the detector boxes when slit 2 is closed and only slit 1 is open. Column P_2 shows a similar distribution obtained with slit 1 closed and slit 2 open. As can be seen, the maximum number of bullets appears in the boxes directly in line with the slit that is left open. The result obtained with both slits open is shown in the column labelled P_{12}. It is now a matter of chance through which slit a bullet will come and this is shown by the scrambled mixture of black and white bullets collected in each box. The important point to notice is that the total obtained in each box when both slits are open is just the sum of the numbers obtained when only one or other of the slits is open. This is obvious in the case of bullets since we know that bullets must pass through one of the slits to reach the detector boxes.

a very large up-and-down motion for the buoy. At other places, a crest from one slit will coincide with a trough from the other so there will be no movement of the buoy at that position. At yet other places, the motion of the buoys will be somewhere between these two extremes. For water waves, it is certainly plausible that the energy of a wave at any given position is related to how big the waves are at that point. In fact, it can be shown that the energy of a wave depends on the square of the maximum height of the wave. Let us call the amount of energy arriving per second the 'intensity' and label this by the symbol I. If we label the

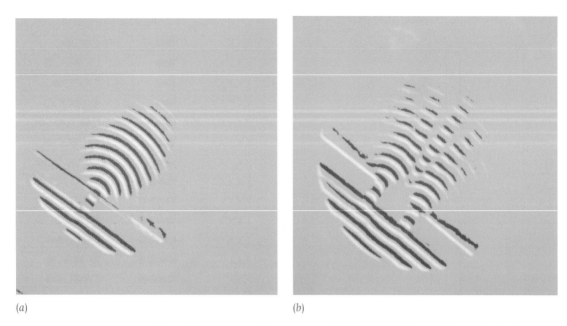

(a)   (b)

Fig. 1.8 Wave patterns with water waves. (a) A wave spreading out from a single slit; (b) the interference obtained with two slits.

maximum height of the wave by h, we can write the relation between I and h as the following equation:

$$I = h^2$$

intensity = height squared

In contrast to our experiment with bullets, we see that the energy of the waves does not arrive at the detector in definite-sized lumps. There, bullets only arrived at one particular position at one particular time. Here, since the height of the resulting wave at the detector varies smoothly from zero up to some maximum value as we move along the detector, we see that the energy of the original wave is spread out. The curve showing how the intensity varies with position along the detector is shown in Fig. 1.9. Since this is the intensity obtained with both slits open, we shall call this curve I_{12}. This intensity pattern has a very simple mathematical explanation. The total disturbance of the water at any position along the detector is given by the sum of the disturbances caused by the waves from slit 1 and slit 2. If we label the height of the wave from slit 1 by h_1, the height from slit 2 by h_2, and the total height obtained when both slits are open as h_{12}, we can write this result as the equation:

$$h_{12} = h_1 + h_2$$

Remember that each of these heights can be positive or negative depending on whether the corresponding wave disturbance raises or lowers the

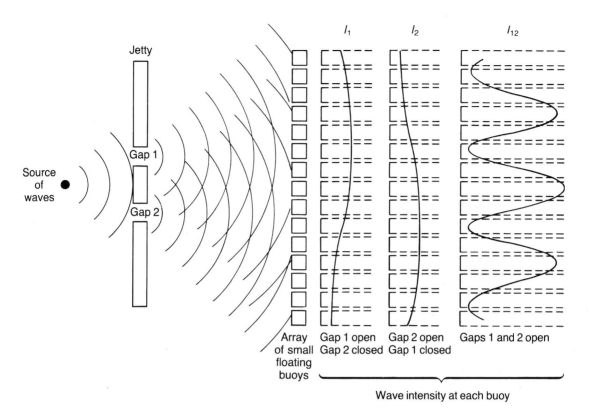

Fig. 1.9 A double-slit experiment with water waves. The detectors are a line of small floating buoys whose jiggling up and down provides a measure of the wave energy. The wave crests spreading out from each slit are shown in the figure and can be compared with Fig. 1.8. The column labelled I_1 shows the smoothly varying wave intensity obtained when only gap 1 is open. Notice that this is very similar to the pattern P_1 obtained with bullets in Fig. 1.7 with only slit 1 open. Again, it is largest at the detector directly in line with gap 1 and the source. The second column shows that a similar pattern, I_2, is obtained when gap 1 is closed and gap 2 is open. The final column, I_{12}, shows the wave intensity pattern obtained with both slits open. It is dramatically different from the pattern obtained for bullets with both slits open. It is not equal to the sum of the patterns I_1 and I_2 obtained with one of the gaps closed. This rapidly varying intensity curve is called an interference pattern.

water level. The resulting intensity is the square of this height or 'wave amplitude'

$$I_{12} = h_{12}^2$$

so that

$$I_{12} = (h_1 + h_2)^2$$

We could now repeat the experiment with one of the gaps closed. In this case we find the results shown in Fig. 1.9. We label the corresponding intensity pattern I_1, since it is the intensity obtained with slit 1 open and slit 2 closed. The curve I_1 is just given by the square of the disturbance caused by the wave from slit 1

$$I_1 = h_1^2$$

Similarly, the curve I_2 is the result obtained with slit 2 open and slit 1 closed, and, in the same way as before, we have the result

$$I_2 = h_2^2$$

It is clear that these two curves are much less wiggly than the pattern I_{12}. Furthermore, the pattern I_{12}, for both slits open, cannot be obtained just by adding up the two intensity patterns, I_1 and I_2, that are each obtained with one of the slits closed. Mathematically, we can see this from our equations as follows:

$$I_{12} = (h_1 + h_2)^2$$
$$= (h_1 + h_2) \times (h_1 + h_2)$$

This may be expanded to read

$$I_{12} = h_1^2 + 2h_1h_2 + h_2^2$$

which is clearly **not** equal to the sum of I_1 and I_2

$$I_1 + I_2 = h_1^2 + h_2^2$$

For wave motion, we say there is *interference*. Unlike the case with bullets, you do not obtain the pattern for 'both slits open' by adding the patterns for 'one slit closed'. It was the observation of such interference patterns for light that convinced Thomas Young that light must be a wave motion. In fact, life is not so simple! We will now describe the results of the double-slit experiment performed with electrons, but similar results would be obtained if the experiment were repeated with light.

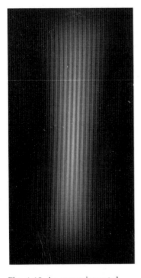

Fig. 1.10 An experimental observation of the two-slit interference pattern obtained with electrons.

With electrons

Source: an electron 'gun', consisting of a heated wire to 'boil off' electrons from the metal, together with an electric potential to accelerate them.

Screen: a thin metal plate with two very narrow slits in it.

Detector: a screen coated with a chemical 'phosphor' that produces a flash of light every time an electron arrives at it.

Results: Flashes of light signal the arrival of electrons at the detector. Electrons arrive singly, in individual 'lumps' of the same size, only at a single place at any one time – just as in the case with bullets. If we turn down the intensity of the electron gun, and thus boil off fewer electrons per minute, we still see the same size flashes at the detector but with fewer electrons arriving per minute. Again, exactly as for bullets, we can count up the number of flashes we see at any given position of the detector during a given interval of time. As for bullets, this allows us to measure how the probability of arrival of electrons varies as we move along the detector. The magic of quantum mechanics will now be revealed! The pattern we see (shown in Fig. 1.11) is the interference pattern characteristic of waves, although, as we have said, the electrons always arrive like bullets!

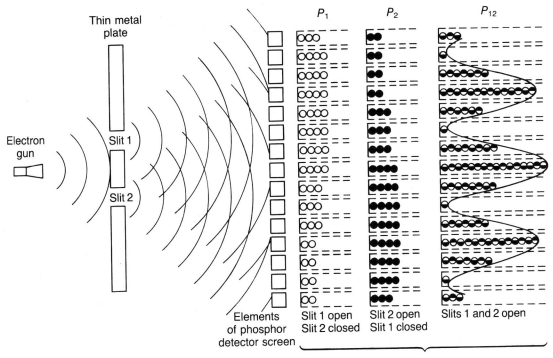

Fig. 1.11 A double-slit experiment with electrons. Electrons always arrive with a flash at the phosphor detector at one point, in the same way that bullets always end up in just one of the detector boxes rather than the energy being spread out, as in a wave. The column marked P_1 shows the pattern obtained with only slit 1 open. Electrons that have gone through slit 1 are represented as open circles, like the bullets of Fig. 1.7. Column P_2 shows the same thing with only slit 2 open and the electrons that have gone through slit 2 indicated by black circles. These two patterns are exactly the same as those obtained with bullets. The difference lies in the column headed P_{12}, which shows the pattern obtained for electrons when both slits are open. This is like the interference pattern obtained with water waves and requires some kind of wave motion arising from each slit as indicated on the figure. It is not the sum of P_1 and P_2 and so we cannot say which slit any electron goes through. We have indicated this lack of knowledge by drawing the electrons, which still arrive like bullets, as half white and half black circles. This fact, that quantum objects such as electrons possess attributes of both wave and particle motion but behave like neither, is the central mystery of quantum mechanics.

This is already very strange – but things become even more mystifying as we look at this result in more detail.

Let us look at a place where the detector observes a dip or 'minimum' of the interference pattern obtained with both slits open. At such positions we find fewer electrons than would be the case if we repeated the experiment with one slit closed! If we do such a 'one slit closed' experiment with electrons, we see the patterns shown in Fig. 1.11, exactly as for waves. But if electrons arrive like bullets, how can this be? Does the electron somehow

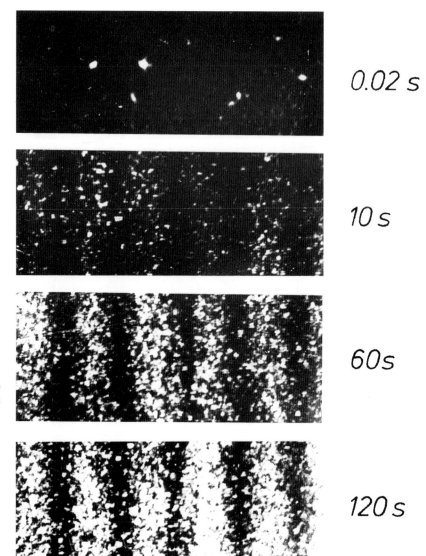

Fig. 1.12 More details of a double-slit experiment performed with electrons. Although interference patterns were once thought of as evidence for wave motion, when looked at in detail it can be seen that the electrons arrive in individual lumps. The top photograph shows a short exposure in which so few electrons have arrived that an almost random pattern of hits is seen. The pictures below show what happens as the exposure becomes longer: more and more electrons arrive until eventually the familiar interference pattern becomes visible.

split up into two and half an electron go through each slit? No! Electrons are never seen in halves – just like bullets they are either all there or not at all. Since the invention of quantum mechanics, many people have struggled to resolve this dilemma. No one has yet succeeded. It is as if the electrons start as particles at the electron gun, and finish as particles when they arrive at the detector, but the arrival pattern of electrons observed at the detector is as if they travelled like waves in between!

We have seen that the mathematics of the interference curve can be summarized in a very simple equation. We also saw, in the case of water waves, that the interference arose from adding the wave heights or 'amplitudes' for waves from the source to go via slit 1 and via slit 2. The intensity or energy of the wave was then related to the square of the sum of these amplitudes. The same mathematics must hold for the electron interference pattern. In the case of electrons, however, we are not measuring the intensity of a real wave motion but rather the probability of arrival of the electrons. From the mathematics of the interference curve, we see that there must be something like the height of a wave in the case of electrons. But what is the meaning of the 'height' of an electron wave? Since the square of this 'height' must give the corresponding probability, it is called a 'quantum probability amplitude'. We shall denote such quantum 'heights' or amplitudes by the symbol a. Thus, our equations for the probability of arrival of electrons will all have exactly the same form as for water waves, except that we shall use the symbol P for probability, instead of I for intensity, and a for quantum amplitude, instead of h for height. With these substitutions, the equations for the probability of the arrival of electrons with 'both slits open' and with 'one slit closed' take the form

$$P_{12} = (a_1 + a_2)^2$$
$$P_1 = a_1^2$$
$$P_2 = a_2^2$$

and, as before, P_{12} is not equal to the sum of P_1 and P_2:

$$P_{12} \neq P_1 + P_2$$

We conclude that electrons show wave-like interference in their arrival pattern despite the fact that they arrive in lumps, just like bullets. It is in this sense that we can say that quantum objects sometimes behave like a wave and sometimes behave like a particle. You may find this all rather mysterious. It is! We cannot do more to explain the magic of quantum mechanics – all we can do is describe the way quantum things behave. This description **is** quantum mechanics.

2 Heisenberg and uncertainty

A philosopher once said 'It is necessary for the very existence of science that the same conditions always produce the same results'. Well, they don't!

Richard Feynman

Watching electrons

We have seen that quantum mechanics does not allow us the comfort of being able to visualize the motion of a quantum particle. In a normal game of billiards we can imagine the paths taken by the individual balls (Fig. 1.1). Figure 2.1 shows the physicist George Gamow's attempt to give some impression of how the same game might look if played with quantum particles. Besides illustrating that the notion of a path is no longer valid in quantum mechanics, this cartoon also illustrates another significant difference between the quantum and classical worlds: the exact position of the white ball is not known. Uncertainty has entered physics and replaced the determinism of Newtonian mechanics.

By the nineteenth century, physicists had been able to explain vast amounts of experimental observations on objects as different as planets and billiard balls. If an observation differed from the predictions of classical physics, they looked for something they had overlooked to explain the deviation. In 1864, physicists' confidence in the whole edifice of classical physics seemed to be spectacularly verified by an analysis of some irregularities in the orbit of Uranus. These were attributed to the existence of a then undiscovered planet – the subsequent discovery of Neptune was a triumph for Newtonian physics. By the turn of the twentieth century, it seemed that all of physics followed from Newton's laws. If one was given a box containing a certain number of particles, all one had to do to be able to predict the motions of every particle any time in the future (or in the past for that matter) was to measure the present positions and speeds of all the particles. By measuring the speeds and positions sufficiently accurately, these predictions could be made as precise as required. This was the deterministic view of Nature encouraged by the success of classical physics. The caveat about 'sufficiently accurately' hardly seemed necessary. After all,

Fig. 2.1 In this illustration George Gamow has Mr Tompkins playing billiards with quantum billiard balls. The original caption is 'The white ball went in all directions!' In such a world, quantum uncertainty would be a familiar experience.

it was 'obvious' that one could measure anything with essentially no limit to the accuracy of the measurement – all one required was a sufficiently sensitive measuring device.

Quantum mechanics does away with this deterministic view of the future once and for all, and an essential element of uncertainty enters the predictions of physics. How does this come about? It arises because the seemingly innocuous belief of the classical physicists, that they could measure both the position and the velocity of a particle as accurately as they wished, is wrong! In quantum mechanics there is a fundamental limit to the accuracy we can achieve, no matter how ingenious or sensitive we make our measuring devices.

To illustrate this point let us return to the double-slit experiment once more. Remember that we talked in terms of the probability of where an electron would hit the screen. This is because we cannot say with certainty where any particular electron will land. We can only predict the relative chance of it landing at any particular position on the screen.

Now recall the experiment with bullets. This was also described in terms of probabilities. But there is a crucial difference between bullets and electrons. In the case of bullets the probability description was used because of our ignorance of the exact initial direction of the bullet – because of the wobbly gun. However, we could make a video of the firing of an individual bullet and then watch the bullet's trajectory to the screen on a slow-motion playback. Even if we only saw a part of the bullet's path, that would be sufficient, according to Newton, to determine the rest of the path. Obviously, the bullet must pass through one of the slits, and we can determine which one by looking at the video replay.

(M. Adams cherchant la planète de M. Leverrier.)

(M. Adams découvrant la nouvelle planète dans le rapport de M. Leverrier.)

(L'Angleterre prenant possession de la nouvelle planète.)

(Suite des découvertes de M. Leverrier.)

(Suite des découvertes de M. Adams.)

Fig. 2.2 The existence of the planet Neptune was predicted by Adams in England and by Le Verrier in France at about the same time. After Neptune's discovery there was a familiar Anglo-French squabble about priority, and this French cartoon depicts Adams stealing Le Verrier's results. Le Verrier tried to repeat his success with Neptune by using an anomaly in the orbit of Mercury to predict the existence of another planet, Vulcan, closer to the Sun. This was, of course, not the correct interpretation of the anomaly, which is now understood using Einstein's theory of general relativity (see our companion book *Einstein's Mirror*).

Why can't we do the same sort of thing for electrons? Let us imagine how we would go about trying to establish through which slit the electron passes. To see the electron just after it has passed through one of the slits we must shine some light on it and observe the reflected light. Let us therefore modify the experimental apparatus by inserting a light source behind the

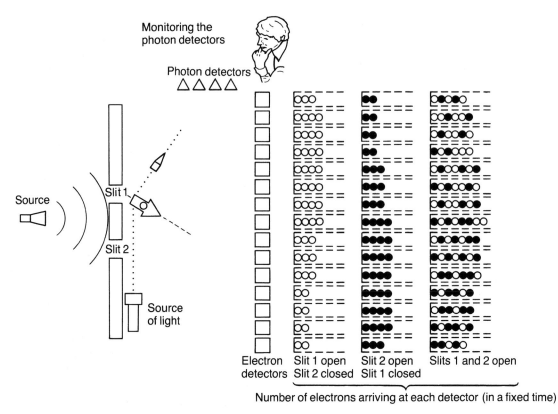

Fig. 2.3 The experimental set-up required to observe through which slit the electron passes in a double-slit experiment. Light, in the form of photons, is directed at the slits. In the figure a photon, represented as a small bullet, has hit an electron behind slit 1. The electron is disturbed slightly in its motion and the scattered photon is observed at the photon detectors. The electron patterns obtained with only one of the slits open are almost the same as before, when we did not observe the electron behind the slits. The surprise occurs with both slits open: there is no interference pattern. The small nudges given to the electrons in their collisions with the photons are always sufficient to wash out the interference pattern completely! We can now say with certainty through which slit the electron went but now the electrons are behaving just like bullets. The observed pattern is just the sum of the patterns for slit 1 and slit 2 separately.

slits (Fig. 2.3). We now arrange things so that if an electron passes through slit 1 we see a flash behind slit 1, and similarly for slit 2. If we now do the experiment what do we see? Well, the first important result is that we never see a half-flash behind both slits simultaneously. There is always a whole flash behind either slit 1 or slit 2. We can now divide up the electrons arriving at the detector into two groups, according to whether they went through slit 1 or slit 2. What's all this quantum nonsense then? The electron obviously either goes through slit 1 or slit 2. So indeed it does, when we watch the electron. But, if we now look at the arrival pattern of electrons at the screen, we see no interference pattern! The result is just the same as we obtained with bullets!

Amazingly, we have a different result depending on whether or not we switch on the light in order to watch the electrons! The resolution of this apparent paradox lies in the quantum nature of light itself. Remember our discussion of the photo-electric effect in chapter 1. When light interacts with matter, it displays its particle-like character. Light, like electrons, arrives in definite chunks of energy called photons. To see an object, therefore, we must bounce at least one photon off it. Now comes the crux of the argument. When we shine light on a bullet, its motion is not noticeably disturbed because the amount of energy in an individual photon is tiny compared with that of the bullet. Electrons, on the other hand, are very delicate quantum objects. Shining light on electrons gives them a jolt that disturbs their motion significantly. A more detailed analysis reveals that this disturbance is always just enough to wash out the interference pattern!

You may think we can turn the light down very low and make the disturbance so small that the interference pattern is not destroyed. This idea ignores the way light works. If we reduce the light intensity, we merely cut down the number of photons emitted per second not the amount of energy in each photon. Now that we only have a few photons around, there is a good chance that an electron can sneak past without being seen. We must therefore make a third category for electrons arriving at the screen. These are the ones that we missed and that we cannot say went definitely either through slit 1 or slit 2. If we look at the arrival pattern of these 'missed' electrons we see the interference pattern once more!

This is what Feynman calls the 'logical tightrope' of quantum mechanical thinking. If we have an experiment that can detect which slit the electron goes through, then one can say with certainty that the electron does go through one or other of the slits. If we have no way of telling which slit the electron went through, then we may not say that the electron goes through either one slit or the other!

Heisenberg's uncertainty principle

It is clear that quantum mechanics is a very cunning and subtle theory. For the double-slit experiment, we have seen that establishing which slit the electron went through destroys the interference pattern. This result is indicative of a very general principle of quantum physics, now named after its discoverer, Werner Heisenberg. It was Heisenberg who first pointed out that the new laws of quantum mechanics imply a fundamental limitation to the accuracy of experimental measurements. In our everyday world we can certainly imagine making measurements sufficiently delicately so that the act of measurement does not cause a perceptible disturbance. In the quantum world this is not the case. Light energy arrives in lumps and making a measurement necessarily gives a significant jolt to the object on which we are making the measurements. Furthermore, there is no way that

Werner Heisenberg (1901–1976) was in his early twenties when he performed his fundamental work on quantum theory. He was awarded the Nobel Prize in 1932 for his discovery of the uncertainty principle.

we can reduce the jolt to zero, even in principle. For objects of microscopic dimensions such jolts are not negligible. This is the essence of Heisenberg's uncertainty principle.

The uncertainty principle can be written down in a precise mathematical form as follows. In our discussion of the deterministic nature of classical physics, we imagined measuring the position and velocity of every particle in a box. We shall often refer to this collection of particles and surrounding box as a 'system' and talk about making measurements on 'the system'. Physicists also usually denote the position of a particle by the symbol x, but instead of speed or velocity, they prefer to talk about a quantity called 'momentum'. Momentum is the mass of the particle multiplied by its velocity, and it is a familiar concept in everyday life. A car moving with a speed of 10 kilometres per hour has more momentum than a football moving with the same speed (and consequently the car will do more damage when it collides with something!). Physicists usually represent momentum by the symbol p. In making measurements on a quantum system, it is not possible to measure the quantities x and p as accurately as we would wish. There is always some minimum error or uncertainty Δx and Δp, associated with their measurement. What Heisenberg discovered, and what actually keeps quantum mechanics from being internally inconsistent, is that the uncertainty in the position measurement, Δx, and the uncertainty in momentum, Δp, are inextricably linked together. If you want to measure the position of a particle very accurately, you inevitably end up by disturbing the system rather a lot, and, consequently, introduce a large uncertainty in the momentum of the particle. Why is this? Well, to determine the position very accurately it is necessary to use light with a very short wavelength – since the wavelength of the light determines the minimum distance within which we can locate the particle. Very short wavelength light has a very high frequency. Here comes catch 22. The energy of a photon depends on its frequency according to a formula first guessed by Max Planck. The formula is very simple: it says that the photon's energy is proportional to its frequency. We can write down this formula relating the energy of the photon, E, to the frequency, f, of the light as follows:

$$E = hf$$

photon energy E = Planck's constant h times frequency f

Max Planck (1858–1947) around 1900. His radical solution to the problem of radiation from hot bodies was the first introduction of quantum ideas. The importance of this work was recognized by the award of the Nobel Prize in 1918, but Planck remained unhappy with the quantum revolution his work had ushered in.

The constant of proportionality, h, is known as Planck's constant. Armed with this result we can now return to the problem of making an accurate measurement of position. We see that in order to locate the particle very precisely we must use high frequency light with a large value of f. But, such high frequency light will arrive in photons with a very large energy and give the quantum system a very large kick. On the other hand, if we want to know the momentum very accurately, we must give the system a very small kick. According to Planck's formula, this means using light of low frequency. Low frequency means long wavelength and this in turn means a large uncertainty in the measurement of position!

Heisenberg's uncertainty principle relates the uncertainties in position and momentum measurements in the following way:

$$(\Delta x) \times (\Delta p) \approx h$$

$$\begin{pmatrix} \text{Uncertainty} \\ \text{in position} \end{pmatrix} \text{ times } \begin{pmatrix} \text{Uncertainty} \\ \text{in momentum} \end{pmatrix} \text{ is approximately } \begin{pmatrix} \text{Planck's} \\ \text{constant} \end{pmatrix}$$
$$\qquad\qquad\qquad\qquad\qquad\qquad\qquad\qquad \text{equal to}$$

This equation puts into mathematical form the correspondence we discussed above. If you want to make the uncertainty in position, Δx, very small, then the uncertainty in momentum, Δp, cannot also be small. If both were small, the product of Δx and Δp would not satisfy Heisenberg's equation, which says that the product of the uncertainties must always be approximately equal to Planck's constant h. Notice that this is the second equation of quantum mechanics that we have come across that contains this mysterious constant.

Planck's constant may be measured in experiments on the photo-electric effect. Its value turns out to be so tiny that Heisenberg's restrictions on the accuracy of measurements have a negligible impact on our everyday observations of cars or billiard balls. For example, suppose a bullet weighing 50 grams has its speed measured to be 300 metres per second with an uncertainty of one hundredth of one per cent. Multiplied by the 'large' mass of the bullet, this uncertainty in the momentum of the bullet leads, via Heisenberg's uncertainty principle, to a fundamental limitation in the accuracy of a position measurement of less than a million million times smaller than the diameter of an atomic nucleus. This is truly a negligible restriction for an object such as a bullet. Contrast this result for a bullet with that for an electron that is measured to be moving with the same speed with the same uncertainty in the measurement. To find the uncertainty in momentum we must multiply the velocity uncertainty by the electron mass. Since the mass of the electron is so much smaller than that of the bullet (see appendix 1), Planck's constant and Heisenberg's uncertainty relation now place real restrictions on the accuracy of a measurement of the electron's position. From Heisenberg's formula we now find that the limit to the accuracy of a position measurement is more than a million times larger than the diameter of an atom: this is clearly a significant restriction at atomic length scales.

There is an amusing postscript to the story of Heisenberg and his uncertainty principle. A few years before his work on the uncertainty principle, Heisenberg was finishing his doctoral thesis in Munich under the supervision of a famous theoretical physicist, Arnold Sommerfeld. At his oral examination, Heisenberg antagonized one of his examiners, a very eminent professor of experimental physics, Wilhelm Wien, by his inability to answer questions of a fairly elementary nature concerning the resolving power of optical instruments. As a result, it was only after special pleading

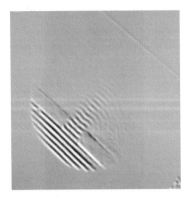 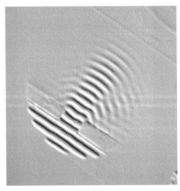 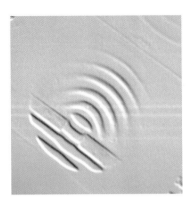

Fig. 2.4 This sequence shows how waves spread out as they pass through a slit. This spreading is known as 'diffraction' and is largest for long wavelengths. This effect limits the amount of detail visible in optical instruments.

by Sommerfeld that Wien was persuaded to pass Heisenberg, and then only with the lowest possible passmark. A few years later, Heisenberg's ignorance of such a basic point in classical optics caught him out. He illustrated his new uncertainty principle by considering a 'gamma-ray microscope' – a hypothetical microscope that could be used to look at electrons with very short wavelength gamma-ray light (see appendix 1). Unfortunately, Heisenberg forgot the lesson of his uncomfortable oral examination and his analysis of this problem did not take the resolving power of the microscope into account! It was left to another great man of physics, Niels Bohr, to point this out gently to Heisenberg, and close this loophole in his argument.

Uncertainty and photography

The probabilistic or statistical nature of quantum processes can be seen not only with electrons but also with light. Imagine looking at a faint star on a dark night. We see the star because light from the star causes chemical changes in the retinal cells of the eye. In order that these chemical reactions take place it is necessary that the light energy arrives in localized chunks – photons. The eye is quite a good light detector: a single photon can stimulate a retinal cell. In general, however, many photons are absorbed by the eye without reaching a light-sensitive cell. For this reason only a few photons in every hundred or so that enter the eye are detected. Obviously, the chemical changes involved in seeing something must be reversible – in fact, the cell reverts to its normal state after about one-tenth of a second. It is this short light storage period that limits the sensitivity of the eye for detecting faint objects. Photography can overcome this limitation of the eye by storing the changes in a permanent way on photographic emulsion.

In the same way as for the eye, single photons can cause chemical changes in the specially prepared photographic material in a roll of film.

What is the active ingredient in the photographic emulsion? If you know this, you can answer one of the questions in the question-and-answer game *Trivial Pursuit*. The question is 'What company is the world's largest user of silver?' and the answer is 'Kodak'. Photographic film consists of lots of individual grains of a silver compound, in which the silver atoms are 'ionized'. A silver ion is an atom of silver which has lost one of its negatively charged electrons. Normally, in a neutral atom, the total charge of all the electrons exactly cancels the positive charge of the nucleus. A silver ion, therefore, has a net positive charge. When a photon is absorbed by the emulsion, an electron is sometimes emitted, in the same way as electrons are knocked out of a metal in the photo-electric effect. This electron can now be attracted by a silver ion to form a neutral atom of silver. Left to itself, the neutral silver atom, surrounded by the ionic silver compound, is unstable, and will eventually eject the electron and revert back to an ion. If, before this happens, other photons have produced several other neutral silver atoms nearby, a stable 'development centre' consisting of a small number of atoms can be formed. In contrast, each grain of the emulsion contains billions of silver ions. When the film is developed, this tiny group of neutral silver atoms induces all the remaining silver ions in the grain to be deposited as opaque metallic silver. How does photography help us to see very faint stars? For such very faint objects the chance of forming a development centre is rather low because of the small number of photons reaching the Earth from the star. But if we wait longer and increase the exposure of the photographic plate, more photons will arrive and the chance of this happening will increase. The photographs shown in Fig. 2.5 are of the Andromeda Galaxy taken at different exposures. The detail of the outer spiral arms is invisible to the eye but is revealed in the long-exposure photographs.

Now consider taking an ordinary photograph with a camera. Figure 2.6 shows several photographs of the same person taken at different exposures. In the top left picture about 3000 photons enter the camera. Most of these photons are absorbed without causing a permanent change in the emulsion. It is evident that 3000 photons are not enough to give us a recognizable image and the photograph appears like a more-or-less random series of dots. As we increase the exposure the number of photons entering the camera increases. The top right picture involves about 10 000 photons and already, although there is no clear image, the blurred impression of a face is beginning to appear. The improvement continues as we increase the number of photons, and the final exposure involves more than 30 000 000 photons. In this last picture, the image intensity seems to vary smoothly from place to place on the photograph. In fact, we know it is built up out of many tiny development centres created by the arrival of individual photons. Furthermore, although in the lowest-exposure photograph the positions of the bright dots that signify the presence of a development centre in that grain of emulsion would seem to be pretty random, we see that they are not. Centres are more likely to develop in places where the

Fig. 2.5 Four photographs of the Andromeda Galaxy showing how the amount of detail visible increases with increasing exposure time.

image will eventually be bright. Thus, even in the commonplace action of taking a photograph, we can see the quantum mechanical, probabilistic nature of light. We cannot predict with certainty where any particular photon will land, or in which grain a development centre will be produced. All we can talk about are probabilities.

As we have seen, photographic emulsions are not sensitive to individual photons – several neutral atoms must be produced to form a development centre. Nowadays in astronomy, a new type of detector has all but replaced photographic plates. This is the so-called 'charge coupled device', or 'CCD', which can detect the arrival of a single photon. It is much more efficient than photography for the detection of very faint stellar objects, as can

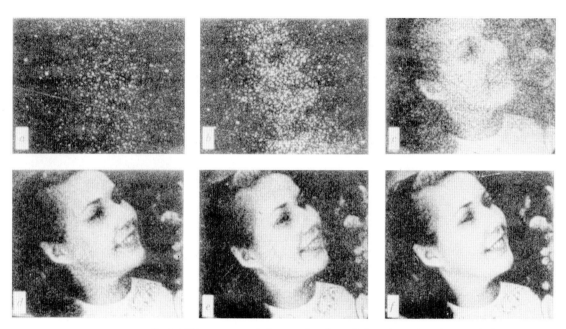

Fig. 2.6 This sequence of photographs of a girl's face shows that photography is a quantum process. The probabilistic nature of quantum effects is evident from the first photographs in which the number of photons is very small. As the number of photons increases the photograph becomes more and more distinct until the optimum exposure is reached. The number of photons involved in these photographs ranges from about 3000 in the lowest exposure to about 30 000 000 in the final exposure.

be seen in Fig. 2.7. A CCD consists of an array of small 'photon detectors' laid out on a silicon chip. As we shall see in a later chapter, silicon is an example of a class of materials that are called 'semiconductors'. Roughly speaking, semiconductors are substances whose electrical properties are halfway between metals, which allow electric current to flow easily, and insulators, which do not allow currents to flow at all. Silicon also has the property of requiring very little energy to release electrons from their parent atoms. By carefully adjusting the temperature at which the CCD is operated, the silicon can be made sensitive to the passage of a single photon. Each 'detector' is in fact just a small region of silicon where the electrons liberated by the arrival of photons can be collected. A measurement of the accumulated charge at each position over the array then corresponds to the pattern of photons striking the CCD. CCDs are beginning to replace film in the new 'digital cameras'. Even this novel CCD technology is being challenged by new electronic photon detectors – Charge Metal Oxide Sensors. These new devices are cheaper to fabricate since they are based on the same silicon processes as modern microprocessors. As yet, these new sensors are not of as high a quality as CCD detectors.

We have now seen how quantum uncertainty manifests itself in something as simple as taking a photograph with film or with a modern

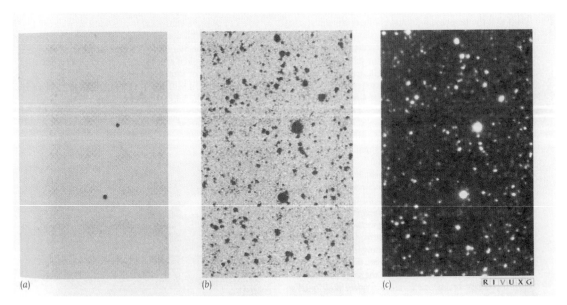

Fig. 2.7 A comparison of a photograph and CCD images of the same part of the sky taken with the same 4m telescope. (a) A negative print with black stars on white sky. (b) A CCD image of the same region showing many more faint stars and galaxies. (c) A coloured view of the region produced by combining CCD images taken using colour filters. This sequence illustrates the dramatic improvement in sensitivity that can be achieved using CCD devices.

electronic image device. Feynman suggested yet another way of looking at this quantum uncertainty. This is in terms of 'classical' and 'quantum' paths for the particle, and this insight has turned out to be of great importance in modern quantum theory.

Feynman's quantum paths

There is another interesting way in which to view the similarities and differences between classical and quantum physics. Consider the double-slit experiment, and suppose that we wish to calculate the probability for an electron to leave the source (S) and arrive at some position at the detector (D) (Fig. 2.8). In order to calculate the observed arrival pattern we found that we had to add the probability amplitudes a_1 and a_2 for the paths 1 and 2

$$a = a_1 + a_2$$

to obtain the complete quantum amplitude a. The probability of arrival at any point is then obtained by squaring this amplitude:

$$P = (a_1 + a_2)^2$$

Richard Feynman (1918–1988) grew up in Far Rockaway, a small town on Long Island, just outside New York. He made important contributions to many areas of theoretical physics. His 'sum over histories' way of looking at quantum amplitudes now plays a central role in modern quantum field theory. During the war, he worked on the Los Alamos 'Manhattan' project, which led to the development of the first atom bomb. One of the great men of quantum mechanics, Niels Bohr, picked on Feynman to try out his new ideas because he was the only person at Los Alamos who would not be in awe of his reputation and would tell him if his ideas were lousy!

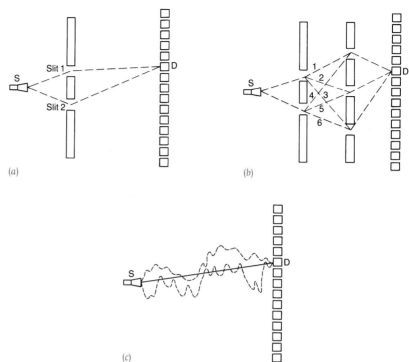

Fig. 2.8 The quantum amplitude may be obtained by adding the amplitudes for all possible paths between source S and detector D. (a) The original double-slit experiment with two possible paths for the electron. (b) With two screens between source and detector and a total of five slits there are now six possible paths. (c) Adding more screens and then cutting more and more slits leads to the situation where there are no screens at all! The quantum amplitude for an electron to travel from S to D may therefore be regarded as a sum of all possible paths. Two of the infinite number of possible quantum paths are shown as dotted lines in this figure. The path taken by a classical particle is shown as a solid line.

This is the quantum mechanical recipe that is required to account for the interference pattern observed by experiment. We will return to the question of exactly how we determine the appropriate amplitudes for each path when we discuss Schrödinger's equation in the next chapter.

For the moment let us just accept this rule and consider what happens if we complicate the experiment by introducing a second screen, with three more slits, as shown in Fig. 2.8b. There are now six possible paths from S to D, and, according to our quantum mechanical rule, we must add up the amplitudes for all these paths to obtain the total probability amplitude:

$$a = a_1 + a_2 + a_3 + a_4 + a_5 + a_6$$

total amplitude = sum of amplitudes for each possible path

Squaring this amplitude gives the probability of arrival. Now imagine what happens if we continue to put more screens with more slits between the source and the detector. To obtain the total probability amplitude we must

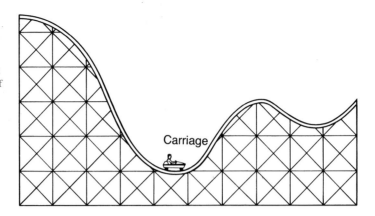

Fig. 2.9 A roller coaster with the carriage at rest in one of the valleys. Heisenberg's uncertainty principle does not allow a quantum roller coaster to remain at rest: instead the quantum carriage would be forever jiggling back and forth across the bottom of the valley.

still add up all the amplitudes for all the many possible paths. If we carry on putting in screens, eventually we will fill up the space between S and D with screens. If we now make more and more slits in each screen, in the end we will have no screens at all! This line of reasoning led Feynman to write down an expression for the total probability amplitude to go from S to D, in the absence of any screens or slits, as the sum of all the amplitudes for every possible path between S and D. In Fig. 2.8*c* we have indicated two such possible 'quantum paths', together with the straight line trajectory that a bullet would follow in going from S to D, in the absence of any screens. In classical physics there is only this one possible path: in quantum physics we must consider all possible paths between S and D to obtain the correct probability of arrival.

We can also see a connection between the sum over all quantum paths and the quantum uncertainty principle. Let us consider first an example of classical motion. We show a section of a 'roller coaster' in Fig. 2.9. If the carriage is placed on the roller coaster at its lowest point, then, according to classical physics, it will remain at rest indefinitely unless we disturb it. To compare with the quantum case that we shall consider next, it is useful to represent the fact that the carriage does not move in a graphical way. Imagine that the position of the carriage is indicated on a horizontal axis representing all possible positions. Similarly, we represent different times by position up and down a vertical axis. The non-moving carriage then corresponds to a straight vertical line on this graph of position against time.

What happens with quantum objects such as electrons? As we describe in detail in the next chapter, it is possible to set up an arrangement of electric fields that act in the same way for the electron as the roller coaster track does for the carriage. What happens if we put an electron onto this electron roller coaster? Heisenberg's uncertainty principle does not allow the electron simply to sit at rest in the bottom of the valley! If it did so, we would know both the position and the momentum of the electron simultaneously and Heisenberg tells us that this is not possible. What happens then? According to quantum mechanics, the electron must be constantly jiggling around near the bottom of the valley and can never be at rest. This incessant motion is called 'zero-point motion'. As a result, the electron cannot have zero energy but must have an associated 'zero-point energy'. What would a graph of the position of the electron against time look like? Obviously it will not be just the simple vertical line of the roller coaster carriage. Instead, it will be some sort of complicated, jagged curve corresponding to the quantum jiggling of the electron. Using Feynman's 'sum-over-paths' way of looking at quantum mechanics, it is possible to generate typical 'quantum paths' for such an electron in a computer simulation. Some examples of such computer-generated paths are shown in Fig. 2.10.

Fractals: a mathematical curiosity

At this point we shall make a diversion from our main theme to include a brief discussion of some rather curious mathematical curves. The quantum paths shown in Fig. 2.10 represent 'snapshots' of the position of the electron over a given period of time. Going from (a) to (b) to (c), we have divided up the same period of time into finer and finer intervals. This is like increasing the magnification at which we look at the curves of the electron's motion. As can be seen, these quantum paths look very jiggly no matter what magnification we choose to look at them. The level of detail displayed in path (b) is twice that in (a) but notice that if half of graph (b) is magnified to the same scale as in (a) it would look very similar. Similarly, the finer graph (c) displays three times the level of detail but if a section of this curve is magnified by three, the resulting curve would look very similar to curve (a). This is what is meant by 'looking the same at all length scales' and is a property that is characteristic of an interesting type of curve studied by mathematicians. We are accustomed to the idea that a line has a definite length associated with it. For example, we can measure the length of a running track in metres. Area is associated with 'length squared' so that a football pitch has an area measured in 'square metres'. We may summarize this by saying that a line or an area has a magnitude expressed in terms of a 'length to the power of dimension D', where $D =$ 1 for a line and $D = 2$ for an area. The curves we have here, however, turn out to be so irregular and jiggly that they can be thought as filling more space than an ordinary line. Such curves are called 'fractals' and can have a dimension greater than one!

A much-loved example that gives some idea of the curious nature of fractals is that of Lewis Richardson's measurement of the length of the coastline of Great Britain. Consider measuring the length of a straight line using a pair of dividers set open so that the points are a certain distance apart, to pace off 'steps'. It is certainly plausible that the value we obtain for the length of the line will not depend on how far apart we have set the dividers. This is not true for a measurement of the length of the coastline

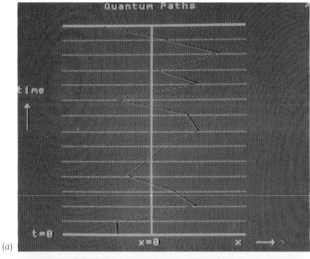

(a)

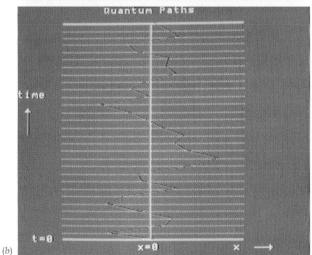

(b)

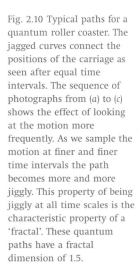

Fig. 2.10 Typical paths for a quantum roller coaster. The jagged curves connect the positions of the carriage as seen after equal time intervals. The sequence of photographs from (a) to (c) shows the effect of looking at the motion more frequently. As we sample the motion at finer and finer time intervals the path becomes more and more jiggly. This property of being jiggly at all time scales is the characteristic property of a 'fractal'. These quantum paths have a fractal dimension of 1.5.

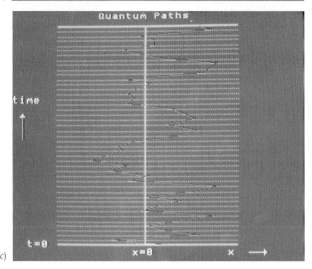

(c)

Fig. 2.11 A computer-generated fractal landscape showing an impressively realistic misty scene. Many natural features can be approximated as fractals and such artificial landscapes are now used extensively in modern science fiction movies.

of Great Britain. If we use a large spacing of the dividers, and go round the coast on a fine scale map, using the dividers as carefully as we can, it is as if we were using a much coarser scale map. The setting of the dividers is too big to follow all the smaller inlets and headlands. We can only follow these details if we use a smaller setting for the dividers. Obviously, the length of coastline between two points is longer than the straight line distance between them, but, equally obviously, the distance we measure will be larger and larger the finer the scale on which we use to measure it. The coastline looks jiggly whatever the scale we choose, and our answer for the length will increase as we look in finer and finer detail. The curve of the coastline is thus a fractal curve whose length depends on the scale at which it is measured. This effect may be part of the reason why different encyclopedias give different values for the length of land frontiers between different countries. Spanish and Portuguese encylopedias, for example give distances for the length of the boundary between the two countries that differ by as much as 20%!

It is clearly unsatisfactory to have a definition of length that depends on the scale at which the measurement is made. The modern theory of fractals was popularised by Benoit Mandelbrot, a scientist in IBM's Research Division. He was inspired by this curious property that the length of the coastline depends on the scale at which it is measured so that finer scales result in greater lengths. The coastline appears as a jagged line at every length scale: looked at from a distance, we can almost imagine the coastline as a 'fuzzy' line with some non-zero thickness. Mandelbrot gave this intuitive idea of fuzziness some mathematical precision. He introduced the notion of a 'fractal dimension' that captures the degree of spikiness in a curve. A smooth curve has fractal dimension $D = 1$, the same as an ordinary dimension, but the more jagged and spiky the curve becomes, the nearer Mandelbrot's fractal dimension approaches $D = 2$, a curve so jaggedy that it fills all of two-dimensional space. From Richardson's data on the west coast of Great Britain, Mandelbrot was able to deduce that the coastline had a fractal dimension of about $D = 1.2$! The jiggly quantum paths shown in Fig. 2.10 turn out to have a fractal dimension $D = 1.5$, in-between a simple curve and a space filling fractal. Mandelbrot produced a wonderful book on fractals that contains many beautiful computer-generated pictures showing the similarity between natural features, such as snowflakes or clouds, and certain types of fractals. Computer-generation of artificial landscapes using fractals is now one of the standard tricks of the trade for modern film-making. A striking example of such a fractal 'lunar landscape' is shown in Fig. 2.11.

3 Schrödinger and matter waves

> Where did we get that [equation] from?
> Nowhere. It is not possible to derive it from
> anything you know. It came out of the mind
> of Schrödinger.
>
> Richard Feynman

De Broglie's matter waves

The early struggles of physicists towards a quantum theory were mostly concerned with attempts to understand the nature of light. The traditional picture of light as a wave motion had been challenged by Planck and Einstein. They had shown that certain experimental results that were impossible to understand in terms of a wave picture could be easily explained if light was thought of as a stream of particles, now called photons. William Bragg, who, with his son, won the 1915 Nobel Prize for studies of crystal structure using X-rays, summarized this dilemma for physics by exclaiming in despair that he was teaching the corpuscular theory of light on Mondays, Wednesdays and Fridays, and the undulatory theory on Tuesdays, Thursdays and Saturdays! Physicists were still wrestling with these apparently contradictory properties of light when, in 1924, Prince Louis de Broglie (pronounced 'de Broy') suggested that all matter, even objects that we usually think of as particles – such as electrons – should also display wavelike behaviour! This revolutionary idea was completely unexpected and, what is more, was included by de Broglie in his Ph.D. thesis. Like most people, physicists are generally reluctant to accept any wild new idea, especially if there is not a shred of evidence to support it. Predictably, de Broglie's examining committee in Paris were distinctly unsure as to what to do about the thesis. One of the examining committee, Professor Langevin, himself an eminent physicist of the day, was described by de Broglie as being '*probablement un peu étonné par la nouveauté de mes idées*' ('probably a bit stunned by the novelty of my ideas'). Nevertheless, Langevin was wise enough to ask de Broglie for an extra copy of the thesis which he sent to Einstein for his opinion. Einstein was impressed and said of de Broglie's work, 'I believe it is a first feeble ray of light on this worst of our physics enigmas'. Fortunately, the examining committee made the right decision and gave de Broglie his doctorate. Only

Prince Louis de Broglie (1892–1987) was descended from a noble French family whose great-great-grandfather died on the guillotine during the French revolution. De Broglie initially obtained a degree in history, but while serving in the French army during the First World War he became interested in science. He was involved with radio communication and was stationed on top of the Eiffel Tower. His proposed mathematical relationship connecting the wave and particle properties of matter earned him the Nobel Prize in 1929. He died in March 1987.

Fig. 3.1 These two photographs show how the interference pattern is altered when the separation of the sources is increased. Larger separations cause the interference bands to move closer.

Erwin Schrödinger (1887–1961) was educated in Vienna, and during the First World War served as an artillery officer. After the war he decided to abandon physics and take up philosophy, but the city in which he had hoped to gain a university position was no longer in Austria. Fortunately, Schrödinger remained a physicist and discovered the central equation of quantum mechanics in 1926. In 1928 he succeeded Max Planck as professor in Berlin. Schrödinger left Germany after Hitler came to power and eventually became professor of theoretical physics at the Institute for Advanced Studies in Dublin, Eire.

a few years later, in 1927, wavelike behaviour of electrons was convincingly demonstrated – by Davisson and Germer in the USA, and by G. P. Thomson in Scotland – and both de Broglie, in 1929, and Davisson and Thomson, in 1937, received Nobel Prizes for their work on matter waves.

If all 'particles' can behave like waves, why did it take physicists so long to observe these matter waves? Why don't we see wavelike behaviour for bullets, billiard balls, or even cars? Again, the answer to these questions lies in the smallness of Planck's constant. According to de Broglie, the wavelength of the matter waves of such everyday objects is very tiny. De Broglie suggested that a particle travelling with a certain momentum p has an associated matter wave of wavelength λ given by the expression

$$\lambda = h/p \quad \text{(de Broglie's relation)}$$

wavelength = Planck's constant divided by momentum

As we saw in our discussion of Heisenberg's uncertainty principle, it is Planck's constant that characterizes the size of all quantum effects. But how does its extreme smallness explain why wavelike behaviour for everyday objects is not observed? Well, what we did not say when we talked about the double-slit experiment was that in order to see wavelike interference effects, the separation of the two slits must be about the same size as the wavelength of the objects – photons or electrons – that are doing the interfering. Since the de Broglie wavelength of a bullet fired from a gun is much smaller than even atomic dimensions, it is impossible to devise an experiment that will show interference with bullets, or indeed any other everyday object. If, on the other hand, we could increase the size of Planck's constant things would look very different, as in Mr Tompkin's nightmares!

Schrödinger's equation

At the time he discovered his now famous equation, Erwin Schrödinger was a moderately successful, middle-aged Austrian physicist working in Zurich. Professor Debye, head of the research group in Zurich, had heard about these peculiar waves of de Broglie and asked Schrödinger

Fig. 3.2 Another view of Mr Tompkins' Wonderland in which Planck's constant is much larger than in our world. The caption read 'Sir Richard was ready to shoot, when the professor stopped him'. The professor went on to explain 'there is very little chance of hitting an animal when it is moving in a diffraction pattern'.

to explain these ideas to the rest of the group. Schrödinger did so, and when he had finished Debye remarked that it all seemed rather childish – to deal properly with waves one should have a wave equation to describe how the wave moves from place to place. Stimulated by this remark Schrödinger went away and discovered the equation that now bears his name. This was a vital breakthrough because it enabled physicists to calculate how quantum probability waves move, and therefore make precise predictions that could be compared with experiment. Just as Newton guessed the simple laws that describe all of classical physics, so Schrödinger guessed the law that describes the motion of quantum objects. Before we write down Schrödinger's equation, in order to make it seem less like a rabbit out of a hat, it is helpful to introduce the principle of conservation of energy. This we can do by looking at the motion of familiar objects in our everyday world.

Imagine yourself back on the roller coaster again (Fig. 3.3). If we start from rest at the top of the hill on the left, as we slide down the track the car

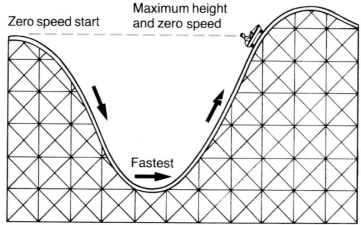

The commemorative plaque on the wall of the original building of the Institute for Advanced Studies in Dublin, Eire. It is said that Eamon De Valera, the President of Eire, found it easier to set up a new institute for Schrödinger than to persuade his Trinity College colleagues to make him a professor. Notice that instead of 'quantum mechanics' the inscription uses the old-fashioned term 'wave mechanics'.

Fig. 3.3 An idealized roller coaster ride illustrating conservation of energy. At the top left-hand side the carriage starts to move from rest. In this position it has zero kinetic energy due to motion and maximum gravitational potential energy due to its height. As the carriage rolls down the track the gravitational energy is converted to kinetic energy so that at the bottom it has zero gravitational energy and maximum speed and kinetic energy. The car then slows as it climbs up the hill on the other side converting back kinetic energy into gravitational energy. On an idealized roller coaster with no energy losses due to heating the track, generating noise, and so on, the carriage will climb up to exactly the same height from which it started.

will go faster and faster until it reaches the bottom. We then start going up the other side, and will gradually slow down until the car comes to a stop. The motion of the car illustrates the principle of conservation of energy. When we start out, high above the valley, we are at rest and have no 'kinetic energy' – energy due to our speed. At the bottom, when we are travelling fastest, we have a lot of kinetic energy. As we travel up the other side, we steadily lose kinetic energy until we are once more at rest. Where has all the kinetic energy gone? As we climb up the slope, we must use up energy in raising the weight of the carriage and its occupants back up to the top of the hill. We describe this by saying that we do work against the pull of gravity and, as we gain height, we are said to gain gravitational 'potential energy'. When we started out, we had no kinetic energy, but because we were high up, we were able to convert our gravitational potential energy into kinetic energy by rolling down the slope and losing height. The total amount of energy is always the same, but its form may change. In principle, the roller coaster will slide down to the bottom of the valley and then coast up the other side to exactly the same height as we started from. Of course, a real roller coaster car will not quite reach the same height as it started from, because some of the initial potential energy will be lost to the surrounding environment in the form of energy used in heating up the tracks, in causing noise and so on. These are all so-called frictional energy losses. To keep things simple for this discussion we shall ignore such losses and our roller coaster should be imagined as being very slippery and shiny. Conservation of energy for our roller coaster example can be summarized by saying that the total energy, for which we use the symbol E, is constant, but this can be made up of varying amounts of kinetic energy, K, and potential energy, traditionally denoted by V. As an equation this reads

$$E = K + V$$

total energy = kinetic energy + potential energy

This equation is true for any position of the roller coaster car on the track and at all times.

Before we leave this example, there is another way of writing the energy equation that will be useful to us later. As we have said in chapter 2, the momentum p of an object is given by its mass m multiplied by its velocity v:

$$p = mv$$

momentum = mass times velocity

From Newton's laws we can show that kinetic energy and momentum are related by the familiar equation:

$$K = p^2/(2m)$$

Our book-keeping relation for energy may therefore be re-written as

$$E = p^2/(2m) + V$$

which is now an equation relating total energy, momentum and potential energy.

What has all this to do with electrons and the Schrödinger equation? In the previous chapter we remarked that it was possible to set up a similar 'roller coaster' for electrons. Quantum objects like electrons still obey the principle of conservation of energy – we are not allowed to create or lose energy, even at the quantum level. But, just like our real roller coaster example, energy can be changed from one form to another. In this case, the relevant form of potential energy is not that due to gravity, but electrical potential energy. Electrons, with their negative electric charge, are attracted to a region of positive charge. We can use a battery and an arrangement of metal plates to set up an electric potential energy curve that has roughly the same shape as the roller coaster track (Fig. 3.4). Electrons moving through this system of plates will be attracted by the positively charged plate and accelerate, gaining kinetic energy as they approach. Again, as for the roller coaster, this gain in kinetic energy is compensated for by a corresponding loss in potential energy, now in the form of electrical potential energy. As before, we can write down an equation describing this book-keeping:

$$E = p^2/(2m) + V$$

where V is now the electrical potential energy.

This equation was Schrödinger's starting point. Using de Broglie's relation between momentum and wavelength, Schrödinger was able to guess at the form of the wave equation for a quantum object travelling in a potential. His equation is displayed in the box below. For an appreciation of the rest of this book it is not necessary for you to understand the mathematics

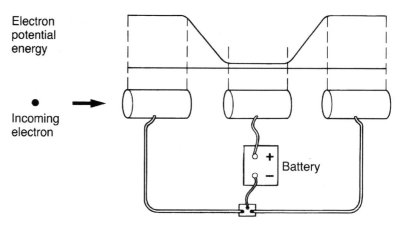

Fig. 3.4 Apparatus to produce an electron roller coaster. A system of metal tubes is connected to a battery to give the electric potential energy curve shown at the top of the figure. An incoming electron from the left is attracted to the positively charged central tube and then overshoots as for our fairground roller coaster example.

of this equation. Our purpose in showing you this equation is not to frighten you. Rather, it is to convince you that there is a precise mathematical basis underlying all the hand-waving discussion of quantum phenomena that you will find in the remaining chapters of this book!

SCHRÖDINGER'S EQUATION

For the motion of a particle with total energy E, moving in one dimension x in a region in which there is a potential V, Schrödinger's equation reads

$$E \psi = -\frac{\hbar^2}{2m} \frac{d^2 \psi}{dx^2} + V \psi$$

It is the convention to represent probability amplitudes by the Greek letter psi (ψ). The mass of the particle is m, and $\hbar$ (pronounced 'aitch bar') is Planck's constant h divided by 2π. Readers familiar with calculus and differentiation can see this equation solved for the problem of an electron in a simple potential in appendix 2.

Electron and neutron optics

When Schrödinger's famous paper was published in 1926, the existence of matter waves had not been experimentally established. Nowadays, the observation of wavelike behaviour for 'particles' is a commonplace occurrence and forms the basis for new ways of revealing the quantum world.

Murray Gell-Mann was born in 1929 and entered Yale University when he was 15. He obtained his Ph.D. degree from MIT when he was 22 and went to Caltech in Pasadena in 1955. Gell-Mann was awarded the Nobel Prize in 1969 for his many contributions to particle physics, not the least of which was the idea of quarks as the fundamental building blocks of matter. In 1984 he helped establish the Santa Fe Institute, where he now works.

George Zweig was a student at Caltech and went to CERN in Geneva, Switzerland, after he completed his Ph.D. It was there that, independently of Gell-Mann, he dreamed up what is now known as the quark model of elementary particles. Zweig is now working on problems in biophysics at the Los Alamos National Laboratory.

One of the most widely used devices that exploits this dual particle and wave nature of matter is the electron microscope. Instead of lenses made of glass, as in ordinary optical microscopes, an arrangement of electric and magnetic fields can be set up that act in the same way for electrons as glass lenses do for light. Why should this be useful? The amount of detail visible on the object being examined depends on the wavelength of the wave that is used in the observation. Roughly speaking, the wavelength we use must be smaller than the size of any detail that we wish to pick out or 'resolve'. The shorter the wavelength, the finer the detail that is revealed. Optical microscopes cannot resolve detail smaller than the wavelength of visible light. Thus, features smaller than about one-millionth of a metre (a micrometre or micron) cannot be resolved with visible light (see appendix 1). For electrons, on the other hand, the wavelength of the associated wave depends on the momentum of the electron, in accordance with de Broglie's relation. Moreover, this wavelength decreases as the momentum is increased. We are therefore able to vary the resolution of the microscope merely by varying the speed to which we accelerate the electrons. A typical electron microscope can operate at wavelengths a million times smaller than optical wavelengths. Such wavelengths are smaller than the size of atoms, and by using careful techniques some atoms can be made visible with electron microscopes. The practical resolution of electron microscopes is limited by technical problems such as defects in the lens systems, vibrations of the apparatus and of the atoms themselves. Nevertheless, electron microscopy provides a spectacular picture of a world completely invisible to optical microscopes. Some dramatic examples of electron microscope photographs are shown in Figs. 3.5–3.7.

Electron microscopes produce an image of the object being examined, and such photographs are relatively easy to interpret. However, electron beams can also be used to probe deep into the inner structure of matter without producing a direct image of the object. Using special-purpose accelerators, electrons can be accelerated to speeds only slightly less than that of light itself. Such very-high-momentum electrons can explore very tiny distance scales. Observing the pattern of electrons scattered by a proton target allows us to infer details about the structure of the proton. Such electron scattering experiments, pioneered at the Stanford Linear Accelerator in California (SLAC, see Figs. 3.8 and 3.9), have enabled us to look deep inside protons. The results were very surprising. Instead of finding all the positive charge of a proton spread out uniformly throughout its volume, these experiments revealed that the charge is concentrated on even smaller constituents within the proton. Moreover, instead of these constituents having the same amount of electric charge as electrons or protons, they are found to have charges 1/3 or 2/3 times this amount! These tiny constituents with their peculiar charges are called quarks. They were first proposed as elementary building blocks of matter by Murray Gell-Mann and George Zweig. Gell-Mann was already a famous physicist, working at Caltech in Pasadena, California. By contrast, Zweig was almost unknown. After completing a

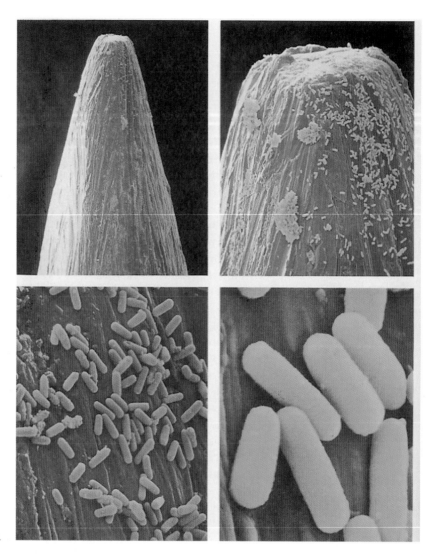

Fig. 3.5 A disturbing
sequence of photographs of
bacteria on the point of a
pin. The magnifications are
×20, ×100, ×500 and ×2500.

Ph.D. at Caltech he had gone to Europe to work at the CERN particle physics
laboratory in Geneva. In his pioneering paper, Zweig originally called these
new fundamental constituents aces. It was Gell-Mann who introduced the
word quark, after a nonsense word in James Joyce's novel, *Finnegan's Wake*.
Since in the quark theory of matter, the proton is made up of three quarks,
the quotation from Joyce, 'Three quarks for Muster Mark!', is very appropriate and Gell-Mann's name has stuck. This is despite the fact that in German
quark means cream-cheese made from skimmed milk and is used colloqui-
ally to mean rubbish! We shall meet quarks again in a later chapter when
we discuss elementary particle physics and the strong nuclear force.

There is another interesting side story about quarks, or rather aces,
which indicates that physicists are not as immune from prejudice as they
would like to think. At the time that Gell-Mann and Zweig suggested this

Fig. 3.6 This electron microscope picture shows a family of dust mites seemingly gently grazing in a field. The magnification is about ×200.

theory involving, as it did, new, even more fundamental particles, another theory was in fashion that could be crudely characterized by the slogan of 'nuclear democracy'. In this rival theory, no particle was any more fundamental than any other, and so deeply were most physicists committed to thinking along these lines that to propose a constituent model involving new fundamental particles seemed like heresy! Gell-Mann realized that there would be great opposition to the quark idea in the USA and made a conscious decision to publish his paper in a European journal, where he felt the prejudice would not be so great. Zweig, on the other hand, was in Europe but, wisely or unwisely, wanted to have what he regarded as his most important discovery published in a US journal. He first had to fight a long battle with the CERN management to be able to send his paper to a US journal. When he finally won his battle, he found that his paper was eventually rejected for publication. Some US physicists went so far as to brand Zweig a charlatan! Fortunately, there was a happy ending. Zweig's paper on quarks became one of the famous unpublished papers of physics. It took almost 20 years before his original paper was finally published in a compilation of influential papers on the quark model. Today, in particle physics, the idea of quarks as fundamental constituents of the proton is accepted as an 'obvious truth' and nuclear democracy is now seen to have been a bold but misguided diversion.

Wavelike properties have now been demonstrated for several other particles besides electrons. In particular, since the late 1970s, a series of beautiful experiments has been performed with neutrons that relies on their wavelike character. Neutrons, as their name implies, are electrically neutral. They weigh about the same as protons, and together with protons

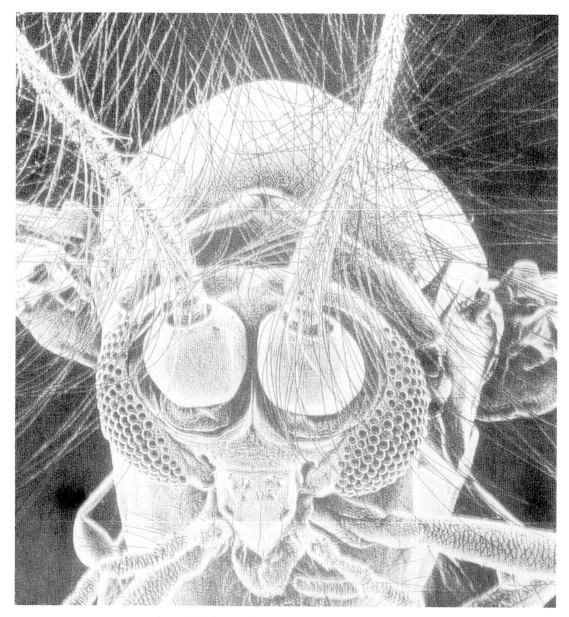

Fig. 3.7 This electron microscope picture shows a common midge at a magnification of about ×500. Large swarms of these can be seen in summer, especially in Scotland, but fortunately their true size is only about 2 mm.

they form the constituents of nuclei. Neutrons are produced in the nuclear reactions that generate power in nuclear reactors, as we shall discuss in a later chapter. All that we need to know here is that beams of neutrons can be produced, and that with such beams it is possible to perform an analogue of the double-slit experiment. For this neutron version of the experiment, the detector is kept in a fixed position and the interference

Fig. 3.8 The SLAC two mile accelerator in Stanford, California runs from the San Andreas hills, under the freeway from San Jose to San Francisco, to the cluster of experimental laboratories at the bottom. The famous earthquake fault runs along the base of the San Andreas hills and there are elaborate safety devices to turn the accelerator off in the event of a serious earth tremor. Electrons and positrons can be accelerated along the two-mile length to speeds close to the velocity of light.

is observed by altering the effective length of one of the two interfering paths. This is done by inserting a gas cell into one of the beams. By varying the density of gas in the cell, the effective path length for neutrons can be changed, causing an interference pattern to be seen in the intensity of neutrons arriving at the detector. Neutron experiments have now become so sensitive that it is possible to observe the effects of a tiny gravitational potential term in the Schrödinger equation for neutrons.

Since the early 1990s, interference experiments with atoms have been performed. The first experiment directed a beam of helium atoms at a tiny gold screen in which two slits had been cut. These slits were separated by only about a millionth of a metre – roughly the wavelength of visible light (see appendix 1). A movable detector observed the arrival of individual helium atoms and gradually the familiar interference fringe pattern emerged. This is exactly analogous to the electron interference experiments discussed in chapter 1. Similar experiments with more massive atoms than helium have also now been performed. Although full experimental

Fig. 3.9 One of the original electron detectors at SLAC. The electron beam entered from the left and collided with protons in a target. The scattered electron is deflected by large magnetic fields and its direction and momentum measured.

vindication of de Broglie's idea of matter waves has taken over 70 years, his Ph.D. examiners clearly made the right decision!

An exciting recent development is the use of laser light (which we discuss in a later chapter) to produce forces on the atoms. In this way, it is possible to construct the equivalent of 'double' or 'multiple' slit devices without using a screen made of matter. The possibility of manipulating and controlling the quantum motion of atoms has laid the foundation for a new field of research in 'atom optics'. We shall return to these developments later in the book.

4 Atoms and nuclei

Atoms are completely impossible from the
classical point of view.

Richard Feynman

Rutherford's nuclear atom

Before quantum mechanics came along, classical physics was unable
to account for either the size or the stability of atoms. Experiments initiated
in 1911 by the famous New Zealand physicist, Ernest Rutherford, had shown
that nearly all the mass and all of the positive charge of an atom are
concentrated in a tiny central core that Rutherford called the 'nucleus'.
Most of the atom is empty space! A table of the relative sizes of atoms, nuclei
and other quantum and classical objects is given in appendix 1. Rutherford
had already won a Nobel Prize earlier, in 1908, for his work on radioactivity.
Radioactivity is now known to be due to the 'decay' of a nucleus of certain
unstable chemical elements: some radiation is given off – in the form of
alpha, beta or gamma rays – and a nucleus of a different element is left
behind (see Fig. 4.1). As you can imagine, it took physicists some time to
disentangle what was going on, and it was Rutherford who showed that
the positively charged, heavy, penetrating alpha rays were, in fact, helium
atoms which had lost two electrons. Beta rays, on the other hand, were
identified as electrons, and gamma rays as high energy photons. At that
time, any work involving the different chemical elements was regarded as
the province of chemists, and Rutherford was somewhat put out at winning
the chemistry Nobel Prize. In his acceptance speech he remarked that he
had observed many transformations in his work on radioactivity but none
as rapid as his own, from physicist to chemist!

How did Rutherford discover the nucleus? He used the traditional
method of physicists, namely that of throwing things at something and seeing what happens. Rutherford, together with his colleagues in Manchester,
UK, fired alpha particles from a radioactive source at a very thin gold foil.
They then watched carefully to see in which directions the particles were
scattered. Most of the time, the alpha particles only changed direction a

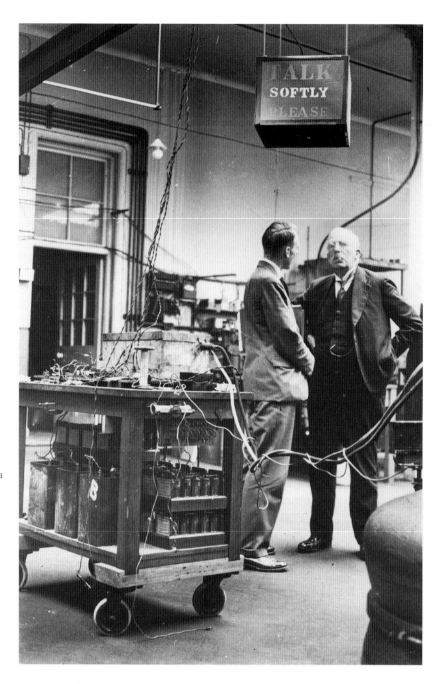

Ernest Rutherford (1871–1937), first Baron Rutherford of Nelson, was born in New Zealand. He is pictured here talking to J. Ratcliffe in the Cavendish Laboratory. Rutherford had a booming voice that could upset delicate experimental instruments and the notice 'TALK SOFTLY PLEASE' was playfully aimed at him. Rutherford was one of the greatest experimental physicists of the twentieth century. As well as his own fundamental research into radioactivity and nuclear physics, he influenced a whole generation of British experimental physicists.

little, but, very occasionally, the particles were deflected through very large angles. Rutherford described his astonishment at the results in graphic terms:

> It was quite the most incredible event that ever happened to me in my life. It was as incredible as if you fired a 15-inch shell at a piece of tissue paper and it came back and hit you!

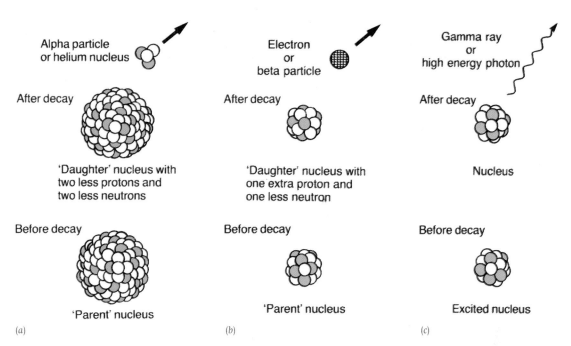

Fig. 4.1 Types of radioactivity. (a) Alpha rays are helium nuclei emitted when an unstable nucleus decays. The final 'daughter' nucleus has two fewer protons and two fewer neutrons than the original 'parent' nucleus. In this figure the blue circles represent protons and the open circles neutrons. (b) A beta particle is an electron ejected from an unstable nucleus as it decays. The new nucleus has one more proton and one less neutron than the parent nucleus. (c) Gamma rays are high-energy photons that are emitted in a transition of a nucleus from an 'excited' state to a lower energy state. The number of protons and neutrons in the nucleus is unchanged.

Rutherford puzzled over these results for some weeks and eventually realized that the alpha particles could only be scattered through such large angles if they had collided with a very dense and small core of matter within the atom – the atomic nucleus (see Fig. 4.2).

We now know that the nucleus of an atom contains particles called protons, with a positive electrical charge, equal and opposite to the charge of the electron, and particles called neutrons that are electrically neutral. Both the proton and the neutron are about 2000 times heavier than an electron, so most of the mass of the atom resides in the nucleus. Different numbers of protons and neutrons in the nucleus then account for the different elements. The protons and neutrons are held within the tiny volume of the nucleus by forces much, much stronger than the electrical repulsion between protons. Furthermore, these 'strong forces' permit only certain numbers of neutrons and protons to combine together to form stable nuclei. The simplest nucleus is that of hydrogen as it consists of just one proton. The next simplest are the alpha particles, which are helium nuclei containing two protons and two neutrons. In a neutral atom, the charge of the nucleus is exactly balanced by the charge of the electrons. Atomic hydrogen has one electron and helium has two. It is the number

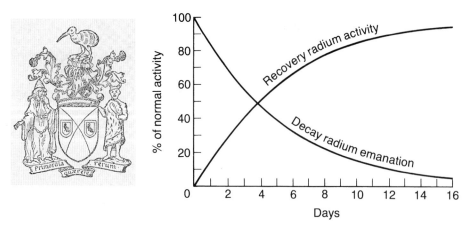

Lord Rutherford of Nelson's coat of arms. Rutherford was made a baron in 1931 and chose to link his name with the town of Nelson, in New Zealand, where he was born. The Kiwi at the top and the Maori holding a club on the right testify to his love of New Zealand. Notice the two curved lines on his shield. These were taken from a graph in a famous paper of his on radioactivity. On the left-hand side of the shield is Hermes Trismegistus (thrice greatest), the name given to the Egyptian God 'Thoth', who was regarded as responsible for mysterious things such as alchemy. This is appropriate since, in a sense, Rutherford received his Nobel Prize for alchemy, albeit in a modern guise! His motto means 'To seek the beginnings of things'.

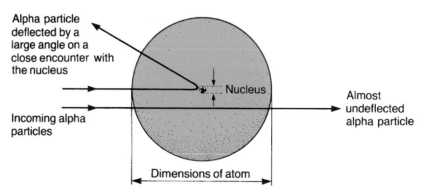

Fig. 4.2 An illustrative picture of Rutherford's alpha scattering experiment. The nucleus is about one hundred thousandth the size of an atom and its size has been greatly exaggerated in this figure. Nonetheless, it is still apparent that most of an atom is empty space! Only if an alpha particle happens to collide with the tiny nucleus will there be scattering through a large angle and this will happen very rarely.

of electrons, or equivalently the number of protons, that determines the chemical character of the different elements. Thus, although the strong nuclear force often allows a nucleus of an element to exist in several varieties, corresponding to different numbers of neutrons, all these 'isotopes' are chemically identical. For example, the gas neon is most commonly found to have a nucleus containing ten protons and ten neutrons but there are two other naturally occurring varieties of neon with nuclei containing eleven

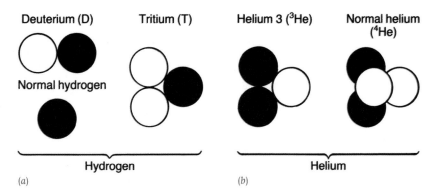

Fig. 4.3 Isotopes of hydrogen and helium. Solid circles represent protons and open circles neutrons. (a) Deuterium, tritium and normal hydrogen; (b) ^{3}He and normal helium, ^{4}He.

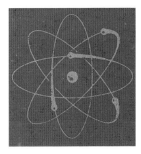

Fig. 4.4 The emblem of the US Atomic Energy Commission shows a schematic picture of the Rutherford–Bohr model of the atom. Electrons are shown orbiting the central nucleus rather like planetary orbits in the solar system.

and twelve neutrons. Since these isotopes of neon have the same number of protons, and therefore the same number of electrons, they have the same chemical properties. Similarly, there are rare isotopes of hydrogen in which the nucleus consists of a proton and one or two neutrons. These isotopes of hydrogen are called 'deuterium' and 'tritium', respectively, and we will see later that they are important in the nuclear reactions of stars and of nuclear weapons. Some isotopes, especially those of heavy elements, are unstable and undergo radioactive decay to more stable elements. We will return to these topics in a later chapter.

Rutherford pictured the atom rather as a miniature solar system, with electrons orbiting the nucleus in the same way as planets circle the Sun (Fig. 4.4). The relatively large orbits of the electrons could then account for the large size of the atom compared with that of the nucleus. The atom as a whole is electrically neutral and the electrons are kept in orbit by the electrical attraction between them and the positively charged nucleus. Unfortunately for classical physics, this whole arrangement is unworkable. To keep in orbit round the nucleus, the electrons cannot travel in a straight line – they must keep changing direction in order to keep in their orbits. In other words, they are always being accelerated towards the nucleus. But according to the well-established laws of electricity and magnetism, a charged particle that is accelerated will radiate light. Classical physics therefore predicts that in a very short time the electrons lose energy by radiation and spiral into the nucleus!

There is no answer to these problems within the framework of classical physics. A young Danish physicist called Niels Bohr then added more fuel to the funeral pyre of nineteenth-century physics. Bohr was at Manchester with Rutherford and was bold enough to recognize that, in spite of all the obvious difficulties, there must be some truth in the planetary model of the atom. He therefore devised a sort of 'recipe book' that gave rules for calculating special stable electron orbits for which the laws of classical physics were 'inoperative'. What gave Bohr the clue to his rules and why did physicists take them seriously? To see this, we must uncover another

Niels Bohr (1885–1962) was very much influenced by Rutherford with whom he worked while he built his model of the atom. Bohr was always concerned to be very clear in his statements on quantum theory. Paradoxically, this made him seem very obscure on first hearing! Nevertheless, he was undoubtedly one of the most influential scientists of the twentieth century. Bohr was regarded as the 'oracle' for questions about the interpretation of quantum mechanics and, in a celebrated debate lasting many years, he and Einstein argued about the philosophical basis for quantum mechanics. Einstein remained unconvinced until the end of his life.

problem for classical physics and introduce a Swiss mathematics teacher named Johann Jakob Balmer.

Physicists had played around making electric sparks in tubes containing various gases. They found that each gas gave off light with a characteristic 'spectrum' – only certain wavelengths were present. These are called line spectra and they can be used to identify different elements. In fact, the element helium was first discovered in light from the Sun: Fig. 4.5 shows a view of the Sun in helium light. Some line spectra are reasonably simple, such as those for hydrogen, helium and the alkali elements, but most are very complicated (Fig. 4.6). Classical physics cannot even account for the stability of the atom let alone explain the details of their spectra. It is here that the rather curious character Johann Jakob Balmer makes the contribution that has immortalized him in all physics textbooks. Balmer was a mathematics teacher who, in his spare time, was obsessed with formulae for numbers. He believed that the whole world was governed by some 'unified harmony' and his aim in life was to find out how to express these harmonic relations numerically. In 1865, when Balmer was forty, he had written a paper explaining the prophet Ezekiel's vision of the temple. Twenty years later, using Anders Jonas Angstrom's measurements of the frequencies of the first four spectral lines of hydrogen, Balmer devised the remarkable formula that now bears his name

$$\lambda = \frac{(364.5)n^2}{(n^2 - 4)}$$

where λ is the wavelength in nanometres and n takes the values 3, 4, 5 and 6. The formula worked with uncanny precision but remained an unexplained curiosity until it was shown to Niels Bohr by his friend and colleague Hans Marius Hansen. Bohr said that after he became aware of this, 'everything became clear'. His allowed electron orbits not only explained Balmer's original formula but also predicted some new line spectra for hydrogen at different wavelengths. These new spectra were given by a formula of a similar form to Balmer's, except that the denominator was changed to $(n^2 - m^2)$ with m set equal to other integers besides the 2 of Balmer's formula. After these new spectral lines had been observed, physicists were forced to take Bohr's model seriously, despite its apparently arbitrary suspension of the laws of physics. Bohr developed his model in 1913: it was not until 1926

Fig. 4.5 The Sun emits light across the whole electromagnetic spectrum. Different parts of the spectrum tell us about different aspects of the processes occurring in the Sun. It is particularly revealing to use a filter that only passes light with a wavelength corresponding to a spectral line of one particular element. In this picture an ultraviolet line from helium has been used to construct a picture of the Sun in helium light. The picture has been coloured by computer with the yellow parts representing the regions where there is most intense emission. This picture was taken by Skylab and picks out a region in the lower atmosphere with a temperature between 10 000 and 20 000 °C. As well as showing the Sun as rather blotchy, the picture also shows a spectacular arch of material propelled away from the Sun by magnetic forces.

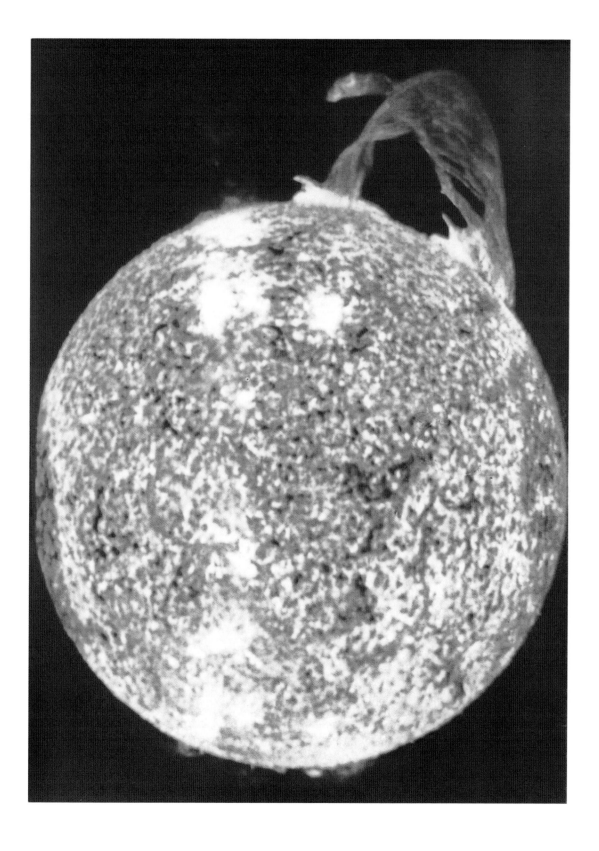

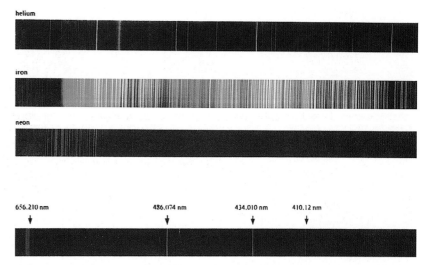

Fig. 4.6 A characteristic spectrum of light is obtained from any element in gaseous form when an electrical discharge is passed through it. This light can be separated out into the different wavelengths by passing it through a prism in a spectrometer. Each element has its own unique series of spectral lines, which can serve as a kind of 'fingerprint' to establish the presence of the element. The spectrum of hydrogen is shown at the bottom. Notice how much simpler this spectrum is than that of iron.

that Schrödinger was able to explain Bohr's rules using the new quantum mechanics.

Quantized energy levels

The key to Bohr's understanding of the line spectrum of hydrogen was the idea that electrons in motion round the nucleus were only allowed in certain particular orbits. If we ignore for the moment the problem that all the orbits are unstable because the electrons should radiate away energy, then each of these orbits will correspond to a specific energy for the electron – we say that the energy is 'quantized'. This is completely at odds with our everyday experience. In our roller coaster example of the preceding chapter, we can set the carriage rocking back and forth across the valley starting at any height we want – equivalent to starting with any value of the total energy (Fig. 3.3). How then does quantum mechanics lead to energy quantization and stable orbits?

The answer to both these problems lies with the wavelike properties of the electrons. According to quantum mechanics, the allowed energies for the electron are found by solving the Schrödinger wave equation with the appropriate potential energy term. Fortunately, we can see how energy quantization comes about without going to the trouble of solving the Schrödinger equation. Imagine a potential like our roller coaster example but where the hills are very high and steep, and the valley is wide and flat, so that we have a sort of box for the electron to sit in. The problem

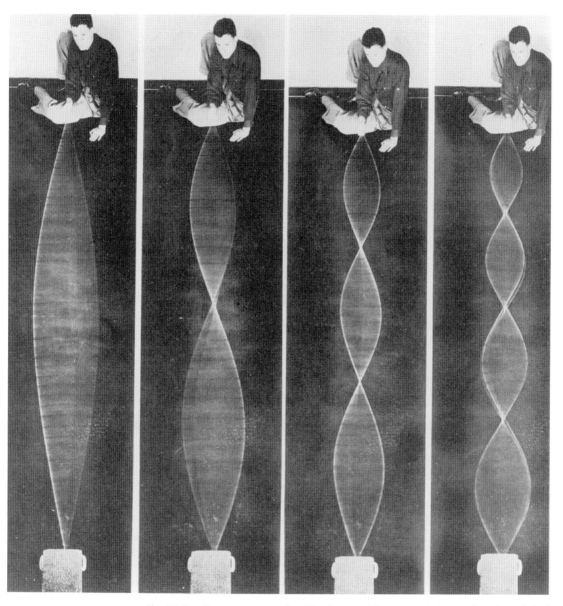

Fig. 4.7 Standing waves on a string. The photographic exposure is longer than the period of vibration so that the photograph records most strongly the positions where the string is moving slowest. Notice that there are some positions where the string is always at rest.

of finding the allowed energies of this quantum system is now very similar to the problem of finding the allowed wave motions of a string that is fixed at both ends (see Fig. 4.7). In the electron case, the steep walls of the potential box act like the fixed end points of the vibrating string and instead of waves on a string we have electron probability waves. For the string, it is clear from Fig. 4.7 that only certain 'wavelengths' can fit in

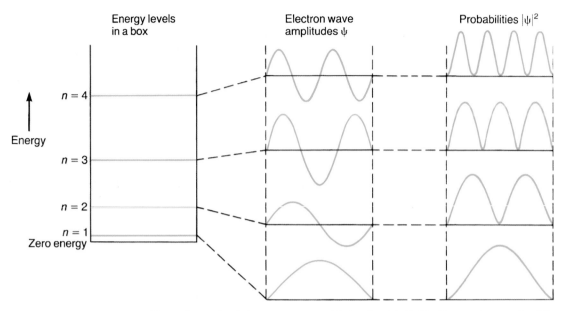

Fig. 4.8 Energy levels for a quantum particle in a box labelled by the quantum number 'n'. The middle picture shows the corresponding wave patterns and the picture on the right the probability patterns for the particle. These are the squares of the wave amplitudes.

between the fixed end points. Furthermore, in the case of the string, it is these wavelengths – the 'fundamental', or longest wavelength, together with the higher 'harmonics', or shorter wavelengths – that determine the sound that we hear. In quantum mechanics, it is the electron probability amplitudes that are required to fit in the box, but now each allowed wavelength corresponds to a definite electron energy and pattern of probability distribution. This is the origin of Bohr's quantized energies. Classically, a ball in a box can be in motion with any energy: quantum mechanically, an electron in a box can only have certain allowed values of energy.

This example of an electron in a box illustrates several general features of quantum mechanics. The mathematical details are contained in appendix 2, but the forms for the electron probability amplitudes can be guessed from our string analogy. In Fig. 4.8 we show these 'wavefunctions' together with a scale showing the corresponding electron energy. The first thing to notice is that the lowest energy of the electron is not zero. Since the uncertainty in position of the electron cannot be greater than the size of the box, Heisenberg's uncertainty principle requires a certain minimum energy. Even when the electron has its lowest possible energy – we say the electron is in the 'ground state' – it cannot sit still but must be forever jiggling around! We saw this effect when we discussed Feynman's quantum paths and Heisenberg's uncertainty principle in the previous chapter. This so-called 'zero-point motion' is a general property of quantum systems: it explains why liquid helium does not freeze solid even at temperatures close to absolute zero. Unlike other gases, helium atoms experience almost no intermolecular forces, as we shall see in chapter 6. It is because of the weakness of these atomic bonds that the quantum mechanical motion of the atoms near absolute zero is sufficient to prevent helium freezing to form a solid.

The next thing to notice is that the probability amplitude for the electron in the box vanishes not only at the ends, but also, for the higher energy states, at places in between. In the case of a vibrating string these 'nodes' – places where there is no movement of the string – cause no surprise. For the electron, however, these nodes are places where there is no chance of finding the electron! The relative probability of finding the electron at different positions within the box is given by the square of the quantum probability amplitude (see Fig. 4.8). From the figure we see that not only are the allowed energies quantized, but also that the probability of finding the electron varies with position within the box and is different for different electron energies. All this is very different from our everyday intuition about particles, but follows when we admit electrons to have wavelike properties.

There is one last moral to take from this example. In order to talk about quantum energy levels and their corresponding wavefunctions it is convenient to label them in some way. Thus, we assign a 'quantum number', $n = 1$, to the ground state, $n = 2$ to the first excited state, and so on. In this example such labelling seems almost a triviality, but the use of quantum numbers to label energies and quantum amplitudes is a general feature of quantum mechanics. Most problems of interest for applications in the real world are not, of course, as simple as this example of an electron in a box, and finding the energies and wavefunctions is often very difficult. Nonetheless, the same general principles hold true, and solving the Schrödinger equation for realistic situations is similar to the problem of finding the vibrational wave patterns for more complicated objects than strings. Some wave patterns for some familiar objects are shown in Figs. 4.9 and 4.10. There is one more lesson that we must learn from these more-complicated vibrations that does not show up in our box example. In Fig. 4.11 we show the vibrational modes of a square drum: the quantum problem of an electron in a two-dimensional box has similar solutions. If we try to label the wavefunctions according to their energy we now find a problem. For the lowest energy state there is just one possible wavefunction, which we can again label by a single quantum number $n = 1$. For the first 'excited state', however, we have a choice. If we label the two directions x and y, we see that we can either excite the x motion to its first harmonic and leave the y motion in the fundamental (see Fig. 4.11, top right), or leave the x motion in the fundamental mode and excite the y motion to the first harmonic (see Fig. 4.11, bottom left). In the case of this square drum, both these possibilities have exactly the same energy. We therefore need another quantum number to distinguish the two possible wavefunctions corresponding to this energy. In the case of the square box, we can label them both $n = 2$ and add a label x or y to specify which direction is excited.

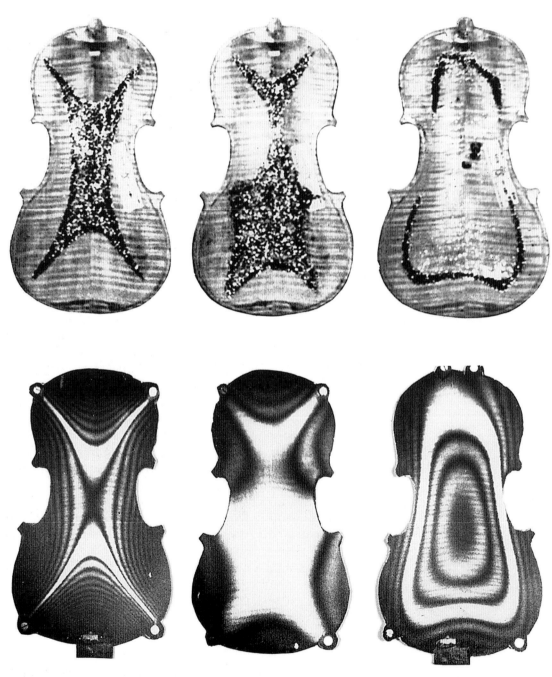

Fig. 4.9 Vibration patterns for a violin. The top three pictures are photographs of the patterns made by a sprinkling of light powder which accumulates on and around regions of little or no vibration. Below these, the same patterns are revealed by a laser interference method. The white areas correspond to regions of little vibration. The interference method is clearly much the more sensitive.

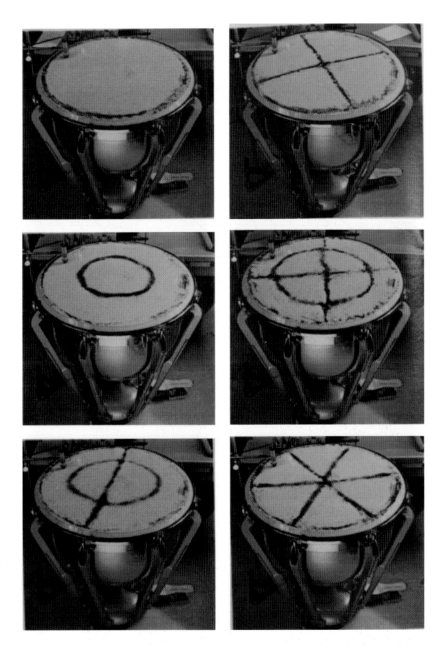

Fig. 4.10 A kettledrum sprinkled with powder reveals six of its many vibrational patterns. The powder collects near the 'nodes' where the vibration is weakest. These patterns are analogous to the quantum probability patterns for electrons in a box.

Thus, we can label the wavefunctions as $2x$ and $2y$. Physicists say that this situation – in which there is more than one possible quantum state with the same energy – is 'degenerate'. This is yet another example of physicists using an everyday word in a special technical sense! We shall find similar degeneracies when we talk about the hydrogen atom wavefunctions. In

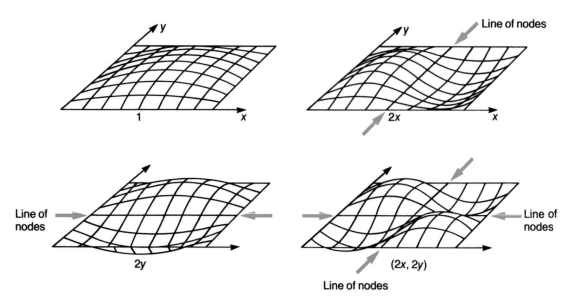

Fig. 4.11 The vibration patterns of a square drum. The surface of the drum has been indicated by a mesh of strings to make clear the correspondence with the modes of the vibrating string.

three dimensions, it should come as no surprise that we shall need at least three quantum numbers to label all the quantum states.

Our analogy between electrons confined to a box and standing waves on strings may seem rather contrived. Remarkably, recent advances in our ability to manipulate atoms now allow us to construct just such boxes for electrons on the surface of materials. With the invention of the scanning tunnelling microscope or STM, which we describe in more detail in chapter 5, physicists are able to move individual atoms on a surface. One of the pioneers of this technology is Don Eigler, a research scientist at the IBM Almaden Research Laboratories in California. Eigler and his group used an STM to construct an 'electron corral' by arranging iron atoms in a circle on a copper surface (Fig. 4.12). The circular box was about was about 7 atoms in diameter and confined the surface electrons in the same way as our square drum example. Using the STM, they were also able to see an image of the electron density within the circular corral. This shows just the standing wave patterns expected for wave patterns on a circular drum (Fig. 4.13). We shall return to recent advances in atomic manipulation and 'nanotechnology' in a later chapter.

The hydrogen atom

One of the reasons for the almost immediate and universal acceptance of Schrödinger's equation was that after a decade or more of fumbling around in the dark, physicists were once more able to calculate

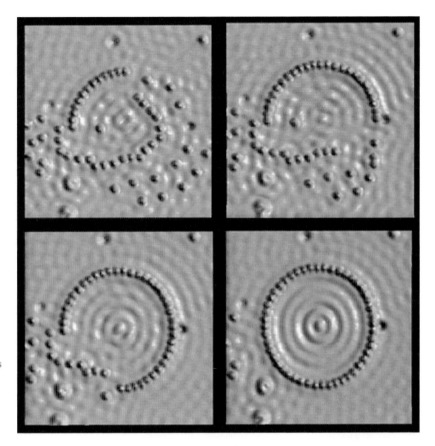

Fig. 4.12 The making of an electron corral. These images show the intermediate steps in the creation of a circular corral of iron atoms on a copper surface by the IBM team led by Don Eigler.

using standard mathematical techniques. Instead of having to follow Bohr's mysterious rules, the energy levels of hydrogen appeared naturally as the allowed frequencies of a wave problem in three dimensions. Indeed, what was astonishing was the remarkable accuracy of these predictions. When giving lectures on the new quantum mechanics, the famous Italian physicist Enrico Fermi used to say 'It has no business to fit so well!'. Nevertheless it does, and we are now in a position to understand both Bohr's quantum rules and the stability of atoms.

The hydrogen atom is a quantum system consisting of a relatively massive proton, with positive charge, accompanied by a very light, negatively charged electron. The electron is attracted to the proton by a force that becomes stronger the closer the electron is to the proton. Classically, there would be nothing to stop the electron lowering its energy as much as possible until it was sitting right on top of the proton. In quantum mechanics we know that Heisenberg's uncertainty principle prevents the electron being stationary. The energy levels are found by solving the Schrödinger equation with an electric potential corresponding to the Coulomb attraction between the proton and the electron. Although the mathematics is more complicated, the resulting energy level spectrum is similar to that of

Fig. 4.13 Electron density waves in a quantum corral. This STM image shows forty-eight iron atoms positioned to form a circle on the copper surface. Copper is a good conductor of electricity and surface electrons are confined by the ring of iron atoms. The STM image shows the standing wave pattern of the electron density inside the corral.

the electron in a box (see Fig. 4.14). With the aid of Planck's famous formula relating the energy of a photon to its frequency

$$E_{\text{photon}} = hf$$

that we talked about in chapter 2, we can now understand Balmer's magic formula. Left to itself, the electron in the hydrogen atom likes to have the lowest possible energy, and the electron therefore occupies the ground state corresponding to the energy quantum number $n = 1$. However, if the atom is disturbed – by collisions with other atoms or by light shining on it – the electron may be excited to a higher energy level with a larger n value. This state of the atom has more energy than usual and after some time the atom will decay back to the ground state. Picturesquely, we can say that the electron jumps down to a lower energy level. To conserve energy, the excess

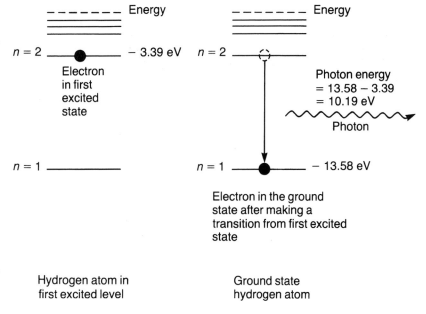

Fig. 4.14 Electron transition from the first excited state of hydrogen ($n = 2$) to the ground state ($n = 1$) with emission of a photon. The other lines in the diagram represent some of the other excited energy levels of the hydrogen atom. The excited energy levels (not shown in the diagram) become closer and closer until the ionization energy is reached (shown as a dashed line). At this energy, an electron has enough energy to overcome the electric attraction of the proton and move away from the atom, leaving a positively charged hydrogen ion or proton.

energy is given off in the form of a light photon whose energy is given by the formula

$$E_{\text{photon}} = E_{\text{initial}} - E_{\text{final}}$$

Since frequency and wavelength of light are related by the classical wave formula

$$c = f \times \lambda$$

velocity of light = frequency times wavelength

we now have a prediction for the wavelengths of spectral lines. Schrödinger's result was exactly the same as that of Balmer and Bohr, namely

$$\frac{1}{\lambda} = R \left(\frac{1}{n_f^2} - \frac{1}{n_i^2} \right)$$

where R is a calculable combination of the mass and charge of the electron, Planck's constant and the nuclear charge. The numbers n_f and n_i are the energy quantum numbers of the final and initial levels, respectively. Fig. 4.15 shows how the various spectra come about. Balmer's original formula had $n_f^2 = 4$ and explained electron jumps to the $n = 2$ energy level. The photon energies involved in these 'photon transitions' correspond to spectral lines in the visible part of the spectrum (see appendix 1). Only this visible 'series' was known at the time of Balmer's discovery. The very large energies involved in photon transitions to the $n = 1$ ground state mean that the corresponding photons have very high energy. The resulting spectral lines are in the ultraviolet part of the spectrum and are known as

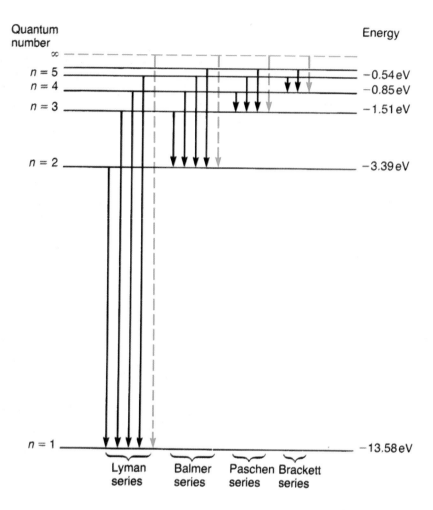

Fig. 4.15 Energy level diagram for hydrogen showing how the various 'series' of spectral lines arise. Each arrowed line represents a possible electron jump and would lead to the emission of a photon with energy equal to the energy difference between the start and end point. Only the Balmer series corresponding to jumps to the $n = 2$ level lead to photons with energies in the visible range of wavelengths. The Lyman series involves electrons making transitions to the $n = 1$ ground state accompanied by high-energy photons corresponding to spectral lines in the ultraviolet. Both the Paschen and Brackett series involve photons of much lower energies and correspond to spectral lines in the infrared. Transitions corresponding to electron capture of a free electron at the ionization energy are shown as dashed lines.

the Lyman series after their discoverer. Similarly, the Paschen and Brackett series correspond to photons with energies in the infrared part of the spectrum arising from electrons making jumps to the $n_f = 3$ and $n_f = 4$ levels, respectively.

The same picture of the energy levels also explains how light is absorbed by atoms. For light to be absorbed, not only does the light photon have to have an energy exactly equal to the difference in energy between two energy levels, but also an electron must be in the right energy level to absorb a photon with this energy. At normal temperatures the energy of collisions between atoms in a gas is usually insufficient to excite many atoms, since there is a large energy difference between the ground state and the first excited state. Hence at room temperatures, most atoms are in the ground state. As we have seen, the energy differences from the ground state to any excited state are so large that the associated photons are in the ultraviolet range of frequencies rather than the visible. Thus, visible light can pass through many gases unabsorbed, since almost all

Fig. 4.16 The Orion nebula, a large glowing cloud of hydrogen, inside which are many newly formed stars. The red colour of the cloud shows the presence of the electron jump from the $n = 3$ to the $n = 2$ level of hydrogen.

the atoms require much larger photon energies for excitation from the ground state. This is the reason why most gases are transparent to visible light.

We can also explain the beautiful colours of nebulae observed in astronomical photographs. Figure 4.16 shows a picture of a large gas cloud in our Galaxy called the Orion nebula. Inside the nebula, hot stars are constantly giving out large numbers of ultraviolet photons. These photons are so energetic that they can knock the electron right out of the hydrogen atom leaving a positively charged 'ion'. As the electrons and protons recombine, the electron loses energy by cascading down through the energy levels giving off photons on the way. The red colour of the cloud corresponds to the electron jump from the $n = 3$ level to the $n = 2$ level of the Balmer series shown in Fig. 4.15.

Wavefunctions and quantum numbers

We have so far only discussed the quantum numbers that arise in very simple situations. For a detailed understanding of the quantum mechanics of the elements that we shall discuss in chapter 6, we need to know more about the wavefunctions and quantum numbers that are needed for the hydrogen atom. Since this involves a rather detailed discussion it may be best to skim this section quickly and not get bogged down! With this warning let us now look at the electron probability amplitudes – hydrogen wavefunctions – that correspond to the energy levels of the hydrogen atom. The wavefunction of the ground state, with energy quantum number $n = 1$, turns out to be very smooth and symmetrical, and looks the same from all directions. The probability of finding an electron is proportional to the square of this wavefunction, and Fig. 4.17 gives an idea of how this probability varies over a slice through the centre of the atom. Moreover, this probability density corresponds to a stationary wave pattern and does not change with time. This solves the problem of why the electron does not radiate in a Bohr orbit. The electron is not described as a particle whirling round the proton but as a stationary pattern of probability in which nothing is being accelerated.

If we look at the excited energy levels in more detail, we see, as in the square drum example, that there is degeneracy – several wavefunctions correspond to the same energy. We therefore expect that we will need several new quantum numbers to label each of these wavefunctions. These extra quantum numbers correspond to the quantization of another classical quantity – 'angular momentum'. In the next few paragraphs we will try to describe these new quantum numbers in a bit more detail: if this proves too heavy going, skip to the next section! Angular momentum is, as its name suggests, related to ordinary momentum, and it is important in problems involving particles in orbit around some centre, such as the motion of the planets round the Sun. Imagine tying a ball to a piece of string and whirling it round and round holding the other end of the string. The angular momentum of the ball is just the ordinary momentum times the length of the string:

$$L = r \times p$$

angular momentum = length of string times ordinary momentum

For a fixed length of string, the faster the ball goes, the more angular momentum the ball has. Angular momentum is important because, like energy, it is conserved in both classical and quantum systems. In the case of our ball on a string, imagine shortening the string as the ball whirls around. Angular momentum is conserved, and therefore as the length, r, of the string decreases, the ordinary momentum, p, of the ball must increase and the ball goes round faster. In the quantum mechanical treatment of the hydrogen atom, angular momentum is conserved just as in this classical example. However, in the quantum case we are not permitted to have any

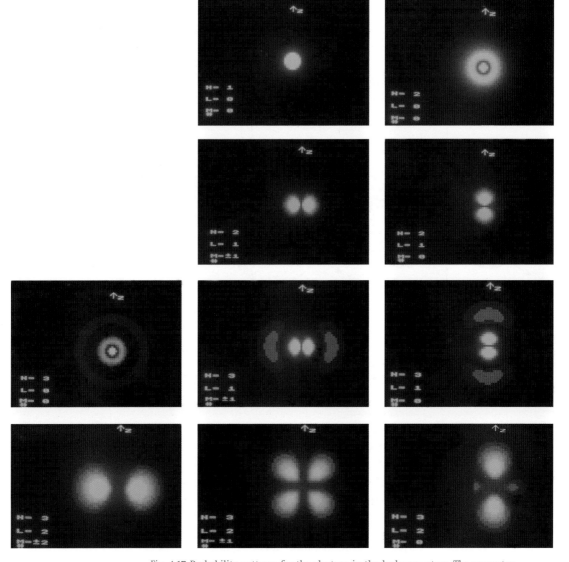

Fig. 4.17 Probability patterns for the electron in the hydrogen atom. The computer-generated pictures show slices through the atom for the lowest energy levels. The pictures are coded so that the bright regions correspond to high probability of finding the electron. They are labelled by the quantum numbers N, L and M and the probability distributions for all states with N = 1, N = 2 and N = 3 are shown. Note that the M = ±1 and the M = ±2 patterns are the same.

value we want for the angular momentum. Quantum mechanical angular momentum is quantized, like the energy.

This was how Bohr guessed his stable orbits: the angular momentum was only allowed to take values which were whole units of Planck's constant h divided by 2π. Curiously, although Bohr had the correct energy

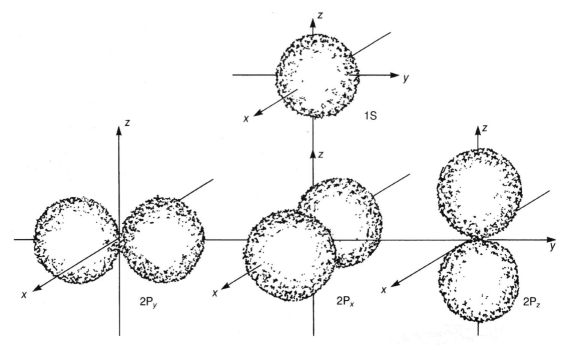

Fig. 4.18 Three-dimensional visualization of equal probability surfaces for the lowest-lying hydrogen energy levels. The 1S state is spherically symmetric and has zero angular momentum ($L = 0$). There are three 2P states all with the same energy and one unit of quantum angular momentum ($L = 1$). The three 'P wave' x, y and z combinations have probability surfaces as shown. These three dumbbell shapes have 'lobes' that are oriented in the x, y or z directions, corresponding to the P_x, P_y and P_z states.

levels, Schrödinger's solution of the hydrogen atom showed that Bohr's guess about angular momentum was not quite right. The ground state of hydrogen actually has zero angular momentum, in contrast to Bohr's prediction! Nevertheless, the angular momentum is indeed quantized and is described by two new quantum numbers, L and M. Thus, at the $n = 2$ energy level we find there are four degenerate wavefunctions which give rise to the probability 'slices' shown in Fig. 4.17. Note that the probability patterns for the $n = 2$, $L = 1$, $M = +1$ and $M = -1$ wavefunctions are the same and are shown in the same picture. The distribution labelled '$n = 2$, $L = 0$, $M = 0$' has angular momentum zero. This slice also has the same circular symmetry as the ground state wavefunction with quantum numbers '$n = 1$, $L = 0$, $M = 0$', and also has zero angular momentum. The other three wavefunctions all have one unit of angular momentum and angular momentum quantum number '$L = 1$'. For $L = 1$ there are three possible values of the second angular momentum quantum number M, namely $M = +1$, 0 and -1. These three possibilities correspond roughly to the three possible directions for the axes of rotation of the electron which we usually refer to as x, y and z. Imagine the ball on a string again and picture it rotating about the vertical 'z axis' marked on the photographs. The

Hans Dehmelt was awarded the Nobel Prize for Physics in 1989 for trapping a single electron and isolating a single atom. He also trapped a single positron – the antiparticle of an electron – and kept it in the trap performing jumps to order for over three months. To counter the commonly held impression that everything in the quantum world is just some vague wave of probability, Dehmelt liked to emphasize the reality and individuality of these trapped elementary particles by giving them names – Astrid the atom and Priscilla the positron! Since winning the Nobel Prize he says 'My life is now a bed of roses, an absolute bed of roses'.

ball will be swinging round in a plane containing the x and y axes. Quantum mechanically this situation corresponds to the $M = +1$ state for which the probability 'lobes' are mainly in the $x-y$ plane. The $M = -1$ state corresponds to the axis of rotation being along the negative z axis and again the probability lobes will be in the $x-y$ plane. The $M = 0$ state with the probability lobes pointing along the z direction, corresponds roughly to having an axis of rotation somewhere in the $x-y$ plane. In order to understand the chemistry of molecules, it turns out to be more convenient to use a slightly different choice of labelling for these three $L = 1$ states. If we take combinations of these M quantum number wavefunctions, we can arrange to have three degenerate wavefunctions whose probability lobes actually point along the x, y and z directions. Instead of using the quantum number M, we can instead refer to the three $L = 1$ states by the labels x, y or z. Figure 4.18 shows an attempt at visualizing the probability distributions for these three $L = 1$ states.

At the next energy level, states with two units of angular momentum, $L = 2$, are present, along with $L = 1$ and $L = 0$ states. There are five possible M values for the $L = 2$ wavefunctions: $M = +2, +1, 0, -1$ and -2, corresponding roughly to the axis of rotation going from the positive z direction round to the negative z direction. It is remarkable that these three quantum numbers provide almost all we need for an understanding of Mendeleev's periodic table of the elements (discussed in chapter 6). However, there is one respect in which hydrogen with its single electron is rather a special case. For atoms with more than one electron, the energies of states with the same value of n but different values of L are not the same: the energies are not degenerate and the energy of the state depends on both n and L.

A last word on notation. The alkali metals lithium, sodium and potassium have line spectra that are similar to those of hydrogen. But unlike hydrogen, the energy levels with the same value of n but different value of L are not the same. The alkali elements have multiple hydrogen-like series corresponding to the different angular momentum values. Because the physicists who first studied these spectra had no idea as to their origin, they labelled these series in a fairly arbitrary way: S for sharp, P for principal, D for diffuse and F for fundamental. We now know that these result from transitions to final states with different values of angular momentum: sharp to $L = 0$, principal to $L = 1$, diffuse to $L = 2$ and fundamental to $L = 3$. Despite understanding the origin of these lines, physicists and chemists still persist in referring to angular momentum states not by their actual angular momentum values, $L = 0$, $L = 1$, $L = 2$ and $L = 3$, but by this incredibly obscure historical labelling scheme, S, P, D and F!

Atom traps and light

In this chapter we have discussed the way in which quantum mechanics describes individual atoms, but so far we have only described

Wolfgang Paul (1913–1993) was born in Saxony in Germany and studied Latin for nine years and Ancient Greek for six at the gymnasium in Munich. He then decided to become a physicist and asked the great German theoretical physicist Arnold Sommerfeld for advice. Sommerfeld suggested he begin an apprenticeship in precision mechanics. In 1989 he was awarded the Nobel Prize for Physics with Hans Dehmelt for his role in the development of the ion trap technique.

observations of the radiation from collections of many atoms. The 1989 Nobel Prize for Physics was awarded to two German-born scientists, Hans Dehmelt and Wolfgang Paul, for developing techniques for observing individual atoms.

In the mid 1950s Hans Dehmelt had the idea of capturing a single electron. His professor, Richard Becker, at the University of Gottingen in the 1940s, had first sowed the seed of this idea when he represented an electron as a little white dot on a blackboard. Since Dehmelt had learnt in his quantum mechanics lectures that no quantum particle could be at rest, this 'mild discrepancy' stayed with him for the next 50 years! It was not until 1973, more than twenty years later, that Dehmelt succeeded in isolating a single quantum particle, using a device called a 'Penning trap' developed by the Dutch physicist Frans Penning in 1936. A Penning trap confines electrons between two negatively charged metal plates enclosed in a vacuum. A magnetic field surrounds the plates and is designed to prevent electrons escaping from the sides of the trap. Electrons can be put into the trap using a negatively charged metal spike and the trapped particles monitored as they oscillate back and forth within the trap. Dehmelt and his team at the University of Washington permitted individual electrons to escape until only a single individual electron remained. At first they were only able to keep the electron in the trap for a few days, then a few weeks and then a few months. Finally they were able to monitor a 'monoelectron oscillator' for almost a year before it escaped from the trap! This allowed them to make measurements of the magnetic properties of an electron with unprecedented accuracy. In 1987, an antimatter version of this experiment was performed using positrons, the antiparticles of electrons (which we describe in chapter 11). In order to emphasize the reality of this trapped individual positron Dehmelt gave her a name – Priscilla Positron! He wrote:

> Here, right now, in a little cylindrical domain, about 30 μm in diameter and 60 μm long, in the center of our Penning trap resides positron (or anti-electron) Priscilla, who has been giving spontaneous and command performances of her quantum jump ballets for the last three months.

When he won the Nobel Prize in 1989, Hans Dehmelt said that he 'felt like dancing'!

Dehmelt was also interested in trapping atoms. In order to trap an atom he used a device invented by Wolfgang Paul of the University of Bonn. When an electron is removed from an atom, the positively charged remainder is called an ion. Paul's ion trap was rather similar to a Penning trap except that it used an oscillating electric field rather than a magnetic field to prevent the particles touching the 'walls'. Dehmelt took his first photograph of a single atom in 1979 while he was at the University of Heidelberg. This was done by illuminating the trapped ion with a laser beam of the correct wavelength for a photon to be absorbed by the ion. The excited ion will then re-emit a photon in any direction. The single ion can be excited

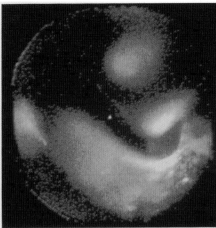

Fig. 4.19 Using a combination of electric and magnetic fields, charged ions can be confined in a 'trap'. By illuminating the ion with laser light of the right frequency an individual ion can be visualized. On the left, an ion trap is shown next to a cent coin to indicate the scale of this device and on the right the bright dot in the centre is a single trapped mercury ion.

and de-excited by the laser beam so that the ion emits hundreds of millions of photons a second thus enabling the ion to be photographed (see Fig. 4.19). In 1980, using a complex system of lasers (see chapter 7) Dehmelt trapped a single atom for several days. Since he wished to emphasize the reality of an individual atom he named the atom Astrid! As he said:

> The well-defined identity of this elementary particle is something fundamentally new, which deserves to be recognized by being given a name, just as pets are given the names of persons.

5 Quantum tunnelling

It is possible in quantum mechanics to sneak quickly across a region which is illegal energetically.

Richard Feynman

Barrier penetration

One of the most startling consequences of de Broglie's wave hypothesis and Schrödinger's equation was the discovery that quantum objects can 'tunnel' through potential energy barriers that classical particles are forbidden to penetrate. To gain some idea of what we mean by an energy barrier, let us go back to our roller coaster and look at a larger section of track, as shown in Fig. 5.1. If we start the carriage from rest, high up on the left, at A, and ignore any small frictional energy losses, we know from the conservation of energy that we shall arrive on the other side at the same height we started from, at C. As we went over the little hill B, at the bottom of the valley, the car slowed down as some of our kinetic energy was changed to potential energy in climbing the hill, but because we started much higher up, we had plenty of energy to spare to get us over the top. However, if we started the carriage from rest at A, we do not have enough energy to climb over the hill D and get to E. This is an example of an 'energy barrier', and we can say that the region from C to E is 'classically forbidden'.

What is remarkable about quantum 'particles' is that they do not behave like these classical objects. An electron travelling on an 'electron roller coaster' of the same form as the roller coaster of Fig. 5.1 can 'tunnel through' the forbidden region and appear on the other side! This 'barrier penetration' or 'quantum tunnelling' is now a commonplace quantum phenomenon. It forms the basis for a number of modern electronic devices such as the tunnel diode and the Josephson junction, of which more later. How can we obtain some understanding of how such tunnelling comes about? One way of thinking about it uses an argument based on Heisenberg's uncertainty principle. In chapter 2 this was discussed in terms of the uncertainties in the measurements of position and momentum. However,

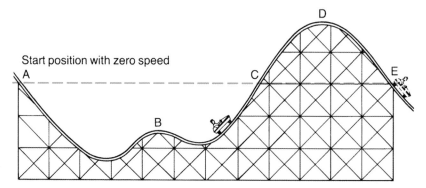

Fig. 5.1 An illustration of what quantum tunnelling would mean for a real roller coaster. If the carriage starts from rest at position A, conservation of energy does not allow it to go higher than position C on the other side of the valley. In quantum theory, however, there is a chance that the carriage can 'tunnel through' the forbidden region between C and E, and emerge on the other side of the hill. For a real roller coaster this tunnelling is extremely unlikely!

Fig. 5.2 In Mr Tompkins' wonderland, where Planck's constant is much larger, his car could tunnel through the wall, 'just like a good old ghost of the middle ages'.

an equivalent relation exists between uncertainties in measurements of time and energy

$$(\Delta E)(\Delta t) \approx h$$

Thus, although classically we can never change the total amount of energy without violating the conservation of energy, in quantum mechanics, if the time uncertainty is Δt, we cannot know the energy more accurately than an uncertainty $\Delta E = h/\Delta t$. Roughly speaking, we can 'borrow' an energy ΔE to get over the barrier so long as we repay it within a time $\Delta t = h/\Delta E$. If the barrier is too high or too wide, tunnelling becomes extremely unlikely and all the electrons will be reflected, just like the roller coaster car. Needless to say, this type of qualitative argument must be checked quantitatively by detailed calculation with the Schrödinger equation, but such arguments do give us insight into the reasons for quantum tunnelling. Another approach is to look at the behaviour of more familiar waves. The phenomenon of tunnelling is then seen to be a general property of wave motion – it only becomes surprising when taken in conjunction with de Broglie's hypothesis that all quantum 'particles' have wavelike properties.

Wave tunnelling

Although both waves on a string and water waves can be made to exhibit 'wave tunnelling', probably the most familiar example involves light in its wavelike guise. Consider what happens when light travels from air into a block of glass. As shown in Fig. 5.3, because light travels more slowly in glass than in air, the wave front slews round and the light changes direction. Such bending of light at a surface is well known and is called 'refraction'. Now consider light travelling from glass to air. Instead of being bent towards the vertical, the light is bent away from it. If we increase the angle at which we shine the light on the glass–air surface, there will be an angle – the 'critical' angle – at which the light emerges in the air just grazing the surface. What happens if we increase the angle still further? What must happen is that all the light is now reflected from the glass–air surface and no light escapes into the air. This phenomenon is called 'total internal reflection'. Such internal reflection of light is the reason why light can be efficiently transmitted down a glass fibre without undue losses and forms the basis of modern fibre optics.

How is all this connected with quantum tunnelling? Well, although no light rays penetrate the air beyond the glass, when the light arrives at the glass surface at an angle larger than the critical angle, there is nonetheless some sort of wave disturbance generated in the air. This is not a wave that carries energy, like ordinary 'travelling' waves, but a 'standing' wave pattern that does not transmit any light energy. Wave patterns on a string that is fixed at both ends are examples of standing waves. The type of standing wave involved here – a so-called 'evanescent' wave – is special

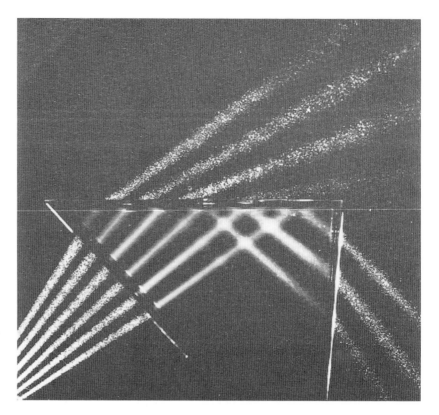

Fig. 5.3 Several rays of light striking a prism at a variety of angles. As can be seen, beyond a certain 'critical' angle the light rays are entirely reflected and no light is transmitted through the prism. The ray on the extreme right is entirely reflected while the other rays show both transmitted and reflected beams.

because the wave disturbance dies away very rapidly the further away we go from the surface. The connection with tunnelling comes about if we place another block of glass parallel to the first one. As we move the two blocks towards each other so that the evanescent wave disturbance begins to penetrate the second block, a transmitted ray of light appears! The closer the two blocks are brought together, the more light energy that reappears as a transmitted ray. This increase in transmitted light when the two blocks are closer together occurs because the amplitude of the standing wave in the 'forbidden' air gap has not decayed away so much. Physicists call this phenomenon 'frustrated total internal reflection' but it is just the optical analogue of quantum tunnelling of de Broglie waves. This effect is used in modern optics as the basis of a 'beam splitter'. The amount of light transmitted or reflected by such a device can be controlled by adjusting the width of the forbidden gap. It is also possible to demonstrate such wave tunnelling with other types of waves. Figure 5.5 shows a ripple tank photograph showing such barrier penetration with water waves.

Applications of quantum tunnelling

There are many devices now in common use that rely on the ability of quantum particles to tunnel through barriers. The example we shall

Fig. 5.4 An optical fibre is wound round a drum carrying light from a helium–neon laser. Light is confined within the fibre by total internal reflection from the sides. The fibre is about 100 m long and has been deliberately badly made so that some light leaks out of the sides so that we see the fibre as red. In a high quality fibre nearly all the light would emerge from the end. In this case, the light coming from the end of the fibre has been directed onto a screen. Pioneering work on optical fibres for telecommunications was performed at Southampton University by Alec Gambling and David Payne.

describe here involves electrons, but there are other examples that we shall come across later in which alpha particles and pairs of electrons are participating in quantum tunnelling. In a metal, the electrons that carry the electric currents are able to move about relatively freely. A simple quantum model of a metal imagines the electrons moving about in an attractive potential well caused by the lattice of positive metal ions. Since it takes energy to knock electrons out of the metal there must be an electrical 'hill' or barrier that keeps them in (see Fig. 5.6a). If we apply a strong positive electric field to the metal, the electric potential will be modified to look like Fig. 5.6b. There is a still a barrier preventing the electrons freely leaving the metal but it is now possible for the electrons to tunnel through the barrier and escape. This quantum mechanical tunnelling process is the basic process at work in an 'electron field emission microscope'. In the last decade, however, these devices have been eclipsed by the amazing power of the revolutionary 'scanning tunnelling microscope' or STM developed by Gerd Binnig and Heinrich Rohrer.

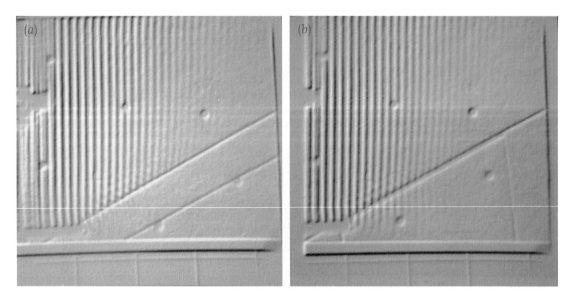

Fig. 5.5 Tunnelling with water waves. (a) The speed of water waves depends on the depth of water. This photograph shows water waves as a row of straight line crests moving left to right across the image. The two sloping lines show the presence of a glass block under the water that leads to a change in the water depth. We see no waves transmitted through this shallow region and the waves undergo 'total internal reflection' at the boundary of the change in water depth. Notice that there is a faint disturbance in the forbidden region beyond the barrier, but this clearly does not correspond to an ordinary water wave. (b) This photograph shows the same scene but with the width of the forbidden region much decreased. One can now clearly see that the water wave can 'jump the gap' and appear on the other side. This is a well-understood wave phenomenon and is the basis for tunnelling in quantum mechanics.

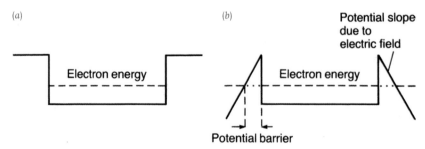

Fig. 5.6 (a) A simplified picture of the potential well for electrons in a metal. The broken line represents the energy of typical 'conduction' electrons – the electrons responsible for electric currents. This energy is below the height of the barrier and is thus insufficient to escape from the well. (b) This diagram shows how the electrical potential is modified in the presence of a large electric field. There is still a barrier but it is now thin enough that conduction electrons can escape from the metal by tunnelling.

Gerd Binnig (left) and Heinrich Rohrer were awarded the Nobel Prize for Physics in 1986 for their discovery of the Scanning Tunneling Microscope or STM. Rohrer was born in Switzerland and studied for his doctorate in physics at ETH in Zurich, with interruptions for his basic training in the Swiss mountain infantry. Because his experimental apparatus was very sensitive to vibrations he learned to work at midnight when the town was asleep. Rohrer joined the IBM Research Laboratory in nearby Ruschlikon in 1963. Binnig was born in Germany and originally thought physics very technical and not at all philosophical and imaginative. After accepting a place at the Ruschlikon IBM Research Laboratory in 1978, his collaboration with Rohrer fully restored his curiosity in physics. The Nobel Prize also helped!

In 1978, Binnig had just been employed by Rohrer as a new researcher at the IBM Research Laboratory in Zurich. After discussions with Rohrer, Binnig came up with the idea of using the 'vacuum tunnelling' of electrons to study the surfaces of materials. The basic idea is very simple. According to quantum mechanics, electrons in a solid have a small but non-vanishing chance of being found just outside the metal surface. As for the evanescent light wave in the previous section, the probability for this to happen is predicted to fall off very rapidly with distance away from the surface. According to quantum mechanics, if we can bring a very sharp, needle-like probe very close to the surface, and apply an electrical voltage between it and the metal, a tunnelling current will flow across the gap even in a vacuum. Since the electron wave function falls off so rapidly, the magnitude of this tunnelling current will be extremely sensitive to the distance of the probe from the metal. If the distance of the tip of the probe from the surface can be controlled very accurately, we can use the magnitude of this current to measure the size of features on the surface. Binnig and Rohrer soon realized that if they could develop an instrument that could scan across the surface of metal accurately and systematically, they would be able to use this effect to reconstruct a contour map of the entire surface. While this was theoretically feasible, there were many experimental obstacles to be overcome before this idea could be transformed into a useful new tool for studying surfaces. First, Binnig and Rohrer had to learn how to make probes whose tip was only a few atoms wide. Then they had to build an apparatus that could reliably position and control the tip to an accuracy of no more than a few atomic diameters from the surface. Binnig described the moment that they first saw the tunnelling effect

Fig. 5.7 A scanning tunnelling microscope consists of a very sharp needle point that surveys the surface of a specimen very accurately. When a high voltage is applied between the surface and the needle, electrons can tunnel from the tip of the probe to the specimen under investigation. This tunnelling current is very sensitive to the height of the probe above the surface. In the microscope, the height of the needle can be adjusted as the needle moves over the surface so that the current remains constant. In this way, the up and down movements of the needle map out the detailed contours of the surface. This image shows a simple representation of the atoms in the needle and surface together with a few stray atoms.

Fig. 5.8 An STM image of the surface of silicon. The rows of atoms are less than 2 nanometres apart and the individual atoms within a row are separated by less than 1 nanometre.

they were looking for:

> Measuring at night and hardly daring to breathe from excitement, but mainly to avoid vibrations, we obtained our first clear-cut exponential dependence of the tunnel current I on the tip-sample separation s characteristic for tunnelling. It was the portentous night of 16 March, 1981.

What was astonishing about the STM was its incredible sensitivity. Binnig and Rohrer reported that 'a change in the distance by an amount equal to the diameter of a single atom causes the tunnelling current to change by a factor of as much as 1,000'. With their new instrument, they reported that 'our microscope enables one to "see" surfaces atom by atom. It can even resolve features that are only about a hundredth the size of the atom.' Perhaps because the device involved such an exotic technology as quantum tunnelling, their invention was not immediately recognized as the revolutionary new tool it has since turned out to be. It was not until 1982 when Binnig and Rohrer solved a long-standing problem concerning the arrangement of atoms on the surface of silicon that the scientific community became convinced of the power of the STM (see Figs. 5.7 and 5.8). The scanning tunnelling microscope is now opening up whole new areas of research at the atomic level with breath-taking new images of the atomic world (Figs. 5.9–5.12). Of his work with the STM Binnig said:

> I could not stop looking at the images. It was like entering a new world. This appeared to me as the unsurpassable highlight of my scientific career and, therefore, in a way, its end.

In 1986 Binnig and Rohrer were awarded the Nobel Prize for physics.

Fig. 5.9 STM image of a Copper 'Landscape'. The resolution is sufficient to show individual atomic steps.

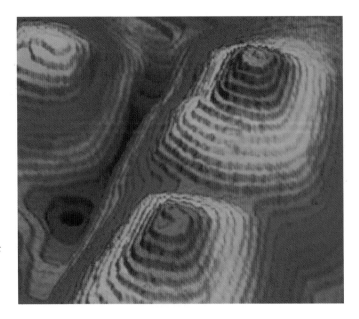

Fig. 5.10 This STM image reveals how spiral gallium antimonide structures have been grown on a gallium arsenide substrate. The ability to grow precise structures at such an atomic level will enable the development of a new generation of electronic and photonic devices.

As Binnig and Rohrer experimented with their STM they noticed that the needle tip occasionally picked up individual atoms. By moving the tip around, they found that they were able to move the atoms about on the surface. This ability to move atoms with the STM has been developed into an exciting new technology by a group at the IBM Research Laboratory in Almaden, California. Don Eigler and Erhard Schweizer first made the headlines when they used an STM to produce an atomic version of the IBM logo (Fig. 5.13). They had started with a clean nickel surface in a vacuum and cooled the system to within 4 degrees of absolute zero using liquid helium to minimize any thermal disturbances. Eigler and Schweizer then introduced a small amount of xenon into the apparatus and located xenon atoms adhering to the nickel surface with the STM. They then dragged 35 atoms into the correct alignment to spell IBM – nine atoms for the 'I', thirteen each for the 'B' and the 'M'. It took them about an hour to construct each letter and they found that apparatus was so stable at these low temperatures that they 'could perform experiments on a single atom for days at time'. Eigler and his colleagues also produced the spectacular quantum corral we saw in the last chapter. We will see other examples of their work when we discuss quantum engineering chapter 9: further examples can be found on the IBM Research website.

There is one more development arising from Binnig and Rohrer's work on the STM that should be mentioned. The image of an STM depends on the electrical properties of the surface; these can be quite complex and the resulting images difficult to interpret. In 1985, Binnig was visiting IBM colleagues in California and together they developed a new

Fig. 5.11 A modern scanning tunnelling microscope (STM).

type of scanning probe microscope – the atomic force microscope or AFM. Instead of using an electrical tunnelling current, the AFM uses a sharp diamond tip mounted on a tiny cantilever arm (see Figs. 5.14 and 5.15). As the diamond tip moves over the surface, weak atomic forces bend the cantilever and this movement can be detected. This bending can be measured by a variety of methods. Binnig himself used an STM to measure the minute motions of the cantilever arm. The AFM has now become a standard tool for investigating surfaces, and is complementary to the STM.

$t = 0$ $t = 46$ s $t = 94$ s

$t = 141$ s $t = 187$ s $t = 234$ s

$t = 281$ s $t = 797$ s $t = 1036$ s

Fig. 5.12 The STM can also be used to track the motion of atoms. In this time sequence (t stands for time and s for seconds) we see successive images of the surface of a germanium crystal. At $t = 0$ we see that there is a site with a missing germanium atom. At room temperature the germanium atoms have enough thermal energy that another atom can hop into the vacant site leaving a new vacancy in its previous position. This remarkable sequence of STM images shows how the position of the vacancy appears to move about the surface.

Nuclear physics and alpha decay

One of the great puzzles in the early days of nuclear physics concerned alpha decay. The puzzle was this. In the radioactive decay of uranium, physicists had measured the energy of the alpha particle that was thrown out of the nucleus and found it to be about 4 MeV. A quick word about units of energy is needed here. An electron-volt, or 'eV', is the amount of energy an electron gains in 'sliding down' a potential hill one volt high. This amount of energy is typical of the electronic energy levels in atoms. For processes involving the nucleus, on the other hand, the energies are much larger, and a more convenient size unit is a million electron-volts, or 'MeV' for short. Now back to the story. Besides discovering the nucleus as

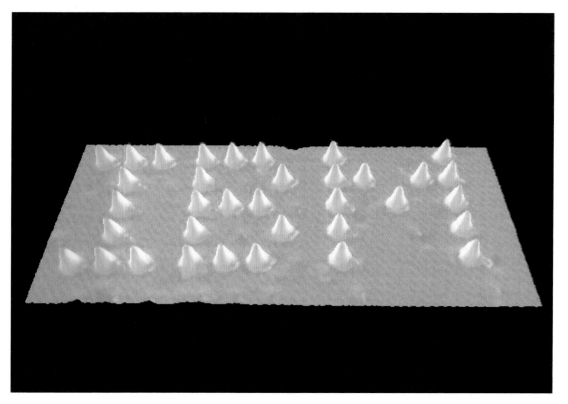

Fig. 5.13 This is the famous atomic version of the IBM Logo. The image was produced by Don Eigler and Erhard Schweizer at the IBM Almaden Research Laboratory in California. They began with a clean nickel surface in a high vacuum and introduced a small quantity of xenon. The system was cooled down to 4 K above absolute zero to minimize any thermal motion and they then set about locating the individual xenon atoms. When Eigler and Schweizer found a xenon atom they dragged it into the correct position for the logo using the tip of the STM. It took them about an hour to form each letter.

Cal Quate, Gerd Binnig and Christoph Gerber enjoy a glass of wine in Ruschlikon in May 2000. In 1986, the three of them wrote a paper on a new type of microscope – the Atomic Force Microscope or AFM. The AFM scans the surface with a probe in close proximity to the surface being studied in the same way as does an STM. In this case, however, the probe is not measuring a tunnelling current but the tiny repulsive or attractive forces between the tip of the probe and the surface. The probe is positioned at the end of a cantilever arm so that small deflections of the cantilever can be measured.

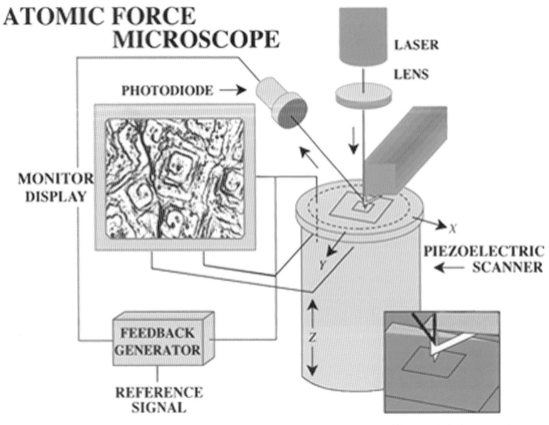

Fig. 5.14 The atomic force microscope, AFM, was invented in 1986. A tip is mounted onto a miniature cantilever arm which is then scanned across a surface in similar fashion to the STM. As the sample is scanned, small forces between the tip and the surface cause the cantilever to deflect. Measuring this deflection then reveals a three-dimensional image of the surface.

we described in chapter 4, Rutherford had done many other experiments shooting alpha particles at atoms. One of the things he had found was that alpha particles with about 9 MeV of energy were strongly repelled by the positive charge of the nucleus. In other words, it seems that for an alpha particle to get inside the nucleus requires much more energy than the 4 MeV of the alpha particles emitted in radioactive decay. To make this problem more graphic, let us look at the analogous situation for our roller coaster. It is as if you were sitting on the track, half-way down the hill, and were suddenly bumped into by a carriage. The only place from which the carriage could have come was from over the top of the hill. But if the carriage had indeed come rocketing down from the top of the hill it would have caused you a nasty injury. Instead, it hit you with just a light bump!

In view of what we have been saying about tunnelling, the answer to the alpha particle paradox is now fairly obvious. But in 1928, when an

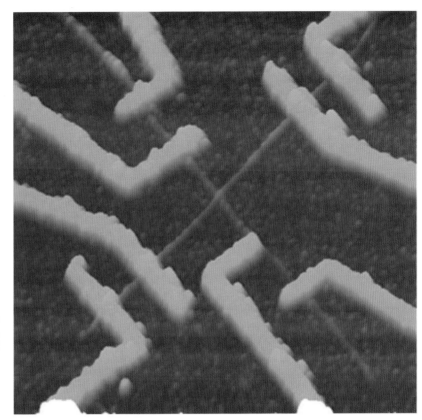

Fig. 5.15 This AFM image shows crossed carbon nanotubes (centre of picture) arranged on a silicon surface. The tubes have a diameter of about one nanometer. Nanotechnologists are experimenting with these systems to develop very small electronic devices.

explanation of alpha decay in terms of quantum tunnelling was suggested by the Russian physicist George Gamow, and by two US physicists, Edward Condon and Ronald Gurney, it was a startlingly new idea and one of the first applications of quantum mechanics to the nucleus. In the nucleus of the common isotope of uranium, ^{238}U, there are 92 protons and 146 neutrons and these are all jostling about in a very small volume. The strong nuclear forces between the 'nucleons' – protons and neutrons – can be thought of as providing an attractive 'potential well' that keeps the nucleons inside the nucleus, rather like our simple model for conduction electrons in a metal (Fig. 5.6a). Inside the nucleus two protons and two neutrons can sometimes get together to form an alpha particle. The resulting potential 'seen' by such an alpha particle is shown in Fig. 5.16. This nuclear potential now looks very similar to that of the electron in a metal in the presence of a large electric field (Fig. 5.6b) Although the height of the barrier is around 30 MeV, the alpha particle can tunnel its way out of the nucleus and appear as a free particle with only 4 MeV of energy. Although we now know far more about nuclear forces and can perform calculations using much more realistic nuclear potentials, the basic explanation in terms of tunnelling remains valid.

George Gamow (1904–1968) was the creator of the intrepid explorer Mr Tompkins. Gamow was the grandson of a Tsarist general and obtained his Ph.D. from Leningrad University. After working at most of the important scientific centres in Europe, he eventually settled in the USA. Besides being one of the best-known popularizers of science, Gamow made important contributions to nuclear physics, cosmology and molecular biology.

There is an interesting 'converse' to this problem that resulted in Cockcroft and Walton winning the Nobel Prize. In 1919, Rutherford, again in his experiments with alpha particles, had seen the first artificially produced nuclear reaction. On firing alpha particles at nitrogen he had observed that protons were occasionally produced. From this fact, Rutherford deduced that he had seen a 'nuclear disintegration'. We would now write this reaction as

$^4_2$He	+	$^{14}_7$N	→	$^{17}_8$O	+	$^1_1$H
helium		nitrogen		oxygen		hydrogen
2 protons		7 protons		8 protons		1 proton
2 neutrons		7 neutrons		9 neutrons		0 neutrons

Figure 5.17 shows an early photograph of this reaction. Rutherford concluded his scientific paper on this observation with a remark that

> If alpha particles – or similar projectiles – of still greater energy were available for experiment, we might expect to break down the nucleus structure of many of the lighter atoms.

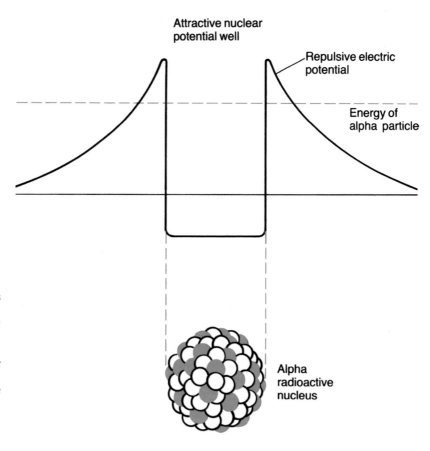

Fig. 5.16 Gamow's original application of quantum tunnelling was to the process of alpha decay. Because alpha particles are unusually stable, we can think of them as existing in a nuclear potential due to all the other particles in the nucleus. It is possible for an alpha particle to tunnel through the barrier and out of the nucleus, causing it to decay.

Cockroft (right) and Walton (left), on either side of Rutherford.

Fig. 5.17 Tracks left by alpha particles in a cloud chamber. The tracks consist of tiny droplets of water that have condensed along the path of an alpha particle. Since the chance of an alpha particle encountering the nucleus of one of the gas atoms is very small, the tracks are almost straight. At the top of the picture, a track appears to cross the other paths and seems to emerge from a small kink in the track of one of the alpha particles. This is, in fact, a nuclear reaction: a proton and an oxygen nucleus are produced in the collision of the helium nucleus with the nitrogen nucleus. The 'sideways' track is the track left by the proton, and the short forward-going track from the kink is left by the oxygen nucleus created in the nuclear reaction.

This possibility was realized with the invention of new types of particle accelerators. In 1932, the same year that the neutron was discovered by Chadwick, a US physicist called Ernest Lawrence had built a machine called a 'cyclotron', which could accelerate particles to energies of several million electron-volts. At the time, it was widely believed that in order for charged particles to penetrate to the core of the nucleus and generate a nuclear reaction, the projectiles would have to have energies of many million electron-volts to surmount the potential barrier. As a result, it was Cockcroft and

Livingstone and Lawrence next to their cyclotron. This was used in 1937 to produce the first artificial element, technetium. Technetium has 43 protons but does not occur naturally because all its isotopes are radioactive with short lifetimes.

Walton, working in Cambridge, UK, with a much more primitive accelerator, who were the first to 'split the atom' using artificially accelerated protons with an energy of less than 1 MeV. Such low-energy protons are occasionally able to tunnel into the nucleus and initiate a reaction. It is said that Cockcroft, in a rare flamboyant gesture, was seen wandering through the streets of Cambridge announcing 'We've split the atom!' to all and sundry. Of course, the name 'atom smashers' belongs to the popular press. What Cockcroft and Walton had seen was the first artificially induced nuclear reaction – a modern version of the old dreams of alchemists about transmutation of the elements. They had converted lithium into helium via the reaction

$$\mathrm{^{1}_{1}H} \quad + \quad \mathrm{^{7}_{3}Li} \quad \rightarrow \quad \mathrm{^{4}_{2}He} \quad + \quad \mathrm{^{4}_{2}He}$$

Their experiment could have been performed by Lawrence almost a year before. He had not considered the attempt worth while because he had thought it necessary to have enough energy to surmount the electric repulsive barrier surrounding the nucleus. Lawrence was on honeymoon on a boat in Connecticut at the time that the Cockcroft and Walton result hit the headlines. At once he sent a telegram to his colleague James Brady back in Berkeley, California: 'Cockcroft and Walton have disintegrated the lithium atom. Get lithium from chemistry department and start preparations to repeat with cyclotron. Will be back shortly.' Brady showed the telegram to his fiancée with the comment 'That's what physicists on their honeymoon think about'.

Fig. 5.18 The electrostatic generator of Cockcroft and Walton. Walton is sitting in the cage under the apparatus.

There is an important postscript to the Cockcroft and Walton experiment. They were able to predict the energies of the two alpha particles they observed in the reaction. To do this they made use of the familiar principle of energy conservation but extended to include differences in nuclear masses by using Einstein's famous equation between mass and energy

$$E = mc^2$$

energy = mass times velocity of light squared

Fig. 5.19 This high-voltage generator was used as a pre-injector for particles in the NIMROD accelerator at the Rutherford Laboratory, near Harwell, UK.

This relation tells us that mass may be regarded as just another form of energy and that the amount of energy, E, that is equivalent to any given mass, m, may be calculated from the formula above. Let us apply this to Cockcroft and Walton's nuclear reaction. To do this we write down the mass energies of the particles on both sides of the reaction equation and add these to their kinetic energies. For a beam of protons incident on a stationary lithium target this gives us a 'mass–energy' equation of the form

proton mass + lithium mass + kinetic energy of proton

= 2 × (helium mass + kinetic energy of helium)

Inserting the masses and kinetic energy of the bombarding proton we predict that the kinetic energy of the two helium nuclei should be 8.5 MeV. Note that this is much larger than the initial proton kinetic energy and turns out to be in good agreement with the experimental measurements. This principle of mass–energy balance is fundamental to the application of energy conservation in the whole of nuclear physics.

Nuclear reactions and Einstein's mass–energy relation

The mass–energy relation used by Cockcroft and Walton to predict the energies of the two alpha particles in the first artificial nuclear reaction leads directly to an understanding of the 'binding energies' of nuclei.

The simplest nucleus after ordinary hydrogen is that of deuterium, a rare isotope of hydrogen that contains one proton and one neutron. Why do these particles stay together in the small confined volume that is the nucleus? We now know that there are attractive strong nuclear forces that bind the neutron and proton together so that they have a lower energy when combined as a deuterium nucleus (the deuteron) than when they exist as separate particles. We can calculate this binding energy, B, using a mass–energy equation as before:

$$B = m_\mathrm{p}c^2 + m_\mathrm{n}c^2 - m_\mathrm{D}c^2$$

binding energy = mass energy + mass energy − mass energy
of proton   of neutron   of deuteron

Using the experimentally measured mass values we find that the binding energy is about 2 MeV. This is the energy that would be liberated if we could take a proton and a neutron and put them together to make a deuteron. Fortunately, nuclear forces have a very short range so that all the protons and neutrons do not rush together to liberate their potential binding energies! By measuring the masses of all the different nuclei we can calculate the binding energy of each nucleus in the same way. If we label the number of protons in a nucleus by Z and the number of neutrons by N we say that the total number of 'nucleons', A, in the nucleus is just the sum of these

$$A = Z + N$$

In Fig. 5.20 we show a plot of the 'average binding energy per nucleon' for all the different elements. We see that the binding energy rises from

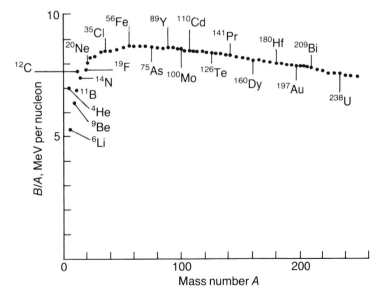

Fig. 5.20 The binding energy per nucleon versus mass number A, the total number of nucleons in the nucleus. The word 'nucleon' means both protons and neutrons. Roughly speaking, the binding energy is the energy required to remove a nucleon from the nucleus. This figure shows that iron is the most stable nucleus and that helium is much more stable than other elements nearby.

about 2 MeV, the value we have just calculated for the deuteron, up to a maximum of around 8.8 MeV per nucleon for iron (Fe) and then falls gradually to about 7.5 MeV for heavy nuclei out to uranium and beyond. Notice that alpha particles (helium nuclei) are especially stable compared with the elements nearby. This is why they are sometimes formed inside heavy nuclei and can tunnel out causing radioactive decay of the nucleus. The position of iron as the most strongly bound nucleus shows that there are two ways of releasing energy from the nucleus. One is the process of 'fusion', in which two nuclei lighter than iron combine to form a heavier nucleus, and the other is 'fission', in which a very heavy nucleus splits into two lighter ones. The binding energy released in these two processes appears as kinetic energy of the final particles.

Let us look at an explicit example of a fusion reaction. By studying all the different binding energies one can find many possible reactions. Probably the best candidate for a fusion nuclear power reactor is the so-called 'D–T' or deuterium–tritium reaction. Tritium is another isotope of hydrogen with one proton and two neutrons. These rare isotopes of hydrogen can interact via the reaction

$$^2\text{H} \quad + \quad ^3\text{H} \quad \rightarrow \quad ^4\text{He} \quad + \quad \text{n}$$
$$\text{deuterium} + \text{tritium} \rightarrow \text{helium} + \text{neutron}$$

which can potentially release 17.6 MeV of energy. But there is a problem with generating power in this way. It is difficult to produce an environment in which a sustained reaction is possible. This is because of the familiar electrical 'Coulomb' barrier that repels the particles as they try to approach each other. It is easy to produce this reaction using beams of deuterons accelerated to energies greater than the repulsive Coulomb barrier, but this is not a commercially viable way of producing large amounts of power. Instead, a very different approach is being followed in the search for cheap fusion power. The idea is to heat up the initial constituents to a very high temperature so that there is enough kinetic energy in the ordinary collisions of the hot gas or 'plasma' to enable this reaction to take place. The generation of such high temperatures and the containment of such a very hot plasma presents us with many formidable technical obstacles. It seems unlikely that these will be overcome for many years. The hope that fusion will provide a cheap, environmentally friendly, inexhaustible source of power is still a long way from becoming a reality!

Given the difficulties of producing energy from fusion in a cost-effective manner, it may seem surprising that such fusion reactions are the basic process by which the stars generate their energy. This is all the more surprising since the temperatures in stars correspond to kinetic energies much lower than the Coulomb barrier. Fusion in the stars is only able to take place at these low temperatures because of quantum tunnelling through the potential barrier. It is not an exaggeration to say that we owe our very existence to this ability of quantum particles to penetrate classically forbidden regions!

Radioactivity, nuclear fission and the atom bomb

Radioactivity is a process by which a nucleus with Z protons and N neutrons can transform itself to one with a different Z and N. Many nuclei are stable and do not decay at all. A plot of all known stable nuclei is shown in Fig. 5.21, together with radioactive nuclei. The three types of radiation emitted by radioactive nuclei were historically called alpha, beta and gamma rays. Alpha particles are now known to be helium nuclei. A nucleus that emits an alpha particle is transformed to the nucleus of a new element with both Z and N decreased by two and with nucleon number, A, decreased by four. In contrast, beta decay of a nucleus does not change the value of A but involves a neutron changing into a proton with the emission of an electron (and an anti-neutrino, as we discuss later). It is also possible for a proton to change into a neutron and for an antielectron – or positron – to be ejected (accompanied by a neutrino). We shall talk about antimatter and beta decay processes in more detail in chapter 11. Finally, the mysterious gamma rays so beloved of comic strips are nothing more than very energetic photons. They arise because some other radioactive decay process – alpha or beta decay – has left the new nucleus in an excited state. These excited states of nuclei are similar to those of the hydrogen atom apart from the energy differences between the energy levels. For atoms, the energy level differences and photon energies are typically measured in units of electron-volts (eV): for nuclei, the corresponding energy differences and photon energies involve millions of electron-volts (MeV).

Why are some nuclei stable and others radioactive? A detailed answer to this question requires an in-depth knowledge of nuclear forces – fortunately, some general features can be understood more easily. The

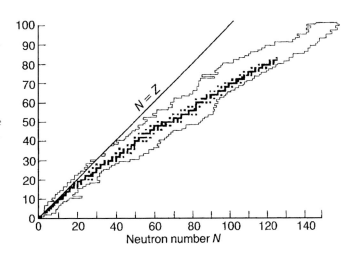

Fig. 5.21 A compilation of all observed nuclei plotted in terms of the number of protons in the nucleus versus the number of neutrons. Stable nuclei are shown in black, and unstable nuclei lie within the marked boundary. More massive nuclei contain more neutrons than protons. The protons are positively charged and produce an electric repulsive energy which tends to make the nucleus unstable.

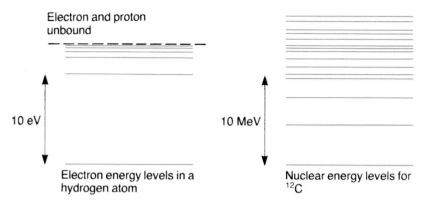

Electron and proton
unbound

10 eV

Electron energy levels in a
hydrogen atom

10 MeV

Nuclear energy levels for
^{12}C

Fig. 5.22 A comparison of atomic energy levels for hydrogen and nuclear energy levels in ^{12}C. The nuclear binding energy is about a million times larger than the energy binding an electron to an atom. When atomic electrons make a transition from one level to another the energy of the photons typically corresponds to that of visible light. In a nucleus there is not the same clear distinction as there is between orbital electrons and the attractive nucleus. Each nucleon can be regarded as moving in an average attractive potential generated by all the other nucleons. When a nucleon makes a transition from an excited state to a lower one, very-high-energy photons, usually called gamma rays, are emitted.

strong attractive nuclear forces are only effective over very short distances, typically smaller than the size of a heavy nucleus. This is why we do not see any direct effects of this enormously strong force on the scale of everyday objects. The electric force, on the other hand, is much weaker than the nuclear force, but its effects are all around us because it is effective over much larger distances than the size of the nucleus. Amongst the neutrons and protons inside a nucleus there is competition between the short range nuclear force tending to bind the nucleons together and the electrical repulsion of the protons tending to break the nucleus apart. For light nuclei, the nuclear force wins out, but for heavy nuclei there is a much more delicate balance between these two opposing forces. If we keep adding protons and neutrons to make heavier and heavier nuclei, the range of the electric force is large enough for all the protons to repel each other. For the nuclear force, on the other hand, the nucleus is now so big, and the nuclear force so short ranged, that any given nucleon only feels strong attractive nuclear forces from nearby nucleons. Consequently, the weaker Coulomb repulsive force, which acts between all the protons in the nucleus, can become comparable or stronger than the attractive nuclear force. This is the reason why stable heavy nuclei with very large A values have more neutrons than protons: the excess neutrons give more attractive binding energy without any Coulomb repulsion. What happens if we start with a radioactive, large A nucleus and change one of the protons into a neutron by beta decay, or eject an alpha particle? We end up with a new nucleus with fewer protons and less Coulomb repulsive energy. The new nucleus is therefore more strongly bound and more likely to be stable. These principles form the basis for the nuclear stability curve.

Lise Meitner and Otto Hahn in their laboratory in Berlin in 1920. After working with Hahn for over 20 years, Meitner left Germany in 1938 when Austria was occupied by Hitler. Her faith in the correctness of Hahn's surprising results led to the discovery of nuclear fission.

This picture of the nucleus also shows that there is the possibility of a heavy nucleus lowering its energy in a more spectacular way than beta decay. Bohr suggested that a heavy nucleus should be imagined more as a 'liquid drop' than as a brittle solid. In a heavy nucleus, the balance between Coulomb repulsion and attractive nuclear forces is very delicate. Perhaps if we add another neutron to the nucleus, the drop may break up into two smaller drops? Looking at the binding energy curve (Fig. 5.20) you can see that it should be possible to gain energy in this way. The neutrons and protons in the two lighter nuclei can have less mass-energy than when they are all together in one very heavy nucleus. Such a possibility was first suggested by a woman chemist called Ida Noddack. She had the temerity to criticize some experiments done by the famous Italian physicist Enrico Fermi and his group in Rome. This group had bombarded uranium with neutrons and believed that they had produced new 'transuranic' elements with Z values greater than the 92 of uranium. Noddack objected that they had not proved that the uranium had not instead split into two large fragments. Unfortunately for Noddack – but perhaps fortunately for the world in view of the imminence of the Second World War – nobody followed up her suggestion. Instead, the famous German chemist Otto Hahn was delighted to have a source of new transuranic elements to study! To their surprise, Hahn and his pupil Fritz Strassman were reluctantly forced to conclude that they could only find some isotopes of barium with a Z value of 56 instead of a new element with a Z value larger than 92. This discovery took place at the end of 1938, on the eve of the Second World War. Because of the prevailing hostility towards Jews, Hahn's collaborator of 30 years, Lise Meitner, whom Einstein once called the German Madame Curie, had been forced to flee to Sweden. Hahn sent a letter to Meitner describing his puzzling results. It was over Christmas of that year, during a visit from her nephew Otto Frisch, that Meitner learnt from Hahn that he had found barium in the reaction products. Lise Meitner was convinced that Hahn was too good a chemist to have made a mistake. It was during a walk in the woods in the snow that she and Frisch arrived at the answer. Using neutron bombardment, Fermi had not made new transuranic elements: instead, the heavy nucleus had 'fissioned' into two lighter elements! The paper describing their conclusions was written by long-distance telephone calls between Stockholm and Copenhagen some days later. Crucial experiments to confirm their idea were performed by Frisch in just two days. The name 'fission' was coined by Frisch from the name biologists use to describe the process by which single cells divide into two.

This process of nuclear fission can be thought of as a tunnelling process, similar to the others we have described in this chapter. The energy of the fissioning nucleus can be pictured as a roller coaster potential like that of Fig. 5.1. This shows two valleys – minima of energy – one of which is lower than the other. Classically, a particle at rest in the upper valley will stay there for ever. Quantum mechanically, such a state is not completely stable – the system has the possibility of tunnelling through to the true

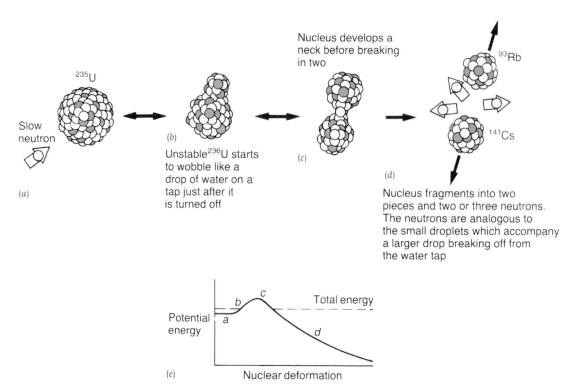

Nucleus develops a
neck before breaking
in two

^{235}U

Slow
neutron

(a)

(b)

Unstable^{236}U starts
to wobble like a
drop of water on a
tap just after it
is turned off

(c)

(d)

^{93}Rb

^{141}Cs

Nucleus fragments into two
pieces and two or three neutrons.
The neutrons are analogous to
the small droplets which accompany
a larger drop breaking off from
the water tap

(e)

Potential energy — Nuclear deformation (graph with points a, b, c, d and Total energy line)

Fig. 5.23 Schematic representation of the liquid drop model of nuclear fission ((a) to (d)) and its relation to quantum tunnelling (e).

lowest energy state. The 'false' minimum is called a 'metastable' state and fission may be imagined as a tunnelling process from a state such as this (Fig. 5.23).

There are two postscripts to this discovery of nuclear fission. One concerns the discovery of true transuranic elements. These were discovered in early 1940 using Lawrence's cyclotron by Ed McMillan. Since the planets Neptune and Pluto lie beyond the planet Uranus, these first transuranic elements, with Z values of 93 and 94, respectively, were called neptunium and plutonium. Both were unstable, but one of them, plutonium, was later to have sinister implications in the production of nuclear weapons.

How nuclear weapons came about brings us to our second postscript. Nuclear reactions, in which the energy liberated by mass conversion is about 100 million times larger than that available in chemical reactions, clearly have enormous potential as an energy source or as a weapon. One vital point about fission that was not mentioned by Meitner and Frisch was the possibility of a 'chain reaction'. A typical fission reaction liberates a few free neutrons in addition to the two large fission fragments:

$$^{235}_{92}\text{U} \;+\; \text{n} \;\rightarrow\; ^{93}_{37}\text{Rb} \;+\; ^{141}_{55}\text{Cs} \;+\; 2\text{n}$$

Fig. 5.24 A painting of the opening ceremony of the world's first nuclear reactor. Fermi's reactor used a controlled and sustained nuclear reaction involving uranium and graphite. Because of a strike by construction workers, the special purpose building for Chicago Pile Number One, as Fermi's reactor was called, was delayed, so Fermi obtained permission to build the reactor in a doubles squash court under the west stands of Stagg Field football stadium. This historic site is now commemorated with a sinister sculpture by Henry Moore.

The rare isotope of uranium, ^{235}U, fissions into rubidium and caesium plus two extra neutrons. The barium observed by Hahn and Strassmann comes from radioactive decay of the unstable caesium isotope. Each of the neutrons produced in this reaction can themselves initiate another fission reaction. The resulting neutrons produced by these reactions in turn generate more fissioning nuclei until an avalanche of fission reactions, a 'chain reaction', is set up. This chain reaction can either release energy in a controlled

Most terrifying weapon in history: Churchill's warning

'THIS revelation of the secrets of nature, long mercifully withheld from man, should arouse the most solemn reflections in the mind and conscience of every human being capable of comprehension. We must indeed pray that these awful agencies will be made to conduce to peace among the nations, and that instead of. wreaking measureless havoc upon the entire globe they may become a perennial fountain of world prosperity.'—*SEE BELOW.*

ATOMIC BOMB: JAPS GIVEN 48 HOURS TO SURRENDER

Radios threaten Tokio: 'You can expect annihilation'

From JAMES BROUGH, Daily Mail Correspondent   New York, Monday Night.

JAPAN is faced with obliteration by the new British-American atomic bomb—mightiest destructive force the world has ever known—unless she surrenders unconditionally in a few days.

Already Japan has felt the terrible effects of one of the bombs. Now she is threatened with being blown out of the Pacific. Soon after President Truman had released the sensational news of the atomic bomb this afternoon, reliable sources here said that a new ultimatum is to be sent to Japan.

This will say : " We will withhold use of the atomic bomb for 48 hours, in which time you can surrender. Otherwise you face the prospect of the entire obliteration of the Japanese nation."

ATOMIC POWER CAN CHANGE THE WORLD—OR DESTROY IT.

By John Langdon-Davies

ZERO HOUR CAME ON JULY 16

Blind girl 'saw' the first big flash

Daily Mail Special Correspondent

New York, Monday.

ZERO hour for the new atomic bomb was 5.30 a.m. on July 16 last, in a remote area on the waters of ...

WE BEAT NAZIS BY FEW MONTHS

Scientists' race

By Daily Mail Reporter

" By God's mercy, British and American science outpaced all German efforts "

THESE words are ...

CHURCHILL TELLS BRITAIN'S PART

Spies and commandos in battle of wits

BRITAIN'S part in the atomic bomb researches, her ceaseless watch on German preparations along similar lines, and her Commando attacks against enemy plants in Norway were described in a statement from 10, Downing-street last night.

JAP TOWN IS WIPED OUT

Guam, Monday. — Japanese industrial town of Paramushi, on Kyushu, was ...

Fig. 5.25 A newspaper headline announcing the atomic bomb ultimatum given to the Japanese.

Fig. 5.26 A soldier on picket duty at Nagasaki was vaporized by the explosion even though he was 3.5 km from the centre of the blast.

Fig. 5.27 An atomic artillery test performed in the Nevada Desert in 1953. The weapon was fired from a 280 mm gun, and was slightly more powerful than the bomb that destroyed Hiroshima.

Fig. 5.28 The B-29 bomber that dropped the first atomic bomb on Hiroshima was piloted by Paul Tibbets. Tibbets named the plane Enola Gay after his mother.

manner, as in a nuclear reactor where the number of neutrons causing fissions can be regulated, or in a catastrophic explosion, as in a fission nuclear bomb. To make a sustained chain reaction you need to have a suitable fissionable material. The most common isotope of uranium is ^{238}U but only

Fig. 5.29 An awe-inspiring photograph of the mushroom cloud from a nuclear test explosion.

the rare isotope ^{235}U is suitable for a nuclear bomb. The transuranic element plutonium, which can be produced from nuclear reactions involving the common ^{238}U uranium isotope, is also suitable. The first 'atom bomb' exploded at the Trinity site in the New Mexico desert was a plutonium bomb. The 'Fat Man' bomb dropped on Nagasaki on August 9th, 1945, was also made from plutonium. 'Little Boy', dropped on Hiroshima on August 6th, 1945, was a uranium bomb made from ^{235}U produced by the Oak Ridge uranium separation plant in Tennessee.

The development of a reactor with a controlled fission chain reaction was vital for the production of plutonium during the Second World War. The first man-made nuclear reactor was built in Chicago by Fermi in 1942. Glen Seaborg, discover of plutonium, element 94, was given the task of extracting plutonium from tons of ^{238}U mixed with other radioactive decay products. At a maximum concentration of around 250 parts per million, Seaborg's job was equivalent to finding a US dime of plutonium in every two tons of uranium reactor products. In fact, Fermi's nuclear reactor was probably not the world's first reactor, Nature had got there first! It is believed that a natural fission reactor operated two thousand million years ago in Africa using natural uranium deposits.

DATING USING RADIOACTIVITY

Almost everything around us is slightly radioactive. The air we breathe, the soil in our gardens, most building materials, and even our own bodies, all contain radioactive elements. Much of this radioactivity originates from naturally occurring uranium and thorium. The Earth's crust has, on average, within one foot of the surface, 8 tons of uranium and 12 tons of thorium per square mile. Both uranium and thorium form one end of complicated decay chains of radioactive elements that eventually end up with stable isotopes of lead. These chains produce radioactive gases that are isotopes of radon and, ingested into the lungs, can be very dangerous and cause cancer. It is said 'that every miner who has worked in the Joachimstal uranium mines in Czechoslovakia for more than ten years has died of lung cancer'. Modern mines now have powerful ventilation systems to flush away these gases. This is one reason not to make our houses too draught-proof!

Given a sample containing a large number of identical radioactive nuclei, we cannot predict when any particular nucleus will decay. However, by measuring the number of decays that occur in the sample in a given time, we can calculate the probability that a nucleus will decay in the next second. We often quantify this probability in terms of a 'half-life' – this is the time taken for half the sample to decay into other elements. Values for half-lives vary over an enormous range: from 4500 million years for ^{238}U, through 1600 years for radium, 3.8 days for radon, to much less than a second for polonium. What this means is that if some material formed 4.5 thousand million years ago contained pure uranium, there would now only be half as much uranium and the rest would have transformed into lead. By measuring the relative amounts of different isotopes in small samples of rock, it is possible to estimate the age of the rock. This technique has been used to date

Moon rocks and meteorites, as well as terrestrial rocks. Using samples of rock from many different parts of the Earth, one arrives at a figure of about 4000 million years for the age of the Earth. The uranium and thorium present at the time of formation of the Earth originated in the huge stellar explosions called supernovae.

There is another important source of radiation, the so-called 'cosmic rays' from outer space. At the surface of the Earth, about 10 000 million neutrinos pass through your fingernail (about one square centimetre) per second. Fortunately, these neutrinos interact so rarely that they do not pose any health hazard! Potentially more harmful are cosmic ray muons – particles like heavy electrons – and about one muon passes through one square centimetre per minute at sea level. These muons are created at the top of the atmosphere by collisions of primary cosmic ray particles (very high energy protons) with the molecules in the atmosphere. Almost all of the primary cosmic rays are absorbed by the atmosphere and only the relatively harmless muons reach the Earth's surface. Nevertheless, this perpetual cosmic ray bombardment is responsible for part of the radioactivity in all living things. This is because humans and all other living things contain carbon, and some of the carbon dioxide in the air we breathe contains the radioactive isotope carbon 14 (^{14}C). Since this has a half-life of 5730 years, it would have decayed away long ago if it were not constantly being replenished by cosmic ray collisions. In all of us there is a small amount of radioactive ^{14}C. When animals or plants die, no new ^{14}C is absorbed and the radioactive carbon decays away without replacement. This is the basis for archaeological radiocarbon dating for time scales of several tens of thousands of years. The method was developed by Willard Libby around 1948 and produced the 'first radiocarbon revolution' since many dates were found to be much earlier than had previously been thought. One problem with the method is that the rate of cosmic ray bombardment of the Earth has not been constant over these time scales. Consequently, the amount of radioactive carbon present at a given time is now calibrated by comparison with tree ring ages of the oldest living trees. These are believed to be the bristlecone pines found in the White Mountains of California. The combination of these two methods produced the 'second radiocarbon revolution', since some dates were found to be even earlier than before. Radiocarbon is useful for archaeological dating from approximately 35 000 years ago. For measuring ages of millions of years the decay of a radioactive isotope of potassium into argon can be used.

Fig. 5.31 The bristlecone pines of the dry and inhospitable White Mountains of California are believed to be the oldest living things. Some trees are over 4000 years old.

Fig. 5.30 A cloud chamber photograph of a cosmic ray shower. The primary particles enter at the top and develop a shower of secondary particles as they pass through a series of brass plates in the chamber.

6 Pauli and the elements

> It is the fact that the electrons cannot all get on top of each other that makes tables and everything else solid.
>
> Richard Feynman

Electron spin and Pauli's exclusion principle

Dimitri Mendeleev (1834–1907) was the youngest of a very large family of between 14 and 17 children! His fame as a chemist allowed him to escape the usual penalties for unorthodox behaviour and his liberal views and support for student causes were tolerated by the Tsarist Russian government. In 1876 he was even able to divorce his wife and marry a young art student without being pursued by the authorities. Perhaps his most endearing eccentricity was that of allowing himself only one haircut a year.

Over a century ago, a Russian chemist, Dimitri Mendeleev, invented a teaching-aid for students struggling with inorganic chemistry. He realized that the properties of the 63 elements then known were repeated 'periodically' as their atomic weight increased. In other words, elements with similar chemical properties were not close together in mass, but instead were found at regular intervals as the mass increased. For example, lithium has a nucleus with three protons and is an alkali metal – a highly reactive soft silvery substance that forms alkaline oxides and hydroxides. The next alkali metal is sodium with eleven protons, then potassium with nineteen, and so on, all with increasing mass. Mendeleev was able to group all the elements into distinct families, and his scheme became known as the 'periodic table' of the elements. All good theories should be able to make predictions and this was no exception. Because of the regularities he had observed, Mendeleev realized that his list of elements must be incomplete. He therefore left gaps in his table corresponding to as yet undiscovered elements, and had the satisfaction of seeing gallium, scandium and germanium discovered during his lifetime. Nevertheless, the explanation of these periodicities remained a mystery for over 50 years until the Austrian physicist Wolfgang Pauli put forward his famous 'exclusion principle'. Pauli's exclusion principle has not only made it possible for physicists to gain an understanding of the different types of solids – metals, insulators and semiconductors – but also enabled nuclear physicists to explain similar 'periodicities' in the properties of nuclei. These nuclear periodicities show up as nuclei that are unusually stable against radioactive decay. These particularly stable nuclei contain 2, 8, 20, 28, 50, 82 or 126 protons or neutrons – the so-called 'nuclear magic numbers'. But before we look at how quantum mechanics enables us to understand Mendeleev's periodicities of

ОПЫТЪ СИСТЕМЫ ЭЛЕМЕНТОВЪ.

ОСНОВАННОЙ НА ИХЪ АТОМНОМЪ ВѢСѢ И ХИМИЧЕСКОМЪ СХОДСТВѢ.

	Ti = 50	Zr = 90	? = 180.	
	V = 51	Nb = 94	Ta = 182.	
	Cr = 52	Mo = 96	W = 186.	
	Mn = 55	Rh = 104,4	Pt = 197,4.	
	Fe = 56	Ru = 104,4	Ir = 198	
	Ni = Co = 59	Pl = 106,6	Os = 199.	
H = 1	Cu = 63,4	Ag = 108	Hg = 200	
Be = 9,4	Mg = 24	Zn = 65,2	Cd = 112	
B = 11	Al = 27,4	? = 68	Ur = 116	Au = 197?
C = 12	Si = 28	? = 70	Sn = 118	
N = 14	P = 31	As = 75	Sb = 122	Bi = 210?
O = 16	S = 32	Se = 79,4	Te = 128?	
F = 19	Cl = 35,5	Br = 80	I = 127	
Li = 7 Na = 23	K = 39	Rb = 85,4	Cs = 133	Tl = 204
	Ca = 40	Sr = 87,6	Ba = 137	Pb = 207.
	? = 45	Ce = 92		
	?Er = 56	La = 94		
	?Yi = 60	Di = 95		
	?In = 75,6	Th = 118?		

Fig. 6.1 'Experimental System of Elements' – a paper sent by Mendeleev to Russian physicists and chemists in 1869. The main difference between this and our modern 'periodic table' of elements (apart from orientation) is because Mendeleev did not know of the existence of the inert gases such as helium and neon.

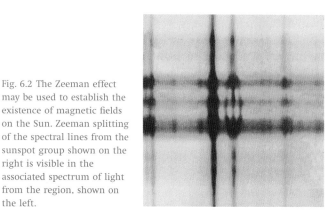 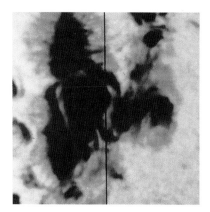

Fig. 6.2 The Zeeman effect may be used to establish the existence of magnetic fields on the Sun. Zeeman splitting of the spectral lines from the sunspot group shown on the right is visible in the associated spectrum of light from the region, shown on the left.

the elements, we must look at another discovery that Pauli had a hand in – this time in a negative way!

In classical electricity and magnetism it is well known that a small loop of wire carrying an electric current acts like a little magnet. Electrons moving in Bohr orbits act like small loops of current and it was therefore possible to calculate their expected magnetic properties. In 1894, long before Bohr's model of the atom, a Dutchman, Pieter Zeeman, had found that the spectral lines of atoms appear split when the radiating atom is put in a magnetic field. Zeeman's first results could be explained in terms of the different angular momentum of electrons in Bohr orbits. But there were later results – too many spectral lines – that could not be explained in this way. In true physicists' jargon this mystery was called the 'anomalous

Wolfgang Pauli (1900–1958) wrote a classic paper on general relativity while still a teenager. He was born in Austria, the son of a chemistry professor. The Pauli exclusion principle, proposed in 1925, explained much of chemistry and made the periodic table of the elements intelligible. Pauli obtained rather belated recognition of this fundamental contribution to quantum mechanics with the award of the Nobel Prize in 1945. He also proposed the existence of the neutrino to explain puzzling features of radioactive decays. By the time its existence was confirmed experimentally, over 20 years after Pauli made his conjecture, most physicists had already accepted its necessity. In the above photograph Pauli and his wife are shown attending the Nobel Prize ceremony in Stockholm.

Zeeman effect'. To illustrate just how bewildering and mysterious all this was at the time, it is appropriate to tell one of the many famous 'Pauli' stories. A friend of Pauli's saw him sitting on a park bench in Copenhagen looking dejected and asked what was making him unhappy. Pauli replied 'How can one avoid despondency if one thinks of the anomalous Zeeman effect?'

The answer to the puzzle was provided by George Uhlenbeck and Sam Goudsmit. They suggested that besides angular momentum due to its orbital motion round the nucleus, the electron also had a 'spin' angular momentum, like the spin of the Earth on its axis as it goes round the Sun. This idea was proposed in 1925, the year before Schrödinger published his wave equation. Up to that time, atomic theory was still a confusing mixture of classical physics combined with Bohr's quantum rules. Uhlenbeck and Goudsmit gave their paper to their professor, Ehrenfest, for his comments, and he suggested they ask the great expert, Lorentz. After thinking the idea over for a week, Lorentz gently pointed out many serious difficulties with a classical picture of a rotating electron. Uhlenbeck and Goudsmit then rushed off to Ehrenfest to withdraw their paper, only to hear him say 'I have already sent your letter in long ago; you are both young enough to allow yourselves some foolishness!' In the end, of course, they were proved to have the right idea, and all the mysterious results on the Zeeman effect are due

Enrico Fermi (1901–1954) was unusual among his generation in that he did brilliant work in both experimental and theoretical physics. In his early experimental work he used the newly discovered neutrons to induce artificial radioactivity. Winning the Nobel Prize in 1938 enabled him to escape Fascist Italy and settle in the USA. As part of the war effort for the atomic bomb project, Fermi built the first nuclear reactor. The establishment of the first self-sustaining chain reaction was announced in a coded telegram sent out by Compton: 'The Italian navigator has entered the new world'.

to the electron having its own spin angular momentum. But, just as with Bohr orbits, a classical picture of a rotating electron cannot be taken too literally as a model for quantum mechanical spin. Uhlenbeck and Goudsmit were luckier than another young physicist named Kronig who had the same idea at about the same time. He had the misfortune to ask Pauli's opinion on the matter – and Pauli convinced him that such a classical idea could not be right!

There is one more discovery about spin angular momentum that we must mention before we can get down to Pauli's exclusion principle and the periodic table. As we discussed in chapter 4, angular momentum is 'quantized' and the axis of rotation can only point in certain directions. It was Pauli who had suggested this 'space quantization' and it was verified in a famous experiment by Stern and Gerlach. For an electron, this has the consequence that there are only two possible directions of rotation: it can either spin clockwise or anticlockwise. We often talk about the electron spinning clockwise as a 'spin up' electron, and anticlockwise as a 'spin down' electron. This property of the electron provided Pauli with the final clue to understanding the structure of atoms.

The essential problem to be explained had been identified by Niels Bohr. If the energies of electrons in atoms are indeed quantized, why is the lowest energy state of an atom not occupied by all the electrons in the atom? It was obvious that the electrons could not all be in the lowest state, since, if they were, all the elements would behave in a similar way. Moreover, as we shall see, it is the shape of the excited state wavefunctions that enables atoms to combine to form molecules. If all electrons were in the symmetrical, lowest energy state there would be no molecules and certainly no life as we know it! Pauli provided the answer with his exclusion principle. This asserts that only one electron is allowed in each quantum state. Consider what this means for electrons in a box (Fig. 6.3) – the quantized energy levels were discussed in chapter 4. With one electron in the box, the lowest energy (ground state) of the system has this electron in the $n = 1$ level, with its spin either up or down. When adding a second electron to the box, we must obey Pauli's exclusion principle. This can also go in the $n = 1$ state, provided its spin is opposite to that of the first electron. However, when a third electron is put in the box it cannot go into the $n = 1$ level since this is now full. To put it there with either spin up or down would mean that two electrons had exactly the same quantum numbers, and this is forbidden by Pauli. It must therefore do the next best thing and go into the lowest available empty energy level – in this case one of the two possible $n = 2$ level spin states. And so on. As Feynman says at the beginning of this chapter, it is the Pauli exclusion principle that makes all things hard and rigid. In essence, all it says is that 'matter' cannot be compressed to nothing but must occupy a certain minimum space. All 'matter-like' quantum particles obey the exclusion principle – such particles are called 'fermions' in honour of Enrico Fermi, who was one of the first to look at the implications of Pauli's principle. In fact, there is another class of quantum 'particles'

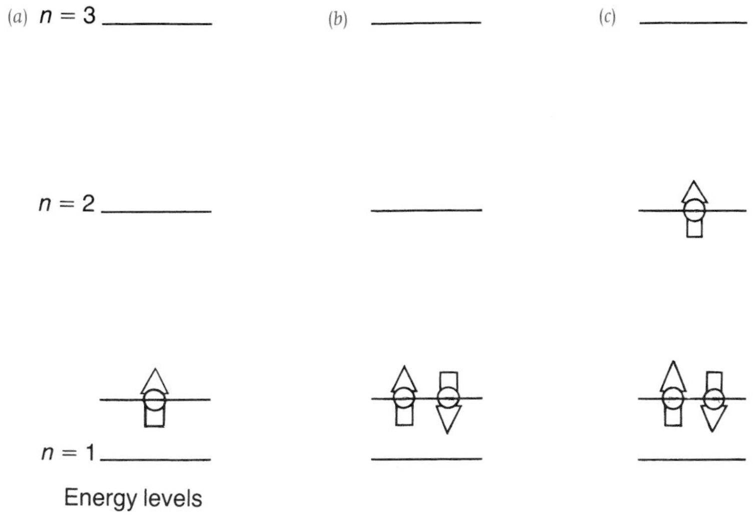

(a) $n = 3$ ————   (b) ————   (c) ————

Fig. 6.3 Electrons in a box. The electrons fill up the energy levels according to the Pauli exclusion principle. (a) One electron in the box. This can have either spin up or down. (b) Two electrons in the box. Both can go into the ground state but their spins must be oppositely directed. (c) Three electrons in a box. The ground state is full so the third electron must go into the first excited state.

$n = 2$ ————

$n = 1$ ————

Energy levels

that we can categorize as 'radiation-like' – photons are an example – which do not obey the Pauli principle. Such particles are known as 'bosons', after the Indian physicist Satyendra Bose, who first considered this possibility. In contrast to fermions, bosons prefer to be together in the lowest energy state if at all possible! We shall examine some observable consequences of such 'bose condensates' in the next chapter.

The elements

We are now in a position to understand not only the variety of elements found in Nature but also their chemical properties. Any detailed understanding of the periodic table requires some knowledge of wavefunctions and quantum numbers. These were discussed in the final section of chapter 4, and, like that section, this one may be rather heavy going at a first attempt. As before, it is probably best to skim quickly through this section and not worry about the details. Only a few references to the results of this section are made in later parts of this book. Here we explain how the Pauli exclusion principle can account for different types of chemical bonding and how the electrons fill up the available energy levels to yield the different elements. For hydrogen, we have seen in chapter 4 how the Schrödinger equation leads to Bohr's quantized energy levels. For a nucleus with Z protons we will need to add Z electrons to make a neutral atom and, according to Pauli, these cannot all go into the lowest energy state. Instead, they will fill up the energy levels starting from the lowest $n = 1$ level, allowing a spin up and a spin down electron in each state labelled by the quantum number n and the two orbital angular momentum quantum numbers L and M. In fact, the energy level diagram that dictates the order of level filling for a many-electron atom will not look quite the same as the level diagram for hydrogen. This is because, in addition to the attractive

Otto Stern (1888–1969) was one of the major experimental physicists of this century. His most important work used molecular beams to demonstrate the quantum properties of atoms. In 1933 Stern was compelled to leave Nazi Germany and he moved to the USA. He won the Nobel Prize in 1943.

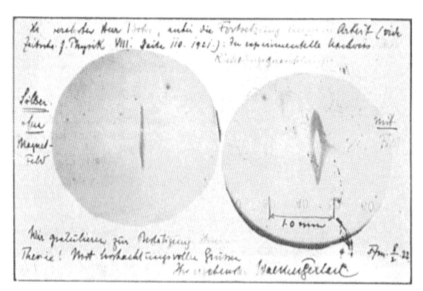

The postcard sent by Gerlach, Stern's collaborator in their famous experiment on 'space quantization', to Niels Bohr, announcing their discovery.

force on an electron from the nucleus, each electron will feel repulsive forces from all the other negatively charged electrons. One effect of this is that an electron in one of the higher n levels – corresponding to a large Bohr orbit – only 'sees' a fraction of the nuclear charge. Some of the positive charge of the nucleus is screened by the negative charges of other electrons closer in. Moreover, the $L = 0$, or S state electrons have probability distributions that are greater near the nucleus than any $L = 1$ (P state) or $L = 2$ (D state) electrons (see Fig. 4.18). The S electrons therefore see more of the nuclear charge and are more tightly bound. Thus, we expect the energy level pattern for many-electron atoms to look something like Fig. 6.4. All we have to do now to explain the periodic table is to fill up these levels

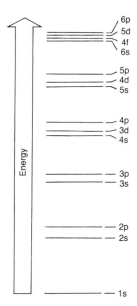

Fig. 6.4 Energy levels for a typical atom. This energy level diagram, along with the Pauli principle, determines the form of the periodic table of the elements.

according to Pauli's housing plan. It was not for nothing that Pauli had the nickname 'atomic housing officer'!

A neutral hydrogen atom has one electron, which is normally found in the $n = 1$ and $L = 0$ ground state: the lowest 1S energy level. The electron can be excited to one of the higher energy levels by collisions, or by shining light on the atom. After a short time, the electron will drop back into the ground state, emitting a photon with an energy corresponding to one of the spectral lines. Perhaps surprisingly, the Pauli principle has a role to play even for hydrogen. What happens if we bring up another hydrogen atom close to the first one? If the two electrons are both in the spin up state, the exclusion principle will prevent the two atoms getting close enough for the wavefunctions of the two electrons to overlap – since this would mean that the two electrons would be in the same quantum state. Conversely, if the two spins are opposite, a close approach is possible, and indeed the two electrons spend most of their time in the region between the two hydrogen nuclei. This results in a binding force between the two hydrogen atoms and allows a stable hydrogen molecule to be formed. This type of chemical bond – in which the two electrons are shared between the two nuclei of the molecule – is called a covalent bond. It is Pauli's exclusion principle that explains why hydrogen is chemically active and why two atoms can form a stable H_2 molecule. Notice that the same principle also forbids a third atom to form a covalent bond with the H_2 molecule since both the two lowest energy spin states are already occupied.

The next simplest element is helium, with two electrons surrounding its nucleus, which has a positive charge twice that of the hydrogen nucleus. These two electrons can both go in the lowest 1S level provided their spins

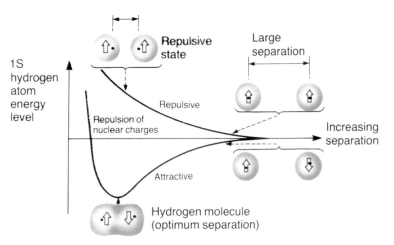

Fig. 6.5 Potential energy curve for two hydrogen atoms plotted against their separation. The atoms repel each other when the two spins are parallel and attract when they point in opposite directions. The two 1S levels of the hydrogen atoms therefore split into two as shown. If the atoms are brought too close together the proton nuclei will start to repel each other significantly. This means that in the case when the two spins are opposite, there is an optimum separation for formation of a hydrogen molecule.

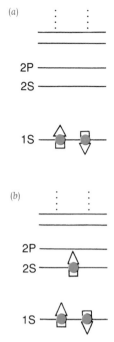

Fig. 6.6 Energy level diagram for (a) helium and (b) lithium.

are opposite (Fig. 6.6a). Since there is now no more room for any other electrons in the 1S state, the Pauli principle will tend to keep other electrons away from the helium atom, just as it does for the H_2 molecule. Thus, we expect helium to be chemically inactive – one of the family of inert gases. The next element, lithium, has three electrons surrounding its nucleus and has an energy level structure similar to that of helium (Fig. 6.6b). The first two electrons can both go into the 1S state, with opposite spins, and this now forms a chemically inactive closed shell, like that of helium. The third electron must go into the lowest unoccupied energy level, which is now the 2S level. Thus, lithium has one electron in an $L = 0$, S level, and this explains why it has similar chemical properties to hydrogen. For example, lithium forms stable Li_2 molecules through covalent bonding in the same way as hydrogen combines to form H_2 molecules.

If we keep adding successive electrons, we begin to fill higher and higher energy levels. Nitrogen, for example, has seven electrons. Two of these electrons fill up the 1S level to form a closed shell. Two more fill up the 2S level to form another closed shell, leaving three electrons to be put into the 2P level. Now, S states have electron probability distributions that are spherically symmetrical – no direction is favoured (see Fig. 4.18). The 2S state is larger than the 1S state, corresponding to the fact that the excited state is less tightly bound than the ground state. However, the P state probability patterns are not spherically symmetrical. As we saw in chapter 4, there are three possible P states, which we can specify by labels x, y or z. These labels tell us whether the lobes of the electron probability pattern are oriented along the x, y or z directions (see Fig. 4.18). In order to get the three electrons as far apart as possible, and thereby minimize the electron repulsion between them, the three electrons in nitrogen go into three different P states, in preference to two of them going into the same P state with opposite spins. The shapes of these P state wavefunctions also allow us to understand the formation of more-complicated molecules. A hydrogen atom can approach a nitrogen atom and attach itself to any of the three P state lobes, provided its spin is opposite to that of the corresponding P electron. Indeed, it is clear that nitrogen can attach a maximum of three hydrogen atoms to itself before the P shell has a full complement of six shared electrons. Fig. 6.7 shows the geometry of the ammonia molecule, NH_3, produced by the covalent bonding of one nitrogen atom and three hydrogen atoms. Nitrogen is clearly a chemically active element able to form many other compounds. Next in the periodic table after nitrogen is oxygen, with four P shell electrons. One of the three P orbitals must now be full, and consequently the oxygen atom can only attach itself to two hydrogen atoms. An impression of the wavefunction of a water molecule, H_2O, formed in this way is shown in Fig. 6.8.

We now move on through the periodic table to neon; this has ten electrons to put into energy levels according to Pauli. The 1S, 2S and 2P shells are now completely filled. We therefore understand why neon, like helium, is chemically inactive, and why the property of inertness has been

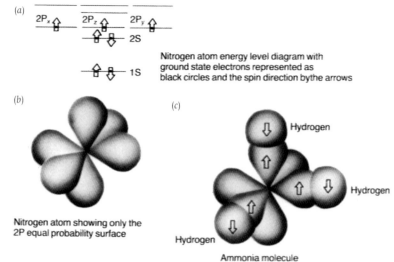

Fig. 6.7 An illustration of the structure of the ammonia molecule NH₃. (a) Energy levels for nitrogen. (b) Sketch of 2P probability surfaces for a nitrogen atom. (c) Arrangement of electron probability surfaces in an ammonia molecule.

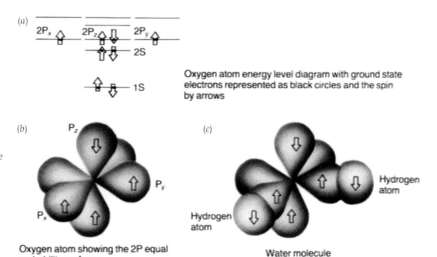

Fig. 6.8 An illustration of the structure of a water molecule. (a) Energy levels for oxygen. (b) Sketch of 2P probability surfaces for an oxygen atom. (c) Arrangement of electron probability surfaces in a water molecule.

repeated. From the energy level pattern shown in Fig. 6.4 we see that another inert element should occur when we have filled up both the 3S and 3P levels. With two electrons in an S shell and six in a P shell this brings us to an element with proton number $Z = 18$: the inert gas argon. The whole of the periodic table can be understood in this way. Elements have similar chemical properties when they have the same number of outer electrons in similar quantum states. Thus, lithium combines with oxygen to form lithium oxide, Li_2O, in exactly the same way that hydrogen combines with oxygen to form a water molecule.

Metals, insulators and semiconductors

One of the great successes of quantum physics has been in understanding the way in which different types of solids conduct electricity. In a solid it is the flow of electrons that gives rise to electric currents. It is a great triumph of quantum mechanics that it can explain what makes different substances metals, insulators or semiconductors. It is no exaggeration to say that it is this quantum mechanical understanding of materials that has led directly to the present technological revolution, with its accompanying avalanche of new and cheap consumer goods ranging from stereo systems and colour TVs to computers and mobile phones. There are many properties of solids that can be understood from a quantum mechanical viewpoint – colour, hardness, texture, and so on – but we will focus on their ability to conduct electric currents. A good conductor, such as copper, must have many conduction electrons that are able to move and carry a current when an electrical potential difference is applied to a wire. An insulator such as glass or polythene apparently has no conduction electrons, since no current flows when a voltage is applied. There is a third category of materials consisting of solids that conduct electricity much better than insulators

Fig. 6.9 A Landsat photograph of San Franscisco Bay with Silicon Valley and San Jose at the bottom right. The San Andreas Lakes that lie along the infamous earthquake fault are clearly visible slanting upwards parallel to the coast in the middle left.

but much worse than metals – such materials are called semiconductors. Germanium and silicon are examples of semiconductors, and their importance for new technology is evident from the naming of the area around San Jose in California as Silicon Valley.

The properties of a solid depend not only on what it is made of but also on the way the atoms or molecules are stacked together. Many materials have their constituent atoms arranged in a regular array – like the pattern of bricks in a wall. This regular pattern of atoms is called a 'crystal lattice', and substances with such a structure are called 'crystalline solids'. There are other materials that have no crystalline structure, but, just like a pile of bricks, still have a certain strength and hardness. Because of their lack of an underlying crystal structure, the properties of such 'amorphous' solids are much more varied than those of the crystalline solids we shall talk about in this chapter. As we shall see, arranging all the atoms in a regular array has a dramatic effect on the allowed energy levels for the atomic electrons.

We can get an idea of the energy level structure of a regular array of atoms by looking at what happens to the atomic energy levels of two atoms as they are brought together. In the case of hydrogen, we saw that the Pauli exclusion principle leads to binding of the two atoms to form a molecule only when the spins of the two electrons are opposite. If the spins are parallel, the Pauli principle keeps the electrons apart and there is no binding. In terms of energy levels, we see that, in the first case, the two electrons have an energy less than that of two isolated atoms – resulting in covalent binding to form a molecule – and, in the second case, the electrons have more energy than two atoms – hence giving no binding (see Fig. 6.5). A similar effect takes place for the energy level of the outermost

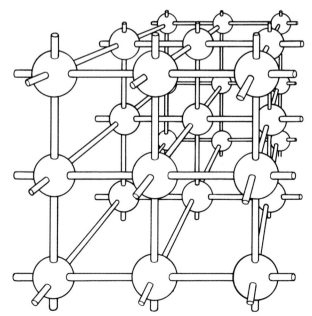

Fig. 6.10 The arrangement of atoms in a cubic crystal lattice. The spheres represent the location of the atoms and the tubes linking them represent the direction of the bond lines holding the atoms in place. In the case of common table salt, the spheres correspond alternately to positively charged sodium ions (sodium atoms that are missing an electron) and negatively charged chlorine ions (chlorine atoms that have gained an electron) separated by about 2.8 angstroms (10^{-10} metres.)

Fig. 6.11 Common table salt magnified 50 times by an electron microscope. The evident cubic structures reflect the underlying lattice structure.

3S electrons of sodium when we bring together two sodium atoms. If we add more and more sodium atoms, we find that the 3S levels splitting increases until all that remains is a 'band' of very closely spaced levels Fig. 6.13. This band is called the 3S band since these levels were formed from the 3S levels of atomic sodium. For N atoms, the 3S band will contain N levels, each of which can hold two electrons, spin up and spin down. The lower-lying atomic energy levels correspond to more-tightly bound electrons with smaller wavefunctions which do not overlap as much as those of the 3S states. The resulting energy level bands are much narrower. Both the 1S and 2S bands can accommodate $2N$ electrons and, for sodium, will be fully occupied. The 2P band can hold $6N$ electrons (three different P states times two different spin states for each of the N atoms) and will also be full. A sodium atom has only one electron in the 3S state so, in a metal containing N atoms, the 3S band will contain only N electrons and will be half full.

Fig. 6.12 The beautiful symmetry of this snowflake reflects the hexagonal bonding structure of the water molecules making up the ice.

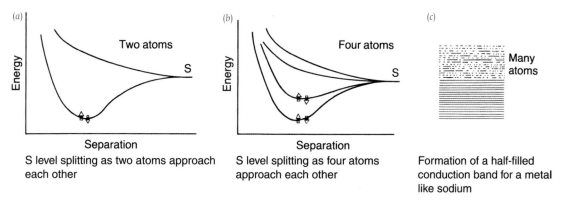

Fig. 6.13 Level splitting for an element such as sodium with a single S state electron in its outer shell. (a) Level splitting as two atoms approach each other. (b) Splitting of levels due to four atoms approaching each other. (c) Formation of a half-filled conduction band as many atoms approach each other.

These 3S electrons are the conduction electrons. If we apply a potential difference to a wire made of sodium metal, these electrons can gain energy and be accelerated in the direction of the potential – we can imagine them jumping to the empty higher energy levels in the 3S band. This picture for the energy levels of conduction electrons in a metal is the reason why many properties of metals can be explained by the simple model of electrons in

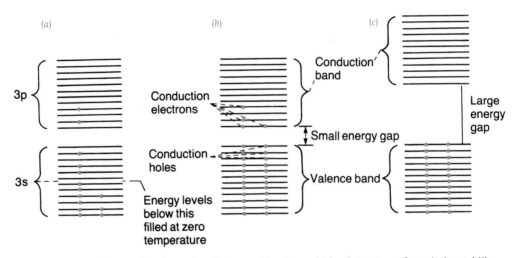

Fig. 6.14 Metals, semiconductors and insulators. (a) Band structure of a typical metal like sodium. There are many unfilled energy levels for the conduction electrons and at normal temperatures a few of the electrons will be excited into the almost unoccupied 3P level. (b) In a semiconductor the valence band is full and there is a small energy gap to the empty energy levels in the conduction band. At normal temperatures some of the electrons have enough energy to jump across the gap. (c) In an insulator the gap between bands is too large for there to be a significant number of electrons able to jump the gap. It therefore conducts electricity very poorly, if at all.

a box that we considered in an earlier chapter. In the covalent binding for the hydrogen molecule, the two electrons are shared between the two hydrogen atoms. In a sense, metals can be thought of as an extreme case of covalent binding in which the conduction electrons are shared by all the atoms in the metal.

The situation we have described above corresponds to the lowest energy state of sodium metal, with the sodium ions fixed in place in a crystal lattice structure. At room temperatures, the lattice ions have some thermal kinetic energy corresponding to vibrations of the ions about their positions in the crystal lattice. Conduction electrons can gain and lose energy in collisions with the lattice ions and with each other. Instead of the conduction electrons filling up exactly the bottom half of the 3S band energy levels, with the top half empty, some of the electrons are thermally excited to these higher levels. This has the effect of leaving some empty levels in the bottom half of the 3S band. Although the energy involved in a typical collision at room temperature is only a fraction of an electron-volt, the gaps between the bands in sodium are small enough for some of the 3S conduction electrons to be excited to the previously empty 3P band.

This complication of thermal excitation of electrons does not significantly change our picture of conduction in metals but it will be crucial for an understanding of insulators and semiconductors. Let us see how this energy band picture, together with the Pauli exclusion principle, can give us an explanation of insulators. Consider what happens if we have a material in which the ground state consists of one band being completely full and the one above completely empty. If there is a large energy gap between the bands, almost no electrons are able to gain enough energy from collisions to jump into the empty band. Thus, when a voltage is applied to the material, there are no empty energy levels close by for the electrons to gain energy – since Pauli does not allow two electrons to occupy the same quantum state. The lower band is already full and there is too large an energy for an electron to jump to the higher empty band. This is the situation in an insulator: there are essentially no free conduction electrons to contribute to an electric current in the upper conduction band. What is a semiconductor? It is a material with a similar band structure to an insulator, but where the energy gap between the bands is much smaller. At ordinary temperatures, a significant number of electrons are excited to the upper conduction band. When a voltage is applied, there are electrons in the upper band with plenty of empty states to move to which allow the electrons to gain energy. There will be some empty states in the lower band which also allow conduction. Thus, semiconductors will conduct currents fairly easily and their conductivity will depend strongly on temperature, in contrast to metals and insulators.

The picture of bands given above is based strongly on the atomic energy levels of the material. One therefore expects metals to have odd numbers of electrons, while insulators and semiconductors should correspond to elements with closed shells. In fact, magnesium, with a filled 3S shell, is a good conductor, while carbon, with only two electrons in its 2P shell, is an insulator! The answer to these puzzles lies in the details of how some bands can overlap and leave no energy gap. In magnesium, the 3S and 3P bands overlap to yield a single band capable of holding $2N + 6N = 8N$ electrons (Fig. 6.15). Since only $2N$ of these levels are filled, magnesium is

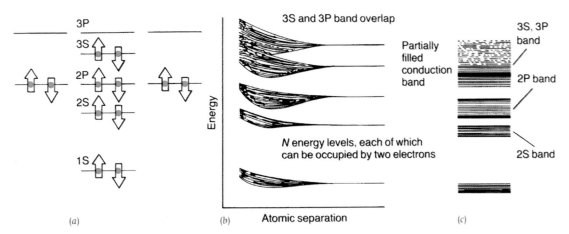

Fig. 6.15 Band overlap in magnesium. (a) Energy levels for an isolated magnesium atom. (b) Variation of the energy level diagram for N magnesium atoms with their separation. (c) Band structure for magnesium.

a good conductor. For carbon, as the N atoms are brought together, the 2S and 2P bands join to form a band with $8N$ states, as for magnesium. However, as the carbon atoms come still closer together, this combined band splits into two bands each with $4N$ states. In carbon, the lower band is full and the upper band completely empty, characteristic of an insulator. The energy levels in germanium and silicon behave similarly, but here the two bands are separated by a much smaller energy gap so that both are semiconductors rather than insulators. The way to understand the detailed band structure of materials was discovered by a Swiss physicist named Felix Bloch. He solved the Schrödinger equation for electrons moving in a potential corresponding to a regular lattice of positive ions. This leads to the energy band structures described above and is the mathematical basis for the quantum mechanical band theory of solids.

Transistors and microelectronics

Pure semiconductors are not in themselves of great practical importance. Only one atom in about a thousand million contributes an electron to conduct electricity: in metals almost every atom contributes one or more electrons. This apparent drawback has the great advantage that the conduction properties of semiconductors can be tailored at will by the introduction of suitable impurity atoms at the level of one in a million. Germanium and silicon both have four valence electrons which fill up most of the $4N$ states of the valence band, which lies below the almost-empty conduction band. If we introduce an impurity such as phosphorous – which has five valence electrons – then, since only four electrons are needed for the covalent bonding of the lattice, there is an electron left over that can easily be detached and contribute to the conductivity. Similarly, if we introduce impurity atoms with only three valence electrons, such as boron, there is now one covalent bond lacking in the lattice, and this tends to capture electrons from the valence band, leaving an empty state which will allow some conductivity. These two situations are represented on the energy level diagram shown in Fig. 6.16. The phosphorus atoms give rise to electron donor states just below the conduction band; these electrons need only gain a small amount of energy to jump into the conduction band. Semiconductors that have been 'doped' in this way are called n-type semiconductors, since there is an extra contribution to the current from the negatively charged electrons from the donor levels. Semiconductors doped with boron are called p-type semiconductors. The boron atoms give rise to acceptor states just above the nearly full valence band, and at room temperatures electrons are readily excited into these levels. Why are they called p-type semiconductors? Well, as we said before, conductivity in the nearly full valence band is only possible because electrons can move into the few unoccupied states. As can be seen from Fig. 6.16, instead of the electron moving, we can equally well think of the 'hole' moving in the opposite

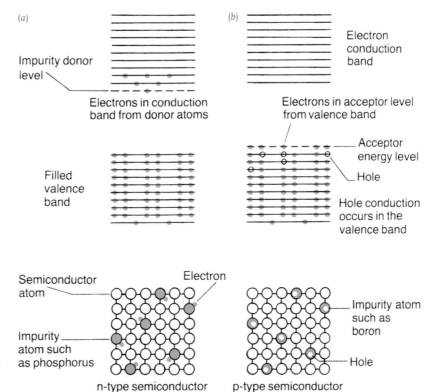

(a)

Impurity donor
level

Electrons in conduction
band from donor atoms

Filled
valence
band

(b)

Electron
conduction
band

Electrons in acceptor level
from valence band

Acceptor
energy level

Hole

Hole conduction
occurs in the
valence band

Semiconductor
atom

Electron

Impurity
atom such
as phosphorus

n-type semiconductor

Impurity atom
such as
boron

Hole

p-type semiconductor

Fig. 6.16 Semiconductors doped with impurity atoms. (a) n-type semiconductor in which the impurity atoms have an extra electron. This results in the effective energy level diagram shown above. (b) p-type semiconductor doped with impurity atoms with one fewer electron, resulting in electron 'holes'. The energy level diagram corresponding to this situation is shown above.

The three inventors of the transistor at about the time they made their invention. From left to right, they are Shockley, Brattain and Bardeen. Bardeen went on to win a second Nobel Prize for his work on super-conductivity. No one else has won two physics Nobel Prizes.

direction to the electron. Since moving a negative charge to the left has the effect of increasing the charge on the right, we can think of the current as positively charged holes moving to the right. In a p-type semiconductor, extra holes have been created in the valence band and we can think of the increased current as due to the positively charged holes we have added to the material.

Why is all this useful? The reason is that p- and n-type semiconductors can be put together to form a switch to control current flow. The simplest semiconductor device is a p-n junction diode. Examination of the energy levels and electron and hole currents across the junction shows that current can only pass in one direction when a voltage is applied. This p–n junction device can therefore convert an alternating current into a unidirectional current – a property called rectification. The development which has affected people's lives most directly is the invention of the transistor by John Bardeen, Walter Brattain and William Shockley at the Bell Telephone Research Laboratories in the USA. The transistor was not discovered by accident but was the culmination of an extensive research programme. As Bardeen described it in his Nobel Prize lecture, 'the general aim of the program was to obtain as complete an understanding as possible of semiconductor phenomena, not in empirical terms, but on the basis of atomic theory'. It seems a long way from de Broglie's probability waves to modern computers! A replica of the first 'point-contact' transistor is shown in Fig. 6.17a. This was in 1947 and it was soon followed in 1951 by the not

(a)

(b)

Fig. 6.17 The first transistors. (a) A replica of the point-contact transistor invented by Bardeen and Brattain. The wedge of semiconductor which forms the base is about 3 cm each side. (b) Shockley's junction transistor looks very unglamorous but was much easier to fabricate reliably.

Jack Kilby was awarded the 2000 Nobel Prize for Physics for his discovery of the integrated circuit. It was this invention that began the silicon revolution and underpins Moore's Law. Kilby was the son of a Kansas electrical engineer and went to the University of Illinois to study engineering after being turned down by MIT. In his first job after the Second World War Kilby set up a small transistor manufacturing line. He joined Texas Instruments in May 1958 and because he had just started in the company and had accumulated no leave, Kilby found himself almost alone in an empty plant during the two week vacation period in July. It was during this period of enforced isolation from his colleagues that Kilby came up with the idea of monolithic semiconductor integrated devices.

very glamorous looking, but more reliable, 'pnp' transistor (see Fig. 6.17*b*). This consists of a thin layer of n-type semiconductor sandwiched between two thicker regions of p-type material. 'Transistor action' is the regulation of the current flowing in an electrode called the collector by a small current applied to an electrode called the base. In the case of the pnp transistor a large current through the high-resistance 'collector–base' p–n junction is controlled by a small current through the low-resistance 'base–emitter' n–p junction. This action can be understood by a detailed consideration of the energy level and electron and hole currents across the two p–n junctions. The word transistor refers to this effect and comes from combining the two words '*trans*fer-re*sistor*'.

Transistors turned out to be ideally suited for the 'on–off' binary logic of computers. Moreover, their reliability and low power consumption, together with a number of incredible engineering advances, have made them the basic ingredient of modern microelectronics. The key idea seems to have been first written down by a British engineer named G. W. A. Dummer, working at the Royal Radar Research Establishment in Malvern, Worcestershire. He was an expert on reliability problems of electronic components and was concerned with the performance of radar equipment under extreme conditions. Dummer eventually realized that it was unnecessary to manufacture all the ingredients for an electronic circuit – transistors, resistors (which impede the flow of current) and capacitors (devices for storing charge) – in separate pieces. The same circuit could be made much smaller and more robust if all these devices were contained in the same piece of semiconductor! In May, 1952, Dummer wrote:

> With the advent of the transistor and the work in semiconductors generally, it seems now possible to envisage electronic equipment in a solid block with no connecting wires. The block may consist of layers of insulating, conducting, rectifying and amplifying materials, the electrical functions being connected directly by cutting out areas of the various layers.

This is an amazingly accurate vision of a modern integrated circuit. In 1952, however, there were still many difficult technical problems to be overcome before Dummer's idea could become a reality. Unfortunately, although Dummer produced a non-working model of such a silicon 'solid circuit' in 1957, the potential of such developments was not appreciated in the UK. Thus, the vital breakthrough in this direction was made in the summer of 1959 by an American, Jack Kilby, working for Texas Instruments. Kilby created the first working integrated circuit or IC (Fig. 6.18). Since ICs are physically made out of tiny chips of silicon, they are popularly known as chips by the trade, or as microchips by the newspapers. The full utility of the IC only became apparent after the invention of a new process for making 'planar' transistors. The planar transistor was discovered late in 1958 by the Swiss-born physicist Jean Hoerni, one of the founder members of Fairchild Semiconductor. With this invention, Robert Noyce, a co-founder

Fig. 6.18 The first 'integrated circuit' or chip. Instead of making the components of a circuit separately, Jack Kilby incorporated a transistor, a capacitor and resistances in the same piece of germanium.

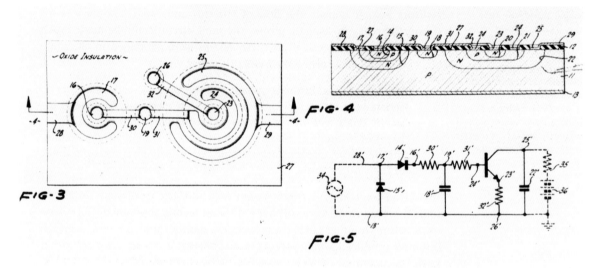

Fig. 6.19 Diagrams from Noyce's patent for integrated circuits made using a planar process. This was a crucial breakthrough in chip manufacture.

of Fairchild, was able to design and produce a truly robust IC that was capable of mass production (Fig. 6.19). Using these ICs, Fairchild were able to market a whole family of logic chips – the decision-making units of computers. That year, 1962, marked the beginning of mass production of ICs. In the same year another technological breakthrough occurred with the discovery of a new type of transistor that could be more easily incorporated in mass-produced chips. This was the MOSFET, the metal-oxide–semiconductor

Fig. 6.20 The first microprocessor. Ted Hoff, an engineer with Intel, had the idea of combining all the elements of a programmable computer on one chip. The chip has dimensions of about 3 mm by 4 mm and it contains over 2000 transistors.

field effect transistor, invented by two young engineers, Steven Hofstein and Frederic Heiman, at RCA's research laboratory in New Jersey. Chip development continued apace, with ever increasing miniaturization and complexity; already by 1967 chips were being produced that incorporated thousands of transistors.

The various stages of computer development may be crudely categorized in terms of generations. The first generation began in the 1950s with the first successful industrial computer, the UNIVAC 1, constructed using electronic valves. The first IBM computer, the IBM 701, was delivered in 1953; by 1956 IBM had already become the largest and most profitable computer manufacturer, building machines by the hundreds! The widespread availability of transistors to replace the costly and unreliable valves saw the beginning of the second generation of computers in about 1959. Along with these 'hardware' developments came 'software' improvements in the art of programming computers – basically, how to get the computer to do what you want it to do! In about 1966, following hard on the heels of second-generation computers, came the third generation,

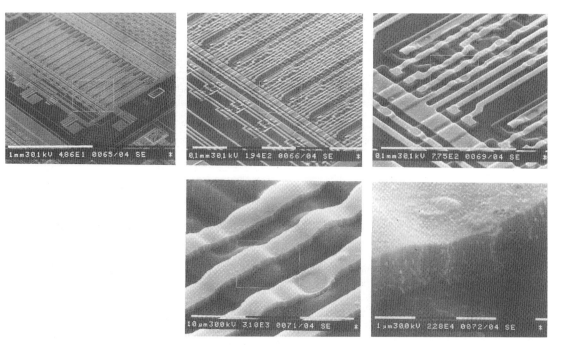

Fig. 6.21 Sequence of electron microscope photographs of a chip with increasing magnification.

Robert Oppenheimer and John von Neumann at the official dedication of the computer built for the Princeton Institute for Advanced Study in 1952. Oppenheimer was head of the Manhattan project at Los Alamos during the Second World War. Von Neumann was a brilliant Hungarian mathematician who moved to the USA before the war. He used the ENIAC computer to perform vital calculations for the design of the implosive lens of the first atomic bombs. He also set down the first specification for a stored-program computer that proved to be of major importance in the development of powerful modern computers. During the McCarthy era Oppenheimer was investigated as a security risk, primarily because he opposed the development of the hydrogen bomb. At the hearings, von Neumann testified to Oppenheimer's loyalty and integrity.

Fig. 6.22 Poster advertising the ENIAC computer. The ENIAC was built to perform ballistic calculations for the US army.

whose main hardware innovation was the incorporation of ICs. This development made third-generation computers smaller, cheaper and far more reliable than previous generations. The most sophisticated ICs now had tens of thousands of transistors and this level of complexity on a chip was known as large- scale integration, or LSI for short. What could come next? Probably the best way to characterize the difference between third- and fourth-generation computers is by the invention of the microprocessor. In 1968 Robert Noyce and Gordon Moore left Fairchild to found Intel (short for *int*egrated *el*ectronics). In response to a contract to develop a set of chips for a new line of electronic calculators, one of Intel's employees, Ted Hoff, Jr, had the bright idea to design a programmable IC chip (Fig. 6.20). Instead

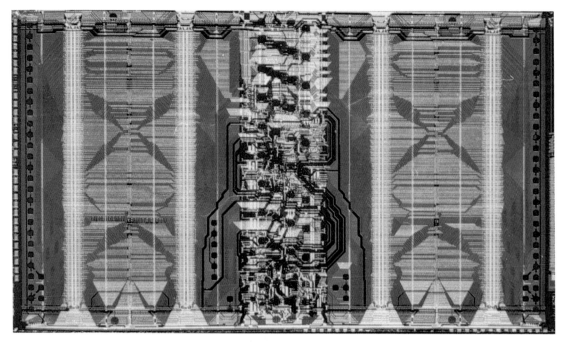

Fig. 6.23 A photograph of a spectacular-looking IBM memory chip.

of having to design a special chip to perform a specific function, such a microprocessor chip could be programmed for the particular application. It took Intel some time to realize that they were sitting on a goldmine. Initially only thought of as a device for calculators and minicomputers, the microprocessor now has applications in all manner of places: washing machines, typewriters, thermostats, video games and personal computers, to name but a few. The first microprocessor was marketed in 1971 and contained 2000 or so transistors; there are now microprocessor chips with several million transistors. Such a high level of miniaturization is abbreviated to VLSI, standing for very-large-scale integration. Are there any limits to this incredible miniaturization of ICs? In chapter 9 we will see that there are, and that it is likely that this period of exponential improvement will come to an end – unless we find some new quantum technologies to replace our present semiconductor technologies.

7 Quantum co-operation and superfluids

> ...there are certain situations in which the
> peculiarities of quantum mechanics can come
> out in a special way on a large scale.
> Richard Feynman

Laser light

Nowadays, everyone has heard of lasers, and laser light displays are
a frequent ingredient of modern rock concerts. Laser light has many applications, ranging from astronomy to hydrogen fusion. What is the special
feature of laser light that makes it so useful? The answer to this question involves a property of wave motion known as 'coherence', with light photons
acting together in a special form of quantum mechanical co-operation. This
type of quantum co-operation will turn out to be vital for an understanding of the peculiar behaviour of quantum 'superfluids'. To understand the
special nature of laser light, however, we must first explain what is meant
by coherence.

Consider the simple wave motion shown in Fig. 7.1. We see that the
pattern repeats itself after one wavelength and the frequency of the wave
corresponds to the number of wavelengths sent out per second. If this wave
is a wave on a string, each point on the string just moves up and down
with a certain amplitude: the maximum distance the point can move out
to before it starts to come back. Up to now, this is really all we have needed
to know about waves. Now consider two waves of the same wavelength but
started at slightly different times, as shown in Fig. 7.2. In the first case,
both waves have their crests and troughs at the same point. In the second
case, the dashed wave has started to fall before the other has reached its
peak. The next figure shows the extreme case when one wave has a trough
when the other has a crest. This is similar to our discussion of interference
in chapter 1. We say that these two wave motions have different phase
differences in these three cases. The phase of a wave is what tells us where
a point on the wave has got to on its up and down motion. If there is a
definite phase difference between two waves, as shown in the figure, the two
waves are said to be coherent and they will display the usual interference

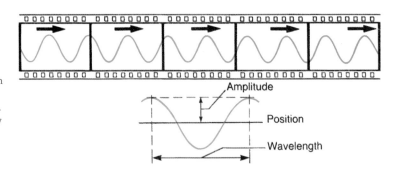

Fig. 7.1 A sequence of 'photographs' of a passing wave. The photographer is standing still – the arrow points over the same crest in all the frames showing how the wave moves to the right. The sketch below shows how the amplitude and wavelength of such a wave motion are defined.

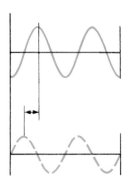
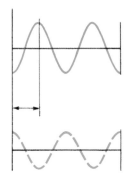

Fig. 7.2 Three pairs of waves with the same wavelength – the top waves are for reference; the bottom waves have different start times.

Charles Townes was born in South Carolina in 1915. During the Second World War he worked for Bell Telephone Laboratories on the design of radar systems. Apparently, one morning in 1951, while waiting for a restaurant to open for breakfast, he had the idea of using molecules rather than electronic circuits to generate microwaves – short wavelength radio waves. By 1953 he had built the 'maser' (microwave amplification by stimulated emission of radiation) using ammonia molecules. Townes went on to speculate about the construction of a similar device for visible radiation.

Theodore Harold Maiman was born the son of an electrical engineer in 1927. He paid his way through college by repairing electrical appliances. While working for Howard Hughes' research laboratories he became interested in the maser invented by Townes and the problem of constructing a similar device for light. Maiman constructed the first laser in 1960.

effects. Two different atomic light sources, on the other hand, do not show interference effects and are said to be incoherent. The reason for the lack of any interference is that the light from the two sources is produced by many different atoms, each emitting photons at different times. Each lamp therefore sends out light that consists of lots of waves with many different phases. Thus, there is no definite phase difference between the light waves coming from the two sources, and all the delicate interference effects are washed out. By contrast, laser light is remarkable in that light from many different atoms is radiated in phase. It is this coherence property of laser light that makes it possible to focus a laser beam on a very small spot and obtain a very high concentration of light energy. A laser beam with less power than an ordinary light bulb can burn a hole in a metal plate.

Laser is an acronym standing for 'light amplification by stimulated emission of radiation'. Stimulated emission is a process of interaction of an atom with light that we have not come across so far. We have seen that if

Fig. 7.3 Welding by laser at the Mirafiori car plant of Fiat in Turin. The laser beam, which is invisible, is delivered from the cone shaped nozzle at the end of the welding head just above the sparks. A 2.5 kW carbon dioxide laser generates the beam.

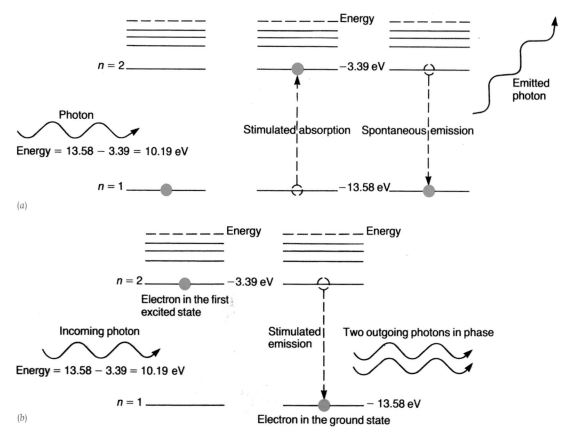

Fig. 7.4 The three possible transition processes for light photons and electrons in an atom. (a) The processes of stimulated absorption and spontaneous emission. In the first case a photon with the right energy can be absorbed by the electron in the ground state and cause it to jump to an excited energy level. After a time this electron will drop back into the lowest energy level giving off a photon with the same energy as the one that was absorbed. This 'decay' process is called spontaneous emission. (b) Stimulated emission of radiation occurs when photons are directed at an atom which is already in an excited state. The stimulating photon and the photon radiated in the transition have the same energy and the same phase.

we shine light on an atom with a photon energy that corresponds exactly to an energy level difference, the electron can be 'stimulated' to jump to the higher state. This process is sometimes called stimulated absorption (Fig. 7.4). We also know that an atom in an excited state can spontaneously emit photon of the right energy so that the electron jumps to the ground state. This process of decay of an excited atom is called spontaneous emission. A third type of process involving photons was discovered by Einstein as early as 1916. In November of that year, he wrote to a life-long friend, Michele Angelo Besso: 'A splendid light has dawned on me about the absorption and emission of radiation'. In Einstein's early days at the Patent Office in Berne, Besso had been Einstein's 'sounding board' and he had been singled out for thanks in the famous paper on special relativity. (When Besso died in 1955 Einstein wrote to the family: 'What I most admired in him as a

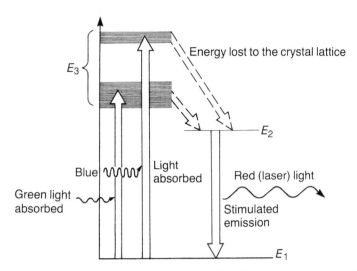

Fig. 7.5 Energy level diagram for a ruby laser. The chromium atoms in the ruby are 'pumped' into the two broad excited energy bands by the absorption of green and blue light. These excited atoms quickly lose energy to the crystal lattice and the electrons drop into the long-lived 'metastable' energy level shown as E_2 in the figure. More electrons will be in this level than in the ground state, so a 'population inversion' exists. Stimulated emission of the transition from this level to the ground state results in red laser light.

human being is the fact that he managed to live for many years not only in peace but also in lasting harmony with a woman – an undertaking in which I twice failed rather disgracefully.') Einstein had realized that if light with the correct photon energy for an electron transition is shone on an atom in an excited state, then the atom could be stimulated to make a transition to the lower energy state. It is natural to call this process stimulated emission of radiation. The excited atom would, of course, have made a transition to the lower state sooner or later – it just makes it sooner in the presence of the stimulus of the radiation. For over 35 years this stimulated emission process gained hardly more than a cursory comment in quantum mechanics textbooks, since it seemed to have no practical application. What had been overlooked, however, was the special nature of the light that is emitted in this way. The photons that are emitted have exactly the same phase as the photons that induce the transition. This is because the varying electric fields of the applied light wave cause the charge distribution of the excited atom to oscillate in phase with this radiation. The emitted photons are all in phase – they are coherent – and, furthermore, they travel in the same direction as the inducing photon.

Such theory is all very well but there are a number of technical problems to be solved before this mechanism can be used to produce intense beams of laser light. At normal temperatures most atoms are in their ground state. We have to look for a way of pumping energy into the laser material so that we manage to get most of the atoms into an excited state. Having more atoms in an excited state than in the ground state is not a normal state of affairs – it is called a population inversion. If we can arrange for such a population inversion, then the stimulated emission process will exceed the stimulated absorption and we will obtain a net amplification of the stimulating light.

The first laser used a ruby crystal, consisting of aluminium oxide in which some of the aluminium atoms had been replaced by impurity chromium atoms. The relevant energy levels of the chromium atoms in this system are shown in Fig. 7.5. By pumping in light with an energy corresponding to the difference between E_1 and E_3 the chromium atoms are excited to the broad short-lived upper state. These excited atoms then decay very quickly to a relatively long-lived state E_2 and a population inversion is obtained. When some of these E_2 states decay spontaneously, the right conditions exist for these photons to cause stimulated emission of the other excited atoms. A diagram of a ruby laser is shown in Fig. 7.6. Photons will be emitted in all directions, but those that do not travel along the length of the ruby rod will escape through the sides before they cause much stimulated emission. Photons traveling along the axis of the rod will be reflected back and forth several times by the mirrors at the ends of the rod. Thus, more

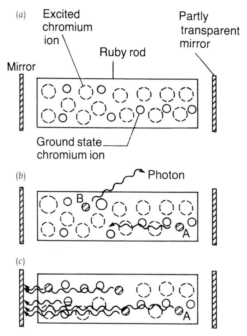

Fig. 7.6 A sequence showing the build up of laser light. (a) The state of affairs after a population inversion has been achieved. The smaller circles represent chromium atoms in their ground state and the larger dashed circles the excited atoms. (b) Two atoms have decayed back to the ground state. In one case the photon leaves the side of the ruby and cannot cause any stimulated emission. In the case where the photon is emitted along the length of the rod, it can generate more decay photons, all with the same phase. (c) The mirrors at the ends help build up a beam parallel to the length of the rod. Several photons are just about to be reflected by the end mirror and will cause more transitions as they pass back along the crystal.

Fig. 7.7 Interior of a ruby laser. The pink cylinder in the top section is the ruby, while the cylinder in the bottom section is the flash tube used to generate the population inversion. Both are cooled by water which enters the apparatus through the pipes visible in the photograph.

and more atoms will be stimulated to emit photons and an intense coherent beam of laser light will emerge from the partially reflecting end of the crystal. In this laser, the pumping required to obtain the crucial population inversion is provided by a flash of light, and a special long-lived metastable state (E_2 in Fig. 7.5) is required to maintain the inversion. Modern lasers can be continuously pumped and do not need the lasing state to be specially long-lived.

The fact that many photons in a laser beam are in the same quantum state is possible only because photons are bosons. For fermions, the Pauli exclusion principle demands that each electron has different quantum numbers but bosons have a tendency to cluster in the same quantum state. We shall discuss this property of bosons in more detail in the next section. We end this section with a brief mention of two very different applications of laser light.

The unique properties of laser light enable us to concentrate light energy in a very intense, short pulse. Using such laser beams, the distance of the Moon from the Earth can be measured with amazing accuracy. Figure 7.8 shows the footprints of the Apollo 14 astronauts near a special reflector they had placed on the Moon's surface. A pulse of laser light is fired through a large telescope at the Moon. By measuring the time of flight of photons reflected from the Moon, the distance to the Moon can be determined to an accuracy of within a few centimetres in a distance of about 400 000 km.

Another interesting application of laser light is three-dimensional photography or 'holography'. Light from a laser is split into two beams by a half-silvered mirror. One beam illuminates the object and the scattered light falls on a photographic plate. The other beam is directed at the photographic plate without scattering off the object. Since the laser light is coherent, the two beams can interfere. The photographic plate records

(a)

(b)

Fig. 7.8 (a) The Moon as seen from the Lure Observatory on the island of Maui, in Hawaii. A pulse of laser light is fired through the telescope at the Moon. By the time the pulse reaches the Moon it has spread out to cover an area more than two miles wide. Part of the beam is then reflected back to Earth by special reflectors left on the surface of the moon by the Apollo 14 astronauts. The return signal can be picked up and the time the light beam has taken to travel from the Earth to the Moon and back can be accurately timed. The travel time is about 2.5 s, and measurements such as this allow us to determine the distance to the Moon to within an accuracy of a few centimetres. (b) The special lunar reflectors.

the interference pattern caused by the recombination of these two beams. This photographic record of the interference pattern is called a hologram, after the Greek word *holos*, meaning whole. This is because a hologram, unlike an ordinary photograph – which just records the intensity of the

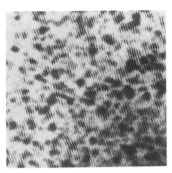

Fig. 7.9 The smudgy looking photograph at the top left is a hologram. The other three photographs show three different views all generated by the same hologram. Not only can one see round objects in different views but also the same pictures, albeit somewhat degraded in clarity, can be produced from just a small piece of the original hologram. This is possible because of the interference mechanism that underlies holography.

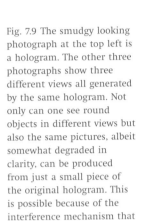

Dennis Gabor (1900–1979) was born in Budapest but received his education in Germany. When Hitler came to power, he came to England and worked as a research engineer for the Thompson–Houston electrical manufacturing company in Rugby. His original paper on holography in 1948 was in the context of electron optics, and it was not until the invention of lasers that his idea gained much attention. He was awarded the Nobel Prize in 1971.

light falling on the photographic plate – also contains information about the phase of the scattered light, since it is a record of an interference pattern. It therefore contains all of the optical information coming from the object being photographed. A hologram bears no resemblance to the object being photographed – it looks like an almost random pattern of smudgy dots. However, when the hologram is illuminated by a beam of laser light, a perfect three-dimensional copy of the original object is reconstructed. If you walk around and look at the image from different angles you see the relative positions of the various pieces exactly as you would for the real thing. In particular, things hidden from view in one position can be seen by looking at the holographic image from another direction. Holography was invented by Hungarian-born Dennis Gabor working in Rugby, UK, in 1947, but remained merely a scientific curiosity for about 15 years. It is only since the advent of coherent beams of laser light that holography has become a multi-million dollar industry, with applications ranging from medical diagnosis to tyre testing.

Bose condensation and superfluid helium

We have seen in chapter 6 how Pauli's exclusion principle, applied to atomic electrons, is able to explain Mendeleev's periodic table of the elements. All the basic matter-like particles – electrons, protons and neutrons – obey the Pauli principle. No two identical fermions, as these particles are called, can occupy the same quantum state. Thus, if we consider

Satyendra Bose (1894–1974). After his paper on the quantum theory of light had been rejected for publication, Bose sent a copy to Einstein in desperation. Einstein personally translated Bose's paper from English into German and arranged for it to be published. From being virtually unknown, Bose suddenly became an internationally famous physicist.

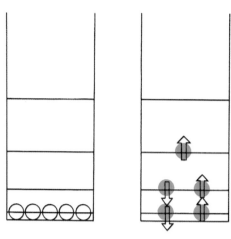

Fig. 7.10 Bosons and fermions in a quantum box. Physical systems tend to have the lowest possible energy. For bosons, this can be achieved by putting all the bosons in the same, ground state energy level. Light photons behave like bosons. Fermions, on the other hand, are particles like electrons that must obey the Pauli exclusion principle. Thus, each energy level can be occupied by at most two fermions, corresponding to the two (spin up and spin down) states of an electron.

putting electrons in a box potential, the electrons cannot all go into the lowest energy level. Instead, they fill up the quantized energy levels in pairs with opposite spin, so that no two electrons have the same quantum numbers. This is what happens for matter-like particles. However, radiation-like particles, such as photons, behave very differently and have a tendency to cluster in the same state! Such particles are called bosons in honour of the Indian physicist Satyendra Bose.

In 1924 Bose was a young Bengali physicist who was virtually unknown in the scientific world. His sixth scientific paper concerned a new derivation of the famous formula with which Planck had introduced both the concept of photons and h, his famous quantum constant. There are many stories in physics about papers that later became famous, after first being rejected for publication. Bose's paper was one of these, but he had the good luck, or foresight, to send a copy to Einstein. Bose asked Einstein if he could arrange for publication in a German journal 'if he thought the work had sufficient merit'. At the time, Einstein was deeply engrossed in his search for a unified theory of all the forces of Nature, but Bose's letter made him deviate temporarily from this, his main line of research. Einstein personally translated Bose's paper into German and sent it off to the journal with a note saying he believed that Bose's work constituted an 'important advance'. Over the course of the next few months, Einstein published several papers extending and clarifying Bose's work. In particular, it was Einstein who first noted the possibility of Bose particles, now called bosons, all 'condensing' into the lowest energy state. We can see what this means by returning to the quantum problem of particles in a box. If, instead of electrons, we put in photons, then the lowest energy state is achieved by all of

Albert Einstein (1879–1955) at the height of his powers in 1916. He had just completed his general theory of relativity as well as his important work on the absorption and emission of light from atoms, which we discuss in this chapter. Einstein received the Nobel Prize in 1921 for his work on the photo-electric effect – another vital contribution to quantum mechanics. In spite of the large part Einstein played in the origin of quantum theory, he remained unhappy with the conventional interpretation of the theory as championed by Heisenberg and Bohr. This is not to say he disputed that quantum mechanics worked, but, rather, that the theory in its present form, with uncertainty playing an essential role, was incomplete. In a letter to Born, who first introduced the probability interpretation of Schrödinger's waves, Einstein made his famous remark that God 'does not play dice' (see chapter 8).

them occupying the lowest energy level (Fig. 7.10). At normal temperatures, enough energy can be transferred in an ordinary collision for most of the bosons to be in excited states. Nonetheless, if we lower the temperature Einstein pointed out that 'from a certain temperature on, the molecules "condense" without attractive forces'. He went on to say: 'The theory is pretty, but is there also some truth to it?' This was in December, 1924; Bose had written to him in June of that year.

Einstein's proposed condensation of bosons – now called Bose or Bose–Einstein condensation – had at first the reputation of having 'a purely imaginary character' and not to represent a real observable physical effect. It was not until 1938 that Fritz London proposed that some strange effects observed with liquid helium could be understood in terms of a Bose condensation of helium atoms. Before we go on to describe these bizarre properties of liquid helium, we must first answer a more fundamental question. As we have said, matter-like particles such as electrons, protons and neutrons are all fermions, so why should helium be considered a boson? The reason is that the usual ^{4}He helium atom contains an even number of fermions: two protons and two neutrons in the nucleus, and two atomic electrons. Experiment tells us that elements with an even number of fermions behave like bosons. Thus, liquid ^{4}He can undergo a Bose condensation at low temperatures, which can explain its remarkable 'superfluid' behaviour. On the other hand, elements with an odd number of fermions are found to obey the Pauli principle and act like fermions. Thus, liquid ^{3}He, with only one neutron in its nucleus, is a fermion and does not undergo a similar condensation to ^{4}He and has very different properties at low temperatures, despite its chemical similarity!

Helium has the lowest boiling point of any gas and was the last to be liquefied. In low-temperature physics, temperatures are usually given in kelvin (symbol K), rather than in degrees Celsius or centigrade. Absolute zero is defined to be the zero of the kelvin scale and corresponds to about -273 °C. There can be no temperature lower than absolute zero. Towards the end of the nineteenth century, physicists in Paris, London and Cracow were vying with each other to produce the lowest temperature. For a long time it had seemed that liquefying hydrogen would be the last step on the road to absolute zero. Sir James Dewar announced the first liquefaction of hydrogen to the Royal Society in London in 1898. In his experiments he had reached about 12 K. By this time, the rare helium gas had been discovered and it had become clear that liquefaction of helium was the real goal. In 1904, Dewar estimated the temperature required to be around 6 K, but it was not until 1908 that the Dutch physicist Kamerlingh Onnes, working in Leiden, finally succeeded in liquefying helium. The boiling point of helium was in fact found to be about 4 K.

Liquid helium is now known to have many remarkable properties. It remains liquid even if cooled as close as possible to absolute zero. This is because of the large zero-point motion of the helium atoms – the necessary quantum jiggling required to satisfy Heisenberg's uncertainty principle.

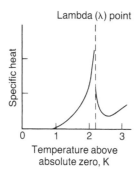

Fig. 7.11 The lambda point for liquid helium. This is a peculiar change that occurs at about 2.2 degrees above absolute zero and is evident in this lambda-like curve formed by these measurements of the specific heat of helium.

Furthermore, at about 2 K, a dramatic change occurs. All boiling ceases and the liquid becomes very still. Other properties also change abruptly. Figure 7.11 shows the variation with temperature of the specific heat: the amount of heat needed to raise the temperature of one gram of helium by one kelvin. Because the shape of this curve looks like the Greek letter lambda, λ, this temperature is known as the lambda point. Below the lambda point, the viscosity, or 'treacliness', of helium suddenly drops by a factor of about one million. Perhaps most surprising of all is the ability, below the lambda point, of liquid helium to 'creep' as a thin film along the walls of its container (Fig. 7.12). If a beaker is lowered into a larger container of liquid helium, a thin film of helium forms over the entire surface of the beaker. This film then acts like a siphon through which helium can flow with almost no viscosity. Thus, no matter what the original difference in levels of the helium inside and outside of the beaker, the helium flows till they are equal! Kurt Mendelssohn recalled the discovery of this 'film transfer' phenomenon in the Clarendon Laboratory in Oxford, UK, with the following words:

> If the beaker is withdrawn from the bath, the level will drop until it
> has reached the level of the bath. If the beaker is pulled out
> completely, the level will still drop, and one can see little drops of
> helium forming at the bottom of the beaker and falling back into the

Fig. 7.12 Liquid helium below the lambda point is a superfluid and shows some remarkable properties. This photograph shows how liquid helium can 'creep' up the sides of a container, flow over the top, flow down the outside and collect in drips at the bottom.

Fig. 7.13 This spectacular photograph of the 'fountain effect' is another example of the strange behaviour of liquid helium.

bath. This is the sort of thing that makes one look twice and rub his eyes and wonder whether it is quite true. I remember well the night when we first observed this film transfer. It was well after dinner, and we looked around the building and finally found two nuclear physicists still at work. When they, too, saw the drops, we were happier.

All these peculiar properties of liquid helium are the result of the helium atoms condensing into the lowest energy state, forming a quantum superfluid. Since essentially all the atoms are in the same quantum state, they behave in a co-operative way and this is what gives the superfluid its unusual properties. As Feynman says at the beginning of this chapter, this is a striking example of the peculiarities of quantum mechanics being observed on a large scale. Without quantum mechanics, de Broglie, Heisenberg, Schrödinger and all, we would have no explanation for all these strange effects!

There is a short postscript to the story of liquid helium that serves as a good lead-in to our discussion of superconductivity. As we have said, liquid ^{3}He is expected to behave quite differently from liquid ^{4}He, since its atoms behave like fermions and cannot undergo Bose condensation. Experiment

Douglas Osheroff (a), Robert Richardson (b) and David Lee (c) were awarded the 1996 Nobel Prize for Physics for their discovery of superfluidity in Helium-3. This was remarkable since an individual Helium-3 atom contains an odd number of particles and normally behaves like a fermion. At temperatures around 0.002 K – a thousand times colder than superfluid Helium-4 – pairs of Helium-3 atoms conspire to act like bosons and make superfluid condensation possible.

bears out this expectation, but a new form of Bose condensation is seen at a very much lower temperature, at about 0.002 K! At this temperature, the weak attractive forces between two ^{3}He atoms are sufficiently strong for the pair to bind together and act like a boson. These pairs of ^{3}He atoms can then undergo a similar Bose condensation to that of the individual ^{4}He atoms. As we shall see, a similar pairing mechanism is responsible for super-conductivity. David Lee, Douglas Osheroff and Robert Richardson of Cornell University were awarded the 1996 Nobel prize for physics for this discovery.

Cold atoms

Superfluid helium involves quantum co-operation among helium atoms. This type of Bose–Einstein condensation occurs for atoms that are already in a liquid state. Can Bose–Einstein condensation take place in a gas before it precipitates into drops of liquid or freezes into a solid? For this to occur the atoms must be sufficiently far apart to prevent ordinary condensation into the liquid state, yet not so distant that Bose–Einstein condensation is inhibited. The key requirement is ultra cold temperatures – less than a millionth of a kelvin above absolute zero. In 1995, Eric Cornell, Carl Wieman and colleagues succeeded in making a dilute gas of atoms so cold that Bose–Einstein condensation took place. All the atoms behaved as a single entity in a collective quantum mechanical fashion. How is it possible to reach such low temperatures that individual atoms move more slowly than a tortoise? One surprising and crucial cooling technique captures atoms between crossed beams of laser light.

Recall that atoms absorb and emit light only when the photon energy is exactly equal to the energy level difference between two allowed electron states. The photon picture of light suggests that the process of light emission is rather like firing a bullet from a gun while that of light absorption is similar to a bullet hitting a target. This image correctly predicts that an atom emitting or absorbing a photon reacts by recoiling. At room temperature a gas consists of a collection of atoms moving in random directions with different speeds. According to the standard model of gases, temperature is a measure of the average speed of the atoms in the gas. If we could cool the gas to absolute zero, the random zero-point motion of the gas atoms would be the minimum required to be consistent with Heisenberg's uncertainty principle. Clearly, to understand the interaction of light with a gas we need to take account of the motion of the atoms.

Imagine an atom moving towards an incoming photon. We are familiar with the Doppler effect for sound in everyday life. For example, if we stand by a railway track the sound of a horn on an express train is raised in pitch as the train approaches us and lowered as it travels away from us. A similar Doppler effect occurs for photons and light. If the atom is moving towards the oncoming photon the optical frequency will be higher because of this optical Doppler effect. Since the atoms in the gas are moving at different speeds, each atom will be tuned to a different photon frequency.

Daniel Kleppner began investigating Bose–Einstein condensates in the 1970s. He was not the first to achieve Bose–Einstein condensation but all the three groups that were first to achieve this feat were led by his former students.

Imagine an atom in the gas moving with exactly the right speed to absorb a photon from an approaching laser beam. When the atom absorbs the photon it will slow slightly from the impact. Of course the photon will ultimately be re-emitted but in a random direction. Since the laser beam contains many photons this process can be repeated many times. The overall effect is rather like the atom moving into a hail of bullets. The net effect is to slow the motion of the atom in the direction of the laser beam and add a small random motion in other directions.

Suppose we tune the frequency of a laser beam to correspond to an energy just below an atomic energy gap of the atom. Atoms moving towards the beam cause a Doppler shift in the photon frequency that will allow photon absorption and slow their motion in the direction of the laser. Since gas atoms move in all directions, if we want to reduce the speed effectively we need to surround the atoms of the gas by six laser beams arranged in three oppositely directed pairs as shown in Fig. 7.14. The resulting configuration of laser beams has been called 'optical molasses' since the atoms feel forces slowing them down in all directions. As the atoms are slowed we must adjust the frequency of the laser beam so that the more slowly moving atoms can continue to absorb photons and be slowed still further. The first demonstration of such laser cooling was made in 1985

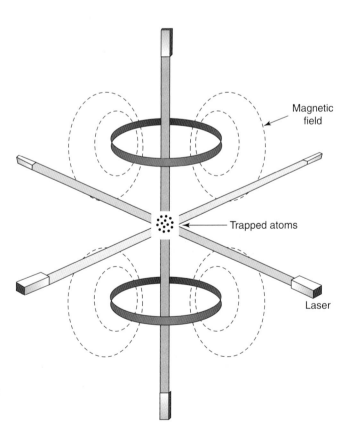

Magnetic field

Trapped atoms

Laser

Fig. 7.14 A schematic diagram of an magneto-optical trap. Six laser beams are used to slow down the trapped atoms. The magnetic fields are designed to keep the atoms trapped.

by Stephen Chu and colleagues at AT&T Bell Laboratories in Holmdel, New Jersey. They were able to cool sodium atoms to a very chilly 240 millionth of a kelvin above absolute zero. This temperature is still much too high for the formation of a gaseous Bose–Einstein condensate. Moreover, gravity causes the cooled atoms to fall out of the optical trap after only about a second. This containment problem was solved by William Phillips and his group at the National Institute of Standards and Technology in Maryland using a combination of magnetic fields.

Many atoms behave like small magnets in a magnetic field. Under a non-uniform magnetic field, a magnet experiences a difference in force on its north and south magnetic poles. Phillips and his team modified the optical molasses set-up by including such magnetic fields above and below the laser atom trap. This modified optical trap was able to keep the atoms for much longer and they successfully cooled atoms to about 40 millionths of a kelvin. This was a puzzling result – since the lowest temperature expected using Doppler cooling was around 240 millionths of a kelvin. It did not take theoretical physicists long to come up with an explanation of how this additional sub-Doppler cooling worked. Claude Cohen-Tannoudji and colleagues in France showed that multiple electronic levels could participate in the absorption and emission of light. Their theory predicted that it should be possible to use laser cooling to slow atoms to a velocity equal to the recoil velocity given to an atom by a single photon. Using their new theoretical understanding of the process, the French team were able to cool helium atoms to 0.18 millionths of a kelvin. In June of 1995 came the breakthrough. A team of physicists at the University of Colorado led by Eric Cornell and Carl Weiman cooled a group of atoms to 20 thousand millionths of a degree above absolute zero and created a new quantum state of matter. About 2000 atoms had formed a Bose–Einstein condensate and no longer behaved like distinct classical particles. In some ways, such a condensate is an atomic version of coherent laser light. The full implications of such condensates are still being explored.

(a)

(b)

(c)

The 1997 Nobel Prize for Physics was won by Steven Chu (a), William Phillips (b) and Claude Cohen-Tannoudji (c) for their development of methods to cool and trap atoms using laser light.

In 1997 Steven Chu, Claude Cohen-Tannoudji and William Phillips were rewarded for their pioneering work on super-cooled atoms with the award of the Nobel Prize for physics. Their work has many possible applications besides Bose–Einstein condensates. The key feature of the technology they created is its ability to manipulate matter with light. This has already led to more accurate atomic clocks, the possibility of atom interference devices and the development of 'optical tweezers'. Optical tweezers use optical forces to control and manipulate objects larger than atoms such as strands of DNA.

(a)

(b)

(c)

In October 2001, the Royal Swedish Academy awarded the 2001 Nobel Prize for Physics to Wolfgang Ketterle of MIT (a), Carl Wieman of the University of Colorado (b) and Eric Cornell of the National Institute of Standards and Technology in Boulder, Colorado (c). The three physicists share the prize of $952 738 between them. Working together in 1995, Cornell and Wieman had succeeded in cooling about 2000 atoms to near absolute zero to create the first 'Bose–Einstein condensate'. This is a strange state of matter in which

Superconductivity

Soon after the discovery of the electron it was realized that many features of the ability of metals to conduct electricity could be explained in terms of the motion of electrons. The resistance to current flow is caused by electrons being scattered by collisions with defects in the crystal lattice of the metal and by interaction with the vibrations of the crystal atoms. As the temperature is decreased, the atoms will vibrate less and less and one would expect the resistance of the metal to approach a constant value. This is indeed what happens for many metals. It was, therefore, all the more surprising to find that the electrical resistance of certain metals suddenly fell to zero when they were cooled below a certain critical temperature. The electrical resistance of normal metals causes energy loss and heating; in these extraordinary materials, by contrast, currents can be set up that persist for years. Such metals are genuinely 'superconductors'.

The phenomenon of superconductivity was discovered by Kamerlingh Onnes, 'the gentleman of absolute zero', at his laboratory in Leiden in 1911. Fig. 7.20 shows a graph of the resistance of mercury taken from his original paper. In 1933, another fascinating property of superconductors was discovered. If a magnetic field is applied to a superconductor, electrical currents are set up in the metal that conspire to exactly cancel the applied magnetic field. This exact cancellation is only possible because electric currents experience no resistance inside the superconductor. This leads to some very striking effects. A small magnet placed over a superconducting dish will float there because of the currents caused by the magnet in the dish (Fig. 7.19). Superconducting levitation has been seriously considered as a method for providing very smooth support for high speed trains.

How can superconductivity be understood? As early as 1935, in Oxford, the brothers Heinz and Fritz London – who did a lot of the early experimental and theoretical work on superconductors – realized that quantum mechanics must be an essential ingredient in any understanding of the effects. However, it was not until 1956 that Leon Cooper came up with the key observation. He showed that although two electrons normally repel each other because of their electric charges, in a metal there is also an indirect attractive force between them, caused by the attraction of the positively

Kamerlingh Onnes (1853–1926) in his cryogenic laboratory in Leiden, The Netherlands. Onnes was the first to liquefy helium, and he received the 1913 Nobel Prize for this achievement. He also was the first to observe the phenomenon of superconductivity – the vanishing of the electrical resistance of some metals at very low temperatures.

charged crystal lattice ions. Roughly speaking, an electron sitting between two positive ions in the lattice brings these ions a little closer together than normal, and another electron will therefore feel a small net attraction. There is therefore the possibility that two electrons will be bound together to form a 'Cooper pair'. These pairs are rather curious in that they consist of electrons with opposite velocities adding up to a net momentum of zero for the pair. Moreover, according to Heisenberg's uncertainty principle, since the momentum of the pair is well defined, the Cooper paired electrons must be very spread out in space. Each pair occupies a region that is several thousand times larger than the size of the individual atoms. This same space is occupied by millions of other overlapping pairs.

Given our discussion of Bose condensation for ^{3}He, it is not too difficult to guess the next step in the argument. The Cooper pairs act like bosons, and condense to form the superconducting state. While this is easy to say, it proved difficult to come up with a quantitative and predictive theory of this condensation. This final step was taken by a trio of physicists now universally known as 'BCS': John Bardeen, Leon Cooper and John Schrieffer. They were working at the University of Illinois, and, because of shortage of space, Bardeen and Cooper were sharing an office. Schrieffer was Bardeen's Ph.D. student and had a desk with other theoretical physics students in a neighbouring building. They were trying to extend Cooper's idea of the formation of a single bound pair of electrons to all the electrons in the superconducting material. Schrieffer later described what they

individual atoms condense into a single quantum state. Four months later, Ketterle used a cloud of sodium atoms to create a Bose–Einstein condensate with many more atoms. A novel arrangement of lasers was used to cool the atoms to within 20 thousand millionths of a degree above absolute zero.

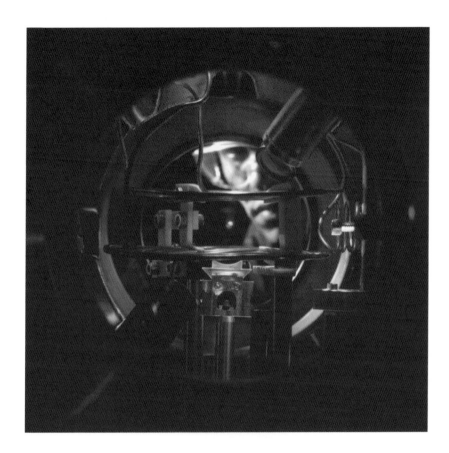

Fig. 7.15 Glowing sodium atoms held in magneto-optical trap. The face of Kristian Helmerson of the National Institute of Standards can be seen through the device.

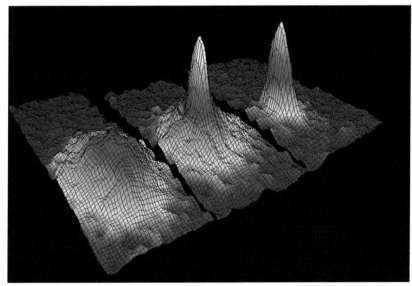

Fig. 7.16 False colour images of the velocity distribution of a cloud of trapped rubidium atoms. The left-most picture shows the cloud just before Bose-Einstein condensation takes place at a temperature of around 400 nK; the centre picture just after condensation at a temperature of around 200 nK; and the right-most picture after further cooling to around 50 nK so that most atoms of the cloud are in the Bose–Einstein condensate. (nK = nano Kelvin or a thousand millionth of a Kelvin).

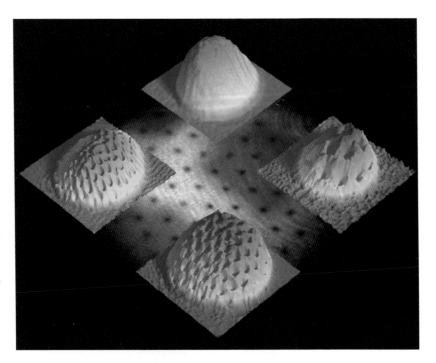

Fig. 7.17 Vortex 'lattices' in a stirred condensate of rubidium atoms. The rotation is in a form such that each atom has a single quantum of rotation. The figures show increasing numbers of vortices with the lowest at the top followed by the right and the other two have more complex patterns corresponding to greater numbers of vortices.

were trying to do as a search for 'a quantum wavefunction to choreograph a dance for more than a million million million couples'. The problem seemed so hard that Schrieffer was thinking of changing his thesis problem to one on magnetism. At this crucial time, Bardeen had to go to Stockholm to receive his share of the Nobel Prize for the invention of the transistor, and before he left he urged Schrieffer to work on the problem for another month. In that month Schrieffer guessed at a manageable form for the wavefunction of the Cooper pair Bose condensate. In the month that followed, B, C and S were able to show that their theory explained all the experimental data. Paradoxically, one finds that metals that are good conductors of electricity at normal temperatures have very little electron–ion interaction and so will not be superconductors at low temperature. Rather, it is the bad conductors at normal temperatures that end up as superconductors.

In the spring of 1986, Johannes Georg Bednorz and Karl Alexander Muller made a remarkable discovery. A ceramic material – lanthanum barium copper oxide – became superconducting at 35 degrees above absolute zero. This may not seem a very significant result but the superconducting transition temperature is more than 10 degrees warmer than those of the classic superconducting materials made from metals or alloys. Indeed, since the original discovery, superconductors based on copper oxide compounds have been discovered with transition temperatures of up to 135 degrees above absolute zero. These so-called high-temperature superconductors hold out the prospect of very different economics and many new applications.

(a)

The atom laser at 200 Hz repetition rate

(b)

0   Density (arbitrary units)   1

(field of view 2.5 mm x 5.0 mm)

(c)

Fig. 7.18 (a) Michael Andrews, Marc-Oliver Mewes and Wolfgang Ketterle (left to right) with the equipment that they used to demonstrate the first atom laser. (b) Atom lasers are essentially moving Bose–Einstein condensates. This first atom laser was powered by gravity. The crescent shapes are clumps of an expanding sodium BEC. (c) Since Bose–Einstein condensates contain atoms occupying a single quantum state, overlapping atom lasers can interfere to produce characteristic interference fringes.

Fig. 7.19 Superconducting levitation. A small magnet floats over a superconducting dish. Superconducting currents flowing in the dish generate forces that repel the magnet and counteract the force of gravity.

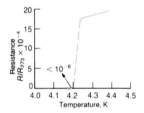

Fig. 7.20 Kamerlingh Onnes' discovery of superconductivity in 1911. A dramatic graph showing how the electrical resistance of mercury suddenly vanishes as the temperature is lowered below about 4.2 K.

The BCS in the BCS theory of superconductivity – John Bardeen (centre), Leon Cooper (right) and John Schrieffer (left). The award of the 1972 Nobel Prize for their work made Bardeen the only person to win two Nobel Prizes in the same subject (with Brattain and Shockley he had discovered the transistor). Bardeen was a student of another famous quantum physicist, Eugene Wigner, who himself won the Nobel Prize in 1963. There are many examples of such 'father–son' Nobel Prize relationships.

Compared with cooling a substance in liquid helium, using liquid nitrogen for cooling is like using milk instead of champagne!

These high-temperature superconductors were discovered by experimenting with materials that had strong electron–ion interactions. This suggested an explanation in terms of conventional Cooper pairs, but recent experiments have shown that the mechanism for high-temperature superconductors is fundamentally different from the classic BCS theory.

(a)

(b)

Alex Bednorz (a) and Johann Muller (b), two physicists working at the IBM Research Laboratory in Rushlikon near Zurich, were awarded the 1987 Nobel Prize for Physics for their discovery of superconductivity in ceramic materials. Their discovery of a new class of materials that became superconducting at much higher temperatures than conventional BCS superconductors came after years of work trying many different materials without success.

Conduction occurs along sheets of copper oxide atomic planes that are sandwiched between insulating layers. In most copper oxide compounds the charge carriers are holes (see chapter 6). It is difficult to explain the formation of Cooper pairs from any conventional hole–interaction. The true mechanism is not yet established.

There are many applications of superconductivity. Superconducting electromagnets are now used to obtain high magnetic fields without the usual power losses of electromagnets constructed using ordinary conductors for the windings in the coil. A problem arises in trying to achieve very high magnetic fields. A magnetic field tends to be induced in the windings of the magnet itself, and too high a magnetic field can destroy the superconductivity of the coil. This problem can be alleviated by using so-called 'Type-II' superconductors (Fig. 7.22). These are superconductors in which magnetic fields are not excluded completely from the metal but are able to penetrate the superconductor in thin 'tubes' of flux. Very high magnetic fields can be produced by electromagnets using such superconducting wire. The property of superconductors in screening out magnetic fields can also be used to make improved electron microscopes.

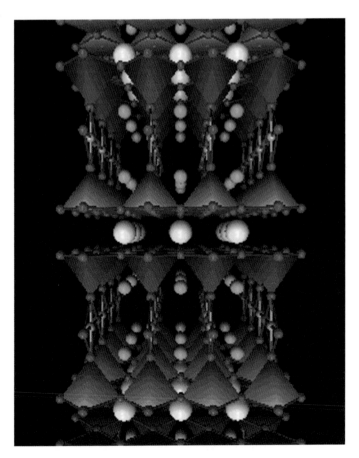

Fig. 7.21 A computer generated image of the complex structure of a high-temperature superconductor.

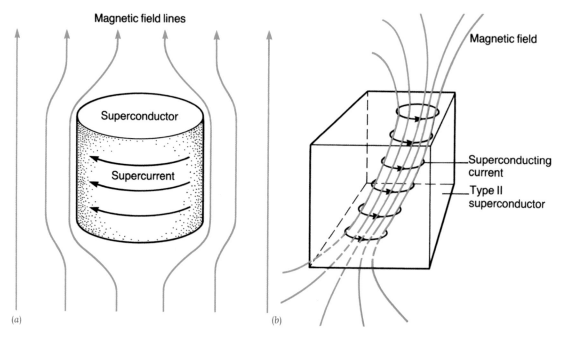

Fig. 7.22 Magnetic fields are excluded from superconductors. (*a*) In a type I superconductor, such as lead or tin, magnetic fields are expelled completely from the metal by the circulating supercurrents induced by the field. (*b*) By contrast, in a type II superconductor, the magnetic field can penetrate the metal in thin tubes.

Perhaps the best-known applications of superconductors are the 'Josephson junction' and a device called a 'SQUID' (Superconducting Quantum Interference Device). Both use a discovery of a British Ph.D. student, Brian Josephson. Philip Anderson, himself a Nobel Prize winner, recalls giving a lecture course on solid-state physics in Cambridge, UK, in 1962, and having Josephson in the audience:

> This was a disconcerting experience for a lecturer, I can assure you, because everything had to be right or he would come up and explain it to me after class.

Josephson was studying the quantum theory of a superconductor–insulator––superconductor sandwich, in which the filling, the insulator, is only a very thin film. He showed that Cooper pairs were able to tunnel through the junction and that this gave rise to some very interesting effects. One prediction was that a current would flow even if no voltage was applied to the junction! He also worked out what would happen in the presence of a magnetic field as well as that of a very-high-frequency oscillating voltage together with a constant voltage. This last arrangement permits the most accurate measurement of the ratio of fundamental constants h/e (Planck's constant divided by the charge of the electron). The Josephson effect has been used to measure incredibly small voltage differences and can also be used as a sensitive radiation detector. By putting one or more Josephson junctions together in an electrical circuit, it is possible to make a device

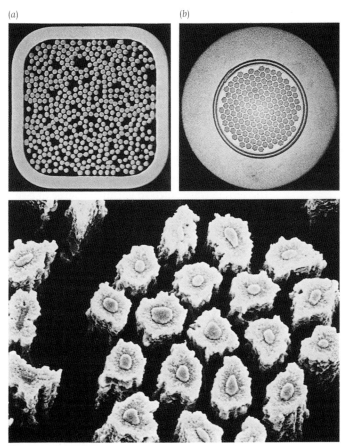

Brian Josephson was only 20 when he made the discovery which was to win him a share of the 1973 Nobel Prize. Josephson was a student attending a course of lectures given in Cambridge, UK, by another great physicist and Nobel Prize winner, Philip Anderson. After class one day Josephson showed Anderson his calculations of tunnelling by super-conducting Cooper pairs of electrons. Josephson's ideas have opened the door to superconducting inter-ferometry, which has many applications in physics and technology.

Fig. 7.23 A montage of photographs of superconducting cable designed to carry very large electric currents. (a) A steel pipe containing hundreds of superconducting wires. Liquid helium is pumped through the cable to maintain the low temperatures necessary for superconductivity to exist in these wires. (b) A magnified photograph of a single wire showing how thousands of superconducting filaments are arranged in hexagonal groups, all encased in a jacket of copper. (c) An electron microscope picture of a single group of these filaments.

for measuring magnetic fields extremely accurately. These devices are the aforementioned SQUIDs, and they are now being used in such diverse fields as medicine and geology. All these applications are possible because the Bose condensation of the Cooper pairs in a superconductor allows us to ob-serve quantum effects on a large scale instead of being restricted to atomic dimensions.

The quantum Hall effect

More Nobel prizes have been associated with superconductivity than any other single topic in physics. We conclude this chapter with another

Klaus von Klitzing was awarded the 1985 Nobel Prize for Physics for his discovery of the quantum Hall effect.

new discovery – the quantum Hall effect – which may have an intimate connection with superconductivity. The classical Hall effect was discovered in the nineteenth century by the American physicist Edwin Hall. He showed that when a magnetic field is applied to a crystal of material carrying a current, the conduction electrons are deflected sideways and a voltage develops across the crystal, at right angles to the direction of the current. As the magnetic field is increased, the Hall voltage will increase.

In 1980, Klaus von Klitzing and his colleagues performed an experiment in which they trapped electrons between two crystalline semiconductors, thus confining the motion of the electrons to a single plane. This is reminiscent of the situation in high-temperature superconductors. When the system was cooled to within a degree or two of absolute zero, they found that the Hall voltage changed discontinuously in discrete steps as the magnetic field was smoothly increased. Moreover, at these Hall voltage steps, the material becomes an almost perfect conductor. This is not technically a superconductor because the magnetic field is not expelled from the conductor – but it does perhaps hint at some relationship with superconductivity. Perhaps more significant was the discovery that the Hall resistance – the Hall voltage divided by the current – is quantized. The unit of quantization is proportional to Planck's constant divided by the square of the charge on the electron. This unit of resistance is related to a fundamental quantity in atomic physics –the so-called fine structure constant.

It was the discovery of this phenomenon at the National Magnetic Field Research Service in Grenoble, France, that led to the award of the Nobel prize to Klaus von Klitzing in 1985. The quantized Hall resistance has now been adopted as the standard for calibrating resistance measuring devices. Since its discovery, the quantum Hall effect has been a very exciting area of research and a variety of new effects has been discovered despite the lack of a full explanation for all aspects of these phenomena.

8 Quantum jumps

> We always have had a great deal of difficulty in understanding the world view that quantum mechanics represents. At least I do, because I'm an old enough man that I haven't got to the point that this stuff is obvious to me.... It has not yet become obvious to me that there is no real problem. I cannot define the real problem, therefore I suspect there is no real problem, but I'm not sure there's no real problem.
>
> Richard Feynman

Max Born and quantum probabilities

In this chapter, we will take a short break from our survey of successful applications of quantum mechanics to take a closer look at the foundations of this great edifice of modern physics. In a sense, we are now disobeying Feynman's warning about the results of the double slit experiment of chapter 1. We shall now ask Feynman's 'forbidden' question 'But how can it be like that?' Nobody disputes that quantum mechanics has been magnificently successful, enabling us to make correct quantitative calculations of atomic and nuclear effects. But there is a great divergence of opinion about the implications of quantum mechanics for the nature of matter, and indeed, of reality itself. To avoid becoming embroiled in a purely philosophical mire, we will focus on two famous paradoxes. The first of these is the 'Einstein–Podolsky-Rosen' paradox, named after its originators Albert Einstein, Boris Podolsky and Nathan Rosen, and usually abbreviated by the initials 'EPR'. The second paradox is named after 'Schrödinger's cat'. Both examples demonstrate the unease that Einstein and Schrödinger, two of the founders of quantum mechanics, had about the foundations of the theory.

Einstein always disliked the fact that quantum mechanics was intrinsically probabilistic and his EPR 'thought' experiment was designed to demonstrate that quantum mechanics could not be the whole story. Nearly 30 years later, the Irish physicist John Bell showed how one might test Einstein's ideas about reality and the EPR 'thought' experiment became a 'real' experiment. Unfortunately for Einstein, the results of such EPR experiments have shown that any attempt to eliminate the non-deterministic, probability aspects from quantum mechanics would require modifications to quantum theory in ways that Einstein would certainly not have liked!

The second paradox concerns the curious case of 'Schrödinger's cat'. Although Schrödinger was stimulated by the EPR example, his 'thought' experiment actually highlights a different problem – 'quantum jumps'. The phrase 'quantum jump' is now widely used in everyday speech. Let us see what it means in its original context of quantum mechanics. We go back to the example of the double slit experiment and arrange things so that only one electron passes through the apparatus at a time. Before we record the arrival of this electron, its position is indefinite and, according to quantum mechanics, all we know is specified as a wave of probability extending over all the detectors. After a flash at a particular detector, we suddenly know the location of the electron. Instead of a spread-out wave function, the probability amplitude has apparently 'collapsed' all of the potential electron positions down to one. This is the famous quantum jump. Although Schrödinger's wave equation accurately describes the spread of the quantum probability wave of the electron it does not predict the quantum jump of the electron to a particular location or quantum state. This is the heart of the so-called 'quantum measurement' problem. Despite the evident importance of this question for a complete understanding of quantum theory, there is no universal consensus amongst physicists about the mechanism by which 'measurement' causes the electron to jump into one particular state. Indeed, there is not even unanimous agreement about exactly what constitutes a quantum measurement! Until recently, most working quantum physicists preferred to close their eyes to such problems. They were apparently happy to accept that quantum mechanics is well enough defined 'For All Practical Purposes' – FAPP – as John Bell used to say. As we will see, new developments in 'quantum computing' are forcing physicists to re-examine these issues.

Before we look at these paradoxes in detail, we should give due acknowledgement to the physicist who first recognised that Schrödinger's wave must be interpreted as a probability wave. Let us go back once more to the double slit experiment for electrons. Here we saw that the mathematical form of the interference pattern for electrons is the square of the sum of the amplitudes for electron waves going through slit 1 and slit 2. Since this interference fringe pattern is directly related to the number of electrons that arrived at a given place, we deduced that the wave itself must represent a 'quantum probability amplitude'. The idea that Schrödinger's waves are probability waves is due to the German physicist Max Born. In spite of the fact that Born wrote some of the earliest papers on quantum mechanics, his role in the interpretation of these quantum waves was strangely neglected in the early textbooks on quantum mechanics. This neglect also extended to the Nobel Prize Committee. Most of the founding fathers of quantum mechanics received their Nobel Prizes only a few years after the significance of their contributions had been recognized. Born had to wait for nearly 30 years until he was awarded the Nobel Prize for his probability interpretation of the quantum mechanical wave function!

The first papers on quantum mechanics by Heisenberg required physicists to understand quantum phenomena in terms of mathematical

Max Born (1882–1970) in the library of his home at Bad Pyrmont in Germany just before he was awarded the Nobel Prize at the age of 72. He led one of the premier theoretical physics groups in Gottingen before leaving Germany in 1933. He became a British National and held the Tait Chair of Theoretical Physics in Edinburgh from 1936 to 1953.

objects called 'matrices'. Matrices are arrays of numbers with the confusing property that matrix A times matrix B need not necessarily be equal to matrix B times matrix A. At that time, although well known to mathematicians, matrix methods were unfamiliar to most physicists. Schrödinger then showed how a wave equation, and the familiar mathematics of differential equations, could explain quantum behaviour. It was not surprising that Schrödinger's equation was welcomed by most of the physics community. But what was the physical significance of Schrödinger's quantum waves? Schrödinger very much wanted to find a real physical interpretation for his quantum waves, but was eventually forced to admit defeat. One problem was that the wave function for a two-electron atom, such as helium, depends on six coordinates – the x, y, z values for both electrons – and it was hard to see how this could correspond to a physical wave motion. Another difficulty was that his equation, unlike the wave equations for classical waves, involved the symbol 'i' representing the square root of -1. The use of such so-called 'complex' numbers is commonplace in physics and is a powerful tool for the solution of many different types of problems. However, quantities measured in experiments are always reassuringly 'real', with no place for complex numbers with an 'imaginary' component involving 'i'. By contrast, Schrödinger's wave functions can be complex numbers and obviously cannot be directly observable quantities. Although communications technology was still relatively primitive in the 1920s – there were no Internet and World Wide Web to broadcast discoveries to the world – developments in the early days of quantum mechanics were surprisingly rapid. Schrödinger's first paper was written in January 1926: by June 1926, Born had put forward his probability interpretation of the quantum wave function. With the benefit of hindsight, we now see that this step by Born represents a fundamental break with classical physics. Probabilities now enter physics as an essential, intrinsic limitation of quantum theory. Of course, probabilities also occur in classical physics, but only as an 'in practice' limitation – not as a fundamental, 'in principle' restriction on what we can ever know about a system. Consider the example of a tossed coin. We usually assume that it will have an equal chance of landing heads or tails. 'In practice', we cannot anticipate which outcome will occur. 'In principle', according to classical physics, if we had a detailed enough knowledge of the exact initial conditions of the coin, and calculated precisely the effects of all the forces acting on the coin, we could predict the result. By contrast, according to quantum physics, we can never escape from probabilities. Einstein was never happy with the introduction of probability into physics and in a letter to Born he made his famous remark 'God does not play dice'. Reportedly, Neils Bohr's rejoinder to Einstein was that it was not the business of physicists 'to prescribe to God how He should run the world.'

Recall that in the double slit experiment, we saw that although electrons 'appeared to travel like waves', they 'arrived in lumps like bullets'. The square of the wave function gives the probability of arrival at any place on the detector array. If we send a large number of electrons through the apparatus, we can predict the statistical distribution of these electrons over the detector array. Alternatively, we can perform the experiment with such a low intensity of electrons that it is extremely unlikely that more than one electron is present in the apparatus at a time. The quantum wave function also predicts the probability of arrival at different positions for this one electron. The position of arrival of a single electron is thus inherently unpredictable: we can only make statements about relative probabilities of arrival for the electron. When the arrival of an electron is detected by a flash at one of the detectors, the previously spread-out probability wave function of this electron obviously collapses down to the region bounded by this detector. How this collapse happens is **not** governed by the Schrödinger equation. This collapse or 'reduction' of the wave function is **the** mystery of quantum mechanics. In order to see just how strange this is, contrast the behaviour of a classical particle described by Newton's Laws. Such a particle follows a classical trajectory all the way to the detector. In principle, just before the particle arrives we could look and see it heading for the chosen detector. Not so in quantum mechanics! Before the electron arrives at a detector we cannot say anything definite about its position, certainly not that it is heading for any particular detector. One of the fundamental difficulties for quantum physics is how classical quantities such as particle trajectories – for example, the track of a particle in a bubble chamber – emerge from Max Born's probabilistic fog.

Photons and polarized light

Let us take a closer look at how quantum mechanics describes quantum 'objects' such as electrons or photons. Instead of discussing electrons and their quantum 'spin' states, in this chapter we will choose light and its description in terms of photons as our fundamental quantum system, in the hope that light and its polarization properties may be more familiar than electrons and their spins. The problems of interpretation are the same for both, or as Feynman once said 'There is one lucky break, however – electrons behave just like light'.

In 1865, James Clerk Maxwell unified the phenomena of electricity and magnetism and summarized all the experimental observations in a set of equations that are now known as 'Maxwell's Equations.' According to Maxwell, light is understood as an electromagnetic wave in which the electric and magnetic fields oscillate in a plane at right angles to its direction of motion. If we concentrate on the behaviour of the electric field, it can oscillate in any direction in this plane (see Fig. 8.1). Ordinary light can be visualized as a collection of oscillating electric fields oriented randomly in all possible directions. Such light is said to be 'unpolarized' – the electric field does not point in any particular direction. Now consider what happens when you put on a pair of polaroid spectacles. Glare caused by light

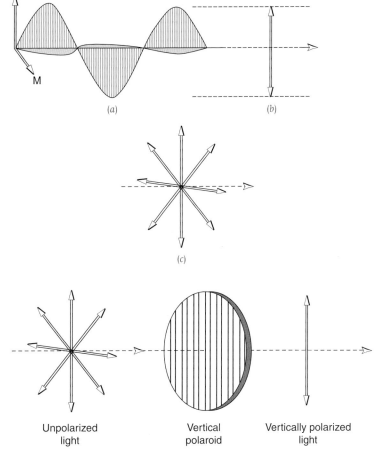

Fig. 8.1 Polarized and unpolarized light. (a) A vertically polarized electro-magnetic wave showing the electric field (E) oscillating in the vertical direction and the magnetic field (M) in the horizontal plane (b) A representation of vertically polarized light referring only to the electric field (c) A pictorial representation of unpolarized light consisting of electromagnetic waves with all directions of polarization.

Fig. 8.2 Unpolarized light arriving at a vertical polaroid to create vertically polarized light.

reflected from the sea or from snow is much reduced. This is because po-laroid glass has the property of only allowing one direction of electric field oscillations to pass through. Figure 8.2 shows a representation of this effect in which unpolarized light passes through a polaroid plate oriented so that light becomes 'polarized' in the vertical direction. The polarization state of light with its electric field oscillating in the vertical direction we denote by 'V'. In picturesque terms, we may regard the polaroid as behaving like a letterbox for which the direction of the 'slot' dictates the orientation in which a letter may be inserted. We can rotate the polaroid so that it passes only light with the electric field polarized in the horizontal, 'H', direction, or indeed in any other direction.

 Some simple experiments using the lenses from a pair of polaroid spectacles will quickly convince the sceptic. For example, from what we have said, if we direct vertically polarized light, V, onto a horizontally ori-ented, H, polaroid, no light should emerge. Using the two polaroid lenses, one can easily verify this statement. If the directions of the 'slots' of the

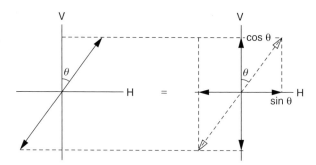

two pieces of polaroid are at right angles to each other, very little light comes through and we can observe how the intensity of the light transmitted increases as the second polaroid is rotated to be parallel to the first. These are the experimental facts: now comes the hard part. In order to explain one of the most important properties of quantum mechanical systems, we must now ask the reader to consider a mathematical description of these experiments. Consider the situation shown in Fig. 8.3. Here, we show how light polarized at an angle θ to the vertical, can be broken down into an equivalent sum of V and H components of polarized light. In terms of the electric fields this equivalence can be written as the 'vector' equation

$$\psi = \cos\theta\,V + \sin\theta\,H$$

where we have used the symbol ψ to represent the original electric field in the θ direction. It is a 'vector' equation in the sense that it has to do with directions in a plane. Physically, the equation means that you can arrive at the same point in the plane **either** by directly travelling unit distance in the θ direction (ψ), **or** by travelling a distance $\cos\theta$ in the vertical direction (V) followed by $\sin\theta$ in the horizontal direction (H). The importance of this equation is that it allows us to calculate the intensity of light transmitted when light polarized at an angle θ passes through polaroids oriented either vertically or horizontally. Since the intensity of light is proportional to the square of the electric field, the amount of light transmitted through a V polaroid varies as $(\cos\theta)^2$. Similarly, the amount transmitted through an H polaroid is given by $(\sin\theta)^2$. Up to now, our discussion of polarized light has used the classical electric field description of the Maxwell equations. But, according to quantum mechanics, at a microscopic level light should be viewed as a stream of tiny 'chunks' of light energy called photons. How do these two descriptions – fields or photons – match up? Our discussion of the variation of the intensity of light transmitted in terms of electric fields must be understood as a prediction for the fraction of the total number of photons that pass through the polaroid.

The essential probabilistic aspect of quantum mechanics becomes apparent when we consider what happens to an individual photon. There is nothing in principle to prevent us lowering the intensity of the light until photons arrive at the polaroid one at a time. Our equation for the electric field ψ at an angle θ now represents the quantum probability wave of a single photon. Let us look carefully at what this means:

$$\psi = \cos\theta\,V + \sin\theta\,H$$

This equation illustrates the essence of the quantum measurement problem. When the photon reaches the vertical polaroid we cannot predict with certainty whether or not it will pass through. All quantum mechanics can say is that the photon has a probability $(\cos\theta)^2$ to pass and a probability $(\sin\theta)^2$ not to pass. In other words, the photon must in some sense be regarded as being in both the V and H states at the same time! The photon is said to be in a quantum 'superposition' of the two states V and H. Superpositions of classical waves like water waves are familiar and unsurprising. Apart from the use of the physicist's term 'interference', what happens when two sets of ripples intersect on the surface of a calm pond is a commonplace event. Similarly, in the double slit experiment, the water waves emerging from each slit are added to obtain the total wave motion. The peculiarities of quantum mechanics emerge as a result of the mysterious 'wave–particle duality'. In the quantum situation, our equation describes not a physical wave but the probability amplitude of an individual photon. This photon is in neither the V or H state but a superposition of the two and 'has to make a choice' at the polaroid and 'jump' into either V or H. The seemingly innocuous phrase, 'has to make a choice', is the heart of the problem. How is this choice made? The photon cannot make the choice – the quantum superposition evolves according to Schrödinger's equation and this does not describe a collapse to one state or the other. It must somehow be the act of observation with the polaroid that causes the collapse into one of the two photon polarization states. But exactly how can some 'classical' measuring apparatus like the piece of polaroid cause this 'collapse of the wavefunction'? After all, any so-called 'classical' measuring apparatus is actually made up of atoms and electrons and these are subject to quantum mechanics and Schrödinger's equation just like the photon. This is the nub of the 'measurement problem' of quantum mechanics and it troubled many of the founders of the theory. It was in response to these challenges that the orthodox 'Copenhagen' interpretation of quantum mechanics was painstakingly pieced together by Niels Bohr and his colleagues at his Institute in Copenhagen. The Copenhagen interpretation offers a very austere and abstract view of the world. Bohr believed that the language of classical physics was inadequate to describe phenomena at the quantum level of reality. Ordinary words are incapable of giving us a satisfactory and unambiguous definition of a quantum superposition. Bohr offers no mechanism to explain the collapse of the wave function on measurement. Instead, to obtain results from quantum theory to compare with experiment, Bohr instructs us to split the experimental system into two parts – one part a classical world containing classical measuring devices and a second part containing the quantum system under observation. This Copenhagen distinction between classical and quantum systems is sometimes called the 'Heisenberg split'. Although such a split is ambiguous in principle, it is clearly sufficiently unambiguous for physicists to use in practice with great success. Nevertheless, there are some physicists, notably the late John Bell, for whom such a 'cookery book' approach to our most fundamental theory of matter was unsatisfactory. Later in this chapter we will see precisely why John Bell detested this 'shifty boundary' between Schrödinger's 'wavy quantum states' on the one hand, and Bohr's 'classical apparatus' on the other. In conversation, Bell went further and insisted that, at its heart, quantum theory was 'rotten'!

There is one more twist to this tale that illustrates another puzzling property of quantum systems and also casts light on what quantum mechanics has to say about the nature of physical reality. Consider a vertically polarized photon approaching a polaroid 'measuring' device. If the polaroid is set to the vertical 'V' position, the photon will certainly pass through unhindered. If the polaroid is set in the horizontal 'H' position, the photon will certainly be absorbed and not transmitted. What happens if we start with a photon polarized at 45 degrees to the vertical? Using the formula we discussed above, the initial photon state can be written down as a superposition of the V and H photon states

$$\psi = \cos(45)\,V + \sin(45)\,H$$

Since $\cos(45)^2$ and $\sin(45)^2$ are both equal to $1/2$, the incoming photon has a 50:50 chance of traversing either a polaroid set in the V direction or a polaroid set in the H direction. However, consider what happens if the directions of the measuring polaroid are chosen to be 'diagonal', either at 45 degrees, 'DV', and or at 135 degrees, 'DH'. The photon now has a 100% chance of passing through a DV polaroid and no chance of passing through a DH polaroid. This seems all very obvious. What can this possibly tell us about the nature of reality? If we choose to measure polarization in the DV direction it looks as though we can say with confidence that the photon really has its polarization aligned at 45 degrees. But if instead we choose to measure in the V or H directions, we know we will see a photon with either a V or an H direction of polarization. In this case, we cannot say that the photon was definitely in either the V or the H state before the measurement. More generally, if we start with a photon in an unknown initial state, we can choose to measure its polarization **either** with a pair of V and H polaroids **or** with a pair of diagonal DV and DH polaroids. Until we decide what direction to set the measuring polaroid it appears that we cannot talk of the photon as having **any** definite polarization. It seems that our choice of the direction of the 'measuring apparatus' has an influence on the polarization of the photon! According to quantum mechanics, until we make a measurement, the direction of the polarization of the photon is apparently not just unknown but really indeterminate. Pascual Jordan, an author of some of the earliest papers on quantum mechanics, went so far as to say that 'observations not only *disturb* what has to be measured, they *produce* it.'

These are some of the uncomfortable questions raised by the probabilistic nature of quantum mechanics. Niels Bohr was well aware of such problems. It was for this reason that his Copenhagen doctrine gave such emphasis to the quantum mechanical predictions for observable quantities rather than speculations as to what quantum mechanics implied about the precise nature of physical reality. Indeed, Bohr declared that quantum mechanics should be regarded rather like a recipe in a cookery book:

> The entire formalism is to be regarded as a tool for deriving predictions, of definite or statistical character, as regards information obtainable under experimental conditions described in classical terms.

However, Aage Petersen, Bohr's assistant, went further and attempted to summarize Bohr's position with the words:

> There is no quantum world. There is only an abstract quantum physical description. It is wrong to think that the task of physics is to find out how Nature **is**. Physics concerns what we can say about Nature.

Heisenberg, who helped Bohr and his colleagues develop this Copenhagen view of the world, also suggested that quantum objects are not 'as real' as everyday objects:

> In the experiments about atomic events we have to do with things and facts, with phenomena which are just as real as any phenomena in daily life. But atoms or elementary particles are not as real; they form a world of potentialities or possibilities rather than one of things or facts.

Such an abstract approach to the microscopic world may enable us to make successful predictions for experiments but appears to contradict all our experience in daily life. The objects we see around us all have a comforting, solid, reality. They do not seem to be conjured into existence only when we choose to look and make a measurement! It is not surprising that Einstein would have none of this. He disliked Bohr's denial of an underlying physical reality and believed passionately that physical objects had real, physical properties whether or not we were there to measure them. In a conversation with Abraham Pais, Einstein highlighted what he felt was the absurdity of the situation by asking Pais: 'Does the Moon only exist when you look at it?' It was to attack Bohr's view of the world that Einstein devised his famous 'thought' experiment with two young colleagues at Princeton, Boris Podolsky and Nathan Rosen. Let us now see how John Bell managed to crystallize such apparently philosophical questions into a form amenable to experimental test.

John Bell and the EPR Paradox

Although Einstein was one of the originators of quantum mechanics, he remained unreconciled to its fundamental probabilistic nature. In a celebrated exchange of arguments with Niels Bohr, known as the Bohr–Einstein debate, Einstein devised a series of challenges to the orthodox Copenhagen view of the world. It was during the famous Solvay conferences in Brussels in 1927 and 1930 that rounds one and two took place. At breakfast with Bohr, Einstein produced several 'thought' experiments that appeared to show that it should be possible to make measurements more accurately than allowed by Heisenberg's Uncertainty Principle. After some sleepless nights, Bohr managed to find a loophole in each of Einstein's arguments and he is generally agreed to be the winner of both rounds one and two. Five years later, Einstein issued a new challenge. Again, his intent was to show that physical quantities, such as the position and momentum of a particle, could, in principle, be known more accurately than allowed by the Uncertainty Principle. In the EPR paper, Einstein's goal was to support his belief in the existence of an 'objective reality', independent of measurement. He was convinced that quantum mechanics must be an incomplete description of the microscopic world and that probabilities only enter quantum physics as a result of our limited knowledge. In essence, Einstein wished the world to be as in our classical coin tossing example. There, if we could measure and calculate everything very carefully, we could in principle predict the result with certainty. By analogy then, perhaps there are some 'hidden variables' in quantum mechanics, which, if we knew them, would enable us to predict the outcome of an experiment with certainty. The EPR paper concludes with words:

> While we have thus shown that the wavefunction does not provide a complete description of reality, we have left open the question of whether or not such a description exists. We believe, however, that such a theory is possible.

Although Einstein was challenging the uncertainties of quantum mechanics in the EPR paradox, his famous thought experiment also illuminates perhaps the strangest aspect of quantum theory – what Einstein once referred to as 'spooky, action-at-a-distance' effects. For certain quantum states – Schrödinger called them 'entangled states' – quantum mechanics seems to require some 'faster than light' influence between separated parts of a quantum system.

We will describe a modern version of the EPR experiment developed by the US physicist David Bohm. In Bohm's arrangement, an atom at rest is stimulated to emit two photons simultaneously (see Fig. 8.4). These photons fly apart in opposite directions and their polarization states can be measured by passing through a pair of polaroid filters. Since the photons travel with the speed of light, the two polaroid polarization detectors can be separated by a large distance. Although originally proposed as a thought

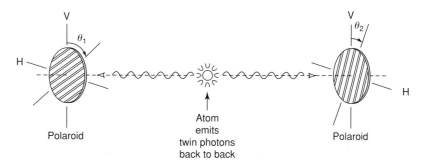

Fig. 8.4 This version of the EPR Experiment is due to US physicist David Bohm. In the centre, an atom at rest decays by emitting two photons back to back. Polaroid detectors in each of the photon paths can be used to measure the photon polarizations.

experiment, advances in photon technology have made it possible to perform the experiment in real life. We obtain the following results:

- If both polaroids are set in the vertical, V, direction, both photons always pass through.
- If both polaroids are set in the horizontal, H, direction, both photons always pass through.
- If one polaroid is set in the V direction and the other in the H direction, we never see both photons passing through together.

In other words, we see VV and HH pairs but not VH or HV photon pairs. Since the two polaroid detectors can be widely separated, we can play all sorts of games in choosing the precise time at which we set the directions of the polaroids. For example, we could choose to set the direction of the second polaroid only *after* the photon has started on its way but *before* a signal could have been sent to tell us what result had been found in the first polaroid. The results are always the same. Although we cannot say whether we will see V or H with certainty, we always see perfect agreement between the polarizations of the two photons. We always see either VV or HH. How can we explain this perfect correlation? Einstein would have wished to interpret these results as evidence that the polarization states of the photons must actually have been pre-determined all along. If one photon is found to be V, it is no surprise if the other one is also V – if these polarizations had been determined at the outset. Any alternative explanation seems to require that some mysterious 'action-at-a-distance' interaction between the photons is able to fix the polarization of the second photon as soon as the first one had been measured to be either V or H.

Action-at-a-distance is not much liked by physicists who prefer to believe in cause and effect. In Maxwell's electromagnetism, forces are transmitted at the speed of light through the intermediary of the electric field. If we have a distribution of electric charges that are widely separated and if we shake some charges in one place, the effects of this shaking are transmitted to the other charges by changes in the electric field that travel with the speed of light. The idea that moving some charges in one place can instantaneously affect other charges at some distant place is not taken seriously. Einstein was profoundly unhappy with this implication of quantum

John Stewart Bell (1928–1991)
was born in Belfast,
Northern Ireland in a family
with no academic traditions.
Despite failing to win a
scholarship to attend
secondary school, he was
able to attend Belfast
Technical College where
he learnt skills such as
bricklaying and carpentry
besides academic subjects.
After graduating from
Queen's University Belfast,
he arrived at CERN after
working on particle
accelerator design in
Malvern, where he met his
wife Mary, and doing
research in the famous
Theoretical Physics group
of Rudolf Peierls in
Birmingham. His famous
paper on his inequality was
written while visiting
Stanford in the USA and
was published in a rather
obscure and short-lived
journal to avoid having to
ask Stanford to pay the
publication charges!

mechanics and wished to avoid such 'spooky' faster-than-light signalling. Bohr's response to Einstein's challenge was unhelpful. In essence, Bohr re-iterated the conventional Copenhagen view that one must consider the quantum system as a whole – the two photons and the two detectors to-gether as a single quantum system – and that it therefore makes no sense to argue about the effect that one measurement has on the other. This is not a very satisfying 'explanation' and certainly did not satisfy Einstein. Within Bohr's explanation is the implication that quantum mechanics must possess peculiar action-at-a-distance properties.

In 1964, nearly 30 years after the EPR paper and 9 years after Einstein's death, John Bell came up with a new twist on the EPR paradox. Bell was an Irishman and was fond of Irish jokes. He often characterized his contribution to the discussion by saying: 'Einstein and Bohr considered correlations (of polarizations) at 0 degrees and 90 degrees – I considered what happened at 37 degrees!' Bell once likened Einstein's two EPR pho-tons to a pair of identical twins. If you separate a pair of identical twins at birth and later observe that one of the twins has red hair, it comes as no surprise that the other twin also develops red hair. Such a correlation is genetically pre-determined – there is no question of any bizarre faster-than-light signalling between the separated twins. David Lindley, in his excellent book *Where Does the Weirdness Go*, discusses the problem in terms of a pair of gloves. If one were to buy a pair of gloves and post one glove to a friend in Hong Kong and the other to a friend in New York, we have no prob-lem understanding what happens when they open their presents. When the friend in Hong Kong finds a left hand glove, she knows immediately that the friend in New York must have the right hand one. There is no mysterious action-at-a-distance happening here. The correlation – left hand in Hong Kong, right hand in New York – is pre-determined from the outset. This type of instantaneous 'knowledge collapse' is familiar from everyday life. John Bell's unique contribution was to show that quantum mechan-ics predicts **more** correlation than such obvious pre-determined 'classical' correlations.

John Bell accomplished this feat by deriving an 'inequality' for the correlations originating from 'common-sense' pre-determined conditions that was violated by the correlations predicted by quantum mechanics. The existence of this inequality then enabled physicists to test whether or not Nature follows quantum mechanics – with its apparent faster than light effects – or deterministic hidden variables as favoured by Einstein. There are many mathematical derivations of 'Bell's inequality' and we do not intend to reproduce these here. Instead, we shall attempt to present a non-mathematical 'intuitive' proof of the inequality that follows an approach outlined by Bell in one of his last 'popular' lectures on the subject. The argument requires some effort and readers willing to take this result on trust can skip to the final paragraph of this section. On the other hand, we suggest that it is worthwhile to go through the argument in detail since this is one of those rare occasions in physics when a fundamental

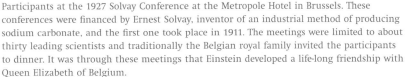

Participants at the 1927 Solvay Conference at the Metropole Hotel in Brussels. These conferences were financed by Ernest Solvay, inventor of an industrial method of producing sodium carbonate, and the first one took place in 1911. The meetings were limited to about thirty leading scientists and traditionally the Belgian royal family invited the participants to dinner. It was through these meetings that Einstein developed a life-long friendship with Queen Elizabeth of Belgium.

Although David Bohm (1917–1992) was the author of a perceptive and careful textbook that espoused the Copenhagen interpretation of quantum mechanics, he later became one of the most outspoken critics of this orthodoxy. It was while he was at Princeton that he had conversations with Einstein that led him to renounce the orthodox view. Bohm later had to leave the USA because of his left-wing political views during the time of the infamous McCarthy hearings. He later developed de Broglie's 'pilot wave' theory of quantum mechanics.

and powerful result can be appreciated without the use of any advanced mathematics.

We begin with Bohm's version of the EPR experiment (see Fig. 8.4) and concentrate on the number of pairs of photons whose polarizations do **not** agree. When the two polaroid polarization detectors are both vertical, we have perfect agreement: there are no cases where one photon of the pair passes through one polaroid and the other photon fails to pass through the second, parallel polaroid detector. So far, we have both polaroid detectors aligned exactly parallel. We are now going to allow these polaroid detectors to be rotated away from the vertical – by an angle θ_1 for polaroid 1 and by an angle θ_2 for polaroid 2. We expect that the number, N, of pairs of photons that disagree – one passes, the other does not – will depend on both these angles. We explicitly allow for this dependence by writing this number as $N(\theta_1, \theta_2)$. When both polaroids are parallel we have $\theta_1 = \theta_2 = 0$ and we found that no pairs disagree. We can summarise this situation by the following equation:

$$N(0, 0) = 0$$

This equation just signifies that the photon pairs are always in perfect agreement.

We shall now attempt to make plausible what we expect to find when we rotate one or both of the detectors away from the vertical. To derive what we expect if the situation is pre-determined in a common-sense way, we shall insist that what happens at one detector does not influence what

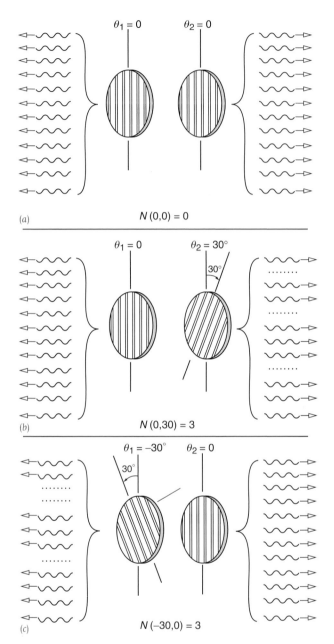

Fig. 8.5 The three experimental configurations to derive the Bell inequality together with sample results for twelve different repetitions of the experiment assuming that the twelve photon pairs are initially vertically polarized. A wavy line is used to indicate that the photon passes through the detector and a dotted line that the photon fails to pass. (a) Both polaroids are set parallel so that $\theta_1 = \theta_2 = 0$; (b) and (c) the situation with one polaroid rotated by ±30 degrees; (d) and (e) possible results obtained with both polaroids rotated by 30 degrees in opposite directions; (d) shows how the maximum possible number of pass/fail disagreements can be obtained and (e) shows how the minumum number of disagreements can be obtained. These common-sense 'results' assume that the photons act independently. Quantum mechanics predicts there will be a greater number of disagreements than those shown in (d).

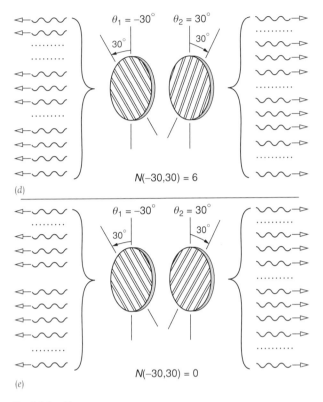

$\theta_1 = -30°$ $\theta_2 = 30°$

$30°$ $30°$

$N(-30,30) = 6$

(d)

$\theta_1 = -30°$ $\theta_2 = 30°$

$30°$ $30°$

$N(-30,30) = 0$

(e)

Fig. 8.5 (cont.)

happens at the other. We are also imagining, like Einstein, that the pair of photons must really be polarized in parallel directions from the outset. When each photon arrives at a detector, we assume quantum mechanics to the extent that we shall use our previous 'vector model' to discuss probabilities of transmission. Let us look at a sample of twelve photon pairs that are created with VV polarization. If we rotate the right hand polaroid by 30 degrees, the number that now pass that detector is given by $(\cos 30)^2$ or $3/4$ (Fig. 8.5). We therefore find that a quarter or three out of twelve of the photon pairs will now 'disagree' – the left photon passes through the left V polaroid but three right photons now fail to pass through the rotated polaroid on the right. We can describe this situation by the following equation:

$$N(0, 30) = 3$$

We now perform another experiment with the right polaroid rotated back to the vertical and the left detector rotated by 30 degrees in the opposite direction. By the same reasoning, we find the result:

$$N(-30, 0) = 3$$

Now consider what we expect to happen when we do the experiment with *both* the polaroids rotated? We are supposing that both the photons start in the V state and that measurements at one detector do not influence measurements at the other. On the right, we must find, as before, that three of the photons fail to pass. Similarly, on the left, three of the photons will fail to pass. Since the pass/fail of any particular photon at each detector is assumed to be probabilistic, these failures may be for photons from different pairs. If these failures occur for six different pairs, we would find that we have a maximum of six disagreements. However, if the failures occur for the same three pairs of photons we would find no disagreements – since nine pairs pass both detectors and three pairs fail both detectors. There can also be situations in between these two extremes. We see that rotating both polaroids together can cause some or all of the expected disagreements to become agreements – in the cases where both photons fail to pass through their respective polaroids. We can summarize this 'common-sense' polarization prediction as an inequality:

$$N(30, -30) \leq N(30, 0) + N(0, -30)$$

This is Bell's inequality. Remarkably, quantum mechanics violates this inequality! In the case we have considered, quantum mechanics does not predetermine the direction of the polarization basis so by using our vector triangle with $\theta = 60$ we predict the left-hand side to be $(\sin 60)^2 = 3/4$. By a similar argument with $\theta = 30$ for both the right-hand terms, we find the right hand side to be $2(\sin 30)^2 = \frac{1}{2}$. Clearly, $3/4$ is not less than or equal to $1/2$ so quantum mechanics violates Bell's inequality.

Bell's inequality makes it possible to put quantum mechanics to the test. A series of beautiful experiments designed to measure these EPR correlations were performed by Alain Aspect and his group in Paris in the late 1970s and early 1980s. These experiments confirmed that Bell's inequality was indeed violated and that the predictions of quantum mechanics agreed with the data. Although these experiments did not perfectly reproduce all the conditions of the EPR thought experiment – real-world photon detectors are less than 100% efficient, for example – most physicists now accept that quantum mechanics has passed this test. What do these experiments tell us about the nature of reality? The observed violation of Bell's inequality means that no hidden variable theory – without some explicit or implicit unpleasant action-at-a-distance property – can agree with experiment. Whilst Einstein would probably have preferred some underlying, deterministic hidden variable explanation for quantum mechanics, he would certainly not have wanted to accept the existence of such 'spooky action-at-a-distance' effects.

Schrödinger's cat

Schrödinger, one of the originators of the theory, remained profoundly unhappy with Bohr's interpretation of quantum mechanics. His paradox of Schrödinger's cat was introduced to highlight the key problem of measurement. According to Bohr, the world can be divided into *quantum systems*, which interact and evolve according to Schrödinger's equation, and *classical systems*, like counters or pointers, which obey the laws of our familiar, everyday world. Unlike classical objects, we have seen that a quantum system can exist in a superposition of several quantum states. It is the process of measurement that is somehow supposed to cause the quantum superposition to collapse down to one definite *classical* state – one that can be determined by the state of a classical counter or the position of a classical pointer. Accordingly, it is apparently the *observer* who decides what properties to measure and when to make the measurement. Only after measurement can we talk about the quantum system as having some definite properties. This is a profound change from classical physics for which the properties of a classical object are assumed to exist independently of any observer or measuring apparatus. Schrödinger was unhappy from the start. As early as 1926, Schrödinger lamented 'If all this damned quantum jumping were really here to stay then I should be sorry I ever got involved with quantum theory'. What is the problem? The problem is that *everything* is quantum mechanical. Despite the apparent solidity of the world around us, everything is made up of atoms and electrons, the same atoms and electrons that are supposed to be peculiar, wave-particle, quantum objects obeying Schrödinger's equation and not Newton's Laws. Why should one need to divide the world into quantum systems and classical measuring apparatus? And how is an observer, who after all is also made of atoms and electrons, able to stand outside the system to make a measurement? These problems are particularly acute if one wishes to contemplate the wavefunction of the universe.

In response to questions such as these, Schrödinger was prepared to sacrifice his (mythical) cat to the cause of science. He described the situation as follows (see Fig. 8.6):

> A cat is penned up in a steel chamber, along with the following diabolical device: in a Geiger counter there is a tiny bit of radioactive substance, so small, that perhaps in the course of one hour one of the atoms decays, but also with equal probability, perhaps none; if it happens, the counter tube discharges and through a relay releases a hammer which shatters a small flask of hydrocyanic acid. If one has left this entire system to itself for an hour, one would say that the cat still lives if meanwhile no atom has decayed. The first atomic decay would have poisoned it. The ψ-function of the entire system would express this by having in it the living and the dead cat (pardon the expression) mixed or smeared out in equal parts.

In other words, until we open the chamber after an hour and observe the cat, quantum mechanics seems to assert that it is in a quantum superposition. We cannot imagine how classical objects can be in a quantum superposition of two different states at once, let alone a living object like a cat. Can it really be that quantum mechanics says that it is the act of observing the cat that causes its wavefunction to collapse to a dead or alive cat? Two of the great quantum mechanics thinkers of the century, John von

Fig. 8.6 The experimental arrangement for Schrödinger's famous thought experiment with his unfortunate cat. The smashing of a glass of hydrocyanic acid will be triggered by the decay of a radioactive nucleus. The experiment is carefully engineered with sufficient radioactive material so that there is exactly a 50% chance for one nucleus to decay in an experiment lasting one hour. According to quantum mechanics, after the hour is up the cat's wavefunction will be an equal superposition of wavefunctions corresponding to an alive cat and a dead cat! When the container is opened we must find the cat dead or alive. The mystery is how such a 'measurement' comes about and the quantum superposition collapses to a definite state of the cat's health.

Neumann and Eugene Wigner worried about this problem. They eventually came round to the view that the consciousness of the observer must play a key role in the collapse of the wavefunction. This immediately takes us into even deeper waters. Can a conscious animal such as the cat cause the collapse of its own wavefunction? What if we construct another steel chamber and attach it to the one containing the cat with an observation window between the two? If we leave 'Wigner's friend' sitting watching the cat for the hour, we can then ask her whether the cat died when the atom decayed or when we opened the chamber to have a look. Or did the measurement and collapse take place earlier because Wigner's friend was watching? If the only non-quantum mechanical aspect of the world is consciousness, why do different observers agree on the same picture of the physical world? No wonder Einstein asked whether the Moon was there when we were not looking at it! Let us look at two attempts to solve this measurement problem – 'many worlds' and 'decoherence'.

The many worlds interpretation of quantum mechanics

The traditional description of the total quantum mechanics of the world by a Monster Wavefunction (which includes all observers)

obeying a Schrödinger equation implies an incredibly complex infinity of amplitudes. If I am gambling in Las Vegas, and am about to put some money into number twenty-two at roulette, and the girl next to me spills her drink because she sees someone she knows, so that I stop before betting, and twenty-two comes up, I can see that the whole course of the universe for me has hung on the fact that some little photon hit the nerve ends of her retina. Thus the whole universe bifurcates at each atomic event. Now some people who insist on taking all quantum mechanics to the letter are satisfied with such a picture; since there is no outside observer for a wavefunction describing the whole universe, they maintain that the proper description of the world includes all the amplitudes that thus bifurcate from each atomic event.
Richard Feynman

Many physicists have proposed different 'interpretations' of quantum mechanics aimed at providing a consistent explanation of the measurement problem and an understanding of the collapse of the wavefunction. Perhaps the most bizarre suggestion was put forward by Hugh Everett in his 1957 Ph.D. thesis. Although Everett was a student of John Wheeler at Princeton University, even Wheeler found the first draft of his thesis 'barely comprehensible'. Although Wheeler was certain there was a very original idea in the thesis, he thought it would be wise to write a companion paper of his own to make Everett's thesis 'more digestible' to his Ph.D. examiners! Everett's idea received little attention until 10 years later when a colleague of Wheeler's, Bryce DeWitt, wrote an article describing Everett's proposal as the 'many worlds interpretation' of quantum mechanics. In the conventional Copenhagen viewpoint, when an observer uses some classical measuring apparatus to make a measurement on a quantum superposition, only one of the possible results is actually realized. The mysterious measurement process somehow collapses all the different possible outcomes down to the one observed outcome. Everett and DeWitt removed this problem in a breathtakingly audacious way – they suggested that *all* possibilities are realized, but each in a different copy of the universe. Furthermore, according to DeWitt, each of these copies of the universe is itself constantly multiplying to allow for all possible outcomes of every measurement. As De Witt says, 'every quantum transition taking place on every star, in every galaxy, in every remote corner of the universe is splitting our local world in myriads of copies of itself'. In this picture there is no collapse of the wavefunction – the universe is replaced by a 'multiverse' of parallel universes.

Despite the considerable appeal of such an idea, there are several problems with such an approach. First of all, if these separate universes cannot interact with each other, it is not at all clear that there is any way to test Everett's proposal. A solution to the measurement problem that makes no new predictions and which cannot be tested seems devoid of content. Even John Wheeler eventually concluded that Everett's viewpoint was only able to offer new insights. There may also be some real problems for a detailed formulation of the theory. Feynman was worried that in each of the different universes there are presumably copies of each of us. Each of us knows which way the world has divided for us and we can follow the track of our past. When we make an observation of our track in the past is the result 'real' in the same way as it would be if the observation was made by an 'outside' observer? Furthermore, although we may consider ourselves to be the outside observer when we look at the rest of the world, the rest of the world includes observers observing us. Will we always agree on what we see? As Feynman says 'These are very wild speculations, and it would be little profit to keep discussing them'.

John Bell was also worried about implications of the many worlds interpretation. Both Everett and Dewitt refer to the branching of the wave-function into many different universes as forming a tree-like structure – the future of a given branch is uncertain but the past is not. Bell believed that, at the microscopic level, the theory 'does not associate a particular branch at the present time with any particular branch in the past any more than with any particular branch in the future'. According to Bell, in Everett's theory there is no association of the particular present with any particular past. There are therefore no trajectories and the configuration of the world, including us, changes in an utterly discontinuous way. How then do we have the illusion that the world changes in a fairly continuous way? In Bell's interpretation of Everett's ideas, this continuity must come about from our memories, which are phenomena of the present. Bell likens this situation to a theory of creation that insisted that the world was created in 4004 BC. Growing knowledge of the structure of the Earth that seemed to indicate that the world had been around for considerably longer than this caused no problem for true believers. It was pointed out that God in 4004 BC would obviously have created a going concern: trees would have annular rings, although the corresponding number of years had not elapsed, and rocks would be typical rocks, some occurring in strata and bearing fossils – of creatures that had never lived. Of Everett's theory Bell concludes: 'if such a theory were taken seriously it would hardly be possible to take anything else seriously'.

In spite of these and other questions, the many worlds interpretation – or multiverse as it is sometimes called – has enduring popular appeal. It is also able to claim the support of several distinguished physicists including David Deutsch and Stephen Hawking. Deutsch proposes a variant of the many worlds interpretation in which the number of worlds, though very large, does not keep increasing. He also believes that the theory can be tested. His test involves a quantum interference experiment in which two quantum states evolve separately for a time and then recombine. An artificial brain, with some sort of microscopic quantum memory, observes the system and splits into two copies in different worlds. The test then 'hinges on observing an interference phenomenon inside the mind of this artificial observer'. We will not follow these speculations further but we shall encounter David Deutsch again in our account of the development of quantum computers.

Decoherence

A less extravagant and rather more mundane attempt to solve the measurement problem goes by the name of 'decoherence'. This approach argues that quantum systems can never be totally isolated from the larger environment and that Schrödinger's equation must be applied not only to the quantum system but also to the coupled quantum environment. In real life, the 'coherence' of a quantum state – the delicate phase relations between the different parts of a quantum superposition – is rapidly affected by interactions with the rest of the world outside the quantum system. Wojciech Zurek is one of the most prominent advocates of this 'decoherence' approach to the measurement problem, and he speaks of the quantum coherence as 'leaking out' into the environment. Zurek claims that recent years have seen a growing consensus that it is interactions of quantum systems with the environment that randomize the phases of quantum superpositions. All we have left is an ordinary non-quantum choice between states with classical probabilities and no funny interference effects. This seems a very prosaic end to the quantum measurement problem! How does this come about? Does decoherence by the environment really supply an answer to all the problems? Let us look at an experiment that claims to see decoherence of 'Schrödinger cat' states in action.

Serge Haroche and Jean-Michel Raimond, working in Paris with their research group, have recently performed some exciting experiments that give support to this decoherence picture. There are three different parts to an experiment that can all interact – the quantum system, the 'classical' measurement apparatus and the environment. In their experiment the quantum system consists of an atom that can be prepared in one of two states. They measure the quantum state of the atom by injecting the atom into a cavity and using the electromagnetic field of the 'cavity' as a classical 'pointer'. What happens if we prepare the atom in a quantum superposition of the two states? If we treat the cavity as a second quantum system in its own right, we find that the supposedly classical pointer is now predicted to be in a 'Schrödinger cat' state – a quantum superposition of two classical states of the 'pointer'. Schrödinger's thought experiment just highlighted the peculiarity of this situation by using his cat as a classical pointer. How do we escape from this apparent paradox? According to the decoherence picture, we must include the unavoidable coupling of the pointer to the environment. The pointer – or cavity – is under a constant bombardment from random photons, air molecules and so on that constitute 'the environment'. Models of this random process as a third quantum system show that all phase information between the two original atomic states with their corresponding pointer positions is very rapidly lost. For the usual classical pointer fields with many photons, this decoherence is predicted to take place in an immeasurably short time. Remarkably, by using pointer cavity fields consisting of only a few photons, Haroche and Raimond have been able to observe and measure the decoherence time of this system. They do this by sending a second atom into the cavity at varying times after the first atom and measuring interference effects that depend on the continued coherence of the wavefunction of the first atom. By observing how fast these interference effects fall off with the time delay between the traversals through the cavity of the first and second atoms, they claim to have 'caught decoherence in the act'!

Einstein's problem with the Moon can be 'explained' using a similar decoherence argument. The Moon is not an inert system – not only are its individual molecules constantly interacting with their neighbours but also its surface is under constant bombardment by particles and radiation, mainly from the Sun. The coherence of any Schrödinger cat state involving the Moon would rapidly be destroyed by these constant interactions. According to such decoherence arguments, we can rest assured that the Moon is really there after all, even when we are not looking at it. Bombardment by solar photons is enough to constitute a measurement and to destroy any quantum coherence.

Would these decoherence arguments have satisfied John Bell as an explanation of the measurement problem? Probably not! We have described not only the quantum system under observation but also the measuring apparatus as a quantum system. The quantum wavefunction for the combined system will be in a superposition of states corresponding to different classical states of the measuring apparatus, as in the experiment of Haroche and Raimond. The decoherence argument says we must include the environment as a third quantum system interacting with our measuring apparatus. As a result, phase randomization rapidly sets in and the quantum superposition is effectively reduced to a sum of different possible outcomes with classical probabilities. Bell had two problems with this approach. Firstly, all quantum states – for system, measuring apparatus and environment – evolve according to the Schrödinger equation. It is mathematically impossible for such evolution to turn a coherent quantum superposition into an incoherent probabilistic sum. Although it is certainly true that the particular measurements one usually chooses to make display little or no quantum coherence, Bell argues that there is nothing *in principle* to stop us considering different types of measurements for which this will not be true. As Bell has said:

> So long as nothing, in principle, forbids consideration of such arbitrarily complicated observables, it is not permitted to speak of wave packet reduction. While for any given observable one can find a time for which the unwanted interference is as small as you like, for any given time one can find an observable for which it is as big as you do **not** like.

In Bell's view, any mechanism for the collapse should also be applicable to small systems and should not be dependent on 'the laws of large numbers'. His second problem, concerned the actual measurement itself. Even if one accepts that decoherence reduces the problem to a probabilistic choice between outcomes, nowhere does decoherence say how any particular out-come is achieved. Bell did not disagree about the practicality of measure-ments in quantum mechanics, but he felt strongly that unless we know 'exactly when and how it [wave function reduction] takes over from the Schrödinger equation, we do not have an exact and unambiguous formula-tion of our most fundamental physical theory.'

There is much more that can be said on the subject of measurement in quantum mechanics. Great physicists from the early years of quantum mechanics, such as John von Neumann and Eugene Wigner, even suggested that the consciousness of the observer was ultimately responsible for the collapse of the wavefunction. Rudolf Peierls advocated an approach based on knowledge and information. Roger Penrose believes that wave packet reduction is caused by quantum gravity. Robert Griffiths, Murray Gell-Mann, James Hartle and Roland Omnes take yet another approach and prefer to describe the problem in terms of 'quantum histories'. In our brief account of the measurement problem, we have only been able to scratch the surface of this debate. Rather than be dismayed by this diversity of views, we hope that the reader may be encouraged by the obvious disagreements of such great physicists! Quantum mechanics is not yet a closed book and there may yet be some surprises in store for us in the twenty-first century. In the next chapter, we shall describe the beginning of a new field – 'quantum engineering' – for which the quantum nature of matter is of paramount importance.

9 Quantum engineering

> What I want to talk about is the problem of manipulating and controlling things on a small scale.... It is a staggeringly small world that is below. In the year 2000, when they look back at this age, they will wonder why it was not until the year 1960 that anybody began seriously to move in this direction.
>
> Richard Feynman

Richard Feynman and nanotechnology

In 1959, in an after dinner speech at a meeting of the American Physical Society in Pasadena, Richard Feynman set out a remarkable vision of the future in a talk entitled 'There's plenty of room at the bottom.' The talk was subtitled 'an invitation to enter a new field of physics' and marked the beginning of what is now known as 'nanotechnology'. Nanotechnology is concerned with the manipulation of matter at the scale of a nanometre – a thousand millionth of a metre. Atoms are typically a few tenths of a nanometre in size. Feynman emphasized that such an endeavour does not need new physics:

> I am not inventing anti-gravity, which is possible someday only if the laws are not what we think. I am telling you what could be done if the laws **are** what we think; we are not doing it simply because we haven't yet gotten around to it.

In his talk Feynman offered two prizes of $1000 each – one 'to the first guy who makes an operating electric motor which is only 1/64 inch cube', and a second 'to the first guy who can take the information on the page of a book and put it on an area 1/25000 smaller' He had to pay out on both prizes – the first less than a year later, to Bill McLellan, a Caltech alumnus. McLellan brought a microscope with him to show Feynman the miniature motor that could generate a millionth of a horsepower. Although Feynman paid the prize, the motor was a disappointment to him since it did not require any new technical advances – he had not made the challenge hard enough! In an updated version of his talk given 20 years later, Feynman speculated that with modern technology it would be possible to mass produce motors 1/40 a side smaller than McLellan's original motor. Feynman also envisaged the creation of a chain of 'slave' machines – each producing tools and machines

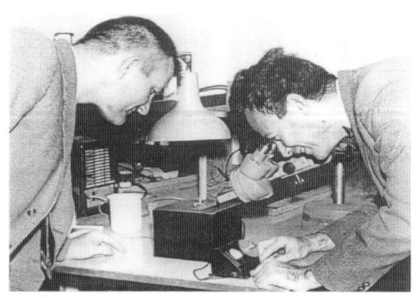

Fig. 9.1 Richard Feynman examining Bill McClellan's tiny motor through a microscope. After offering the prize for a microscopic electric motor, Feynman was inundated with people wanting to show him their efforts. Feynman knew McClellan was different from all the other hopefuls when the first thing he took out of the box was a microscope to be able to see the motor.

at one-fourth of their scale – in order to produce such micromachines. At the time, Feynman was embarassed that he could not conceive of a use for such minute machines and he explored the subject out of purely academic interest. Forty years on from Feynman's original talk we are on the verge of seeing – or rather not seeing – microsystems in all sorts of applications ranging from medical sensors to arrays of tiny optical mirrors. We will describe some of these applications in the next section.

It was not until 26 years after his first talk that Feynman had to pay out on the second prize. The scale of the challenge was equivalent to writing all twenty-four volumes of the *Encyclopedia Brittanica* on the head of a pin. In 1985, Tom Newman was a Stanford graduate student who was using a technique called 'electron beam lithography' to engrave the patterns on silicon to make integrated circuits. A friend showed Newman a copy of Feynman's 1959 talk and pointed out the section offering a prize for 'writing small'. Newman calculated he would have to reduce individual letters down to a scale only 50 atoms wide. By programming his electron beam machine he thought that this should be possible. To check that the prize was still on offer after all this time, Newman sent a telegram to Feynman. He was very surprised to receive a telephone call from Feynman in his laboratory in confirmation. Since he was meant to be finishing his thesis, Newman had to wait till his thesis advisor went to Washington DC for a few days before he made his attempt. He then programmed the machine to write the first page of Charles Dickens' novel *A Tale of Two Cities*. It turned out that the major difficulty was finding the tiny page on the surface after it had been written! Newman duly received Feynman's cheque in November 1985.

Fig. 9.2 A photograph of Tom Newman's miniature page. After Newman had performed the task of engraving the first page of Charles Dickens' *A Tale of Two Cities* at a scale where each letter was only 50 atoms wide, the biggest difficulty he had was finding the tiny text on the silicon surface.

In his talk, Feynman also extended his speculations down to the atomic level. He envisaged a time when atoms could be re-arranged to order. No longer would it be necessary to synthesize chemicals in the traditional way:

> ... it would be, in principle, possible (I think) for a physicist to synthesize any chemical substance that the chemist writes down. Give the orders and the physicist synthesizes it. How? Put the atoms down where the chemist says, and so you make the substance.

This dream is now starting to be realized. Don Eigler and colleagues at the IBM Research Center in Almaden have manipulated individual atoms using the scanning tunnelling microscope (STM) invented by their colleagues at IBM Zurich. Besides creating the world's smallest IBM logo (Fig. 5.13), and creating spectacular quantum 'corrals' (Fig. 9.3, see also Fig. 4.13), Eigler and his group have created 'artificial' molecules, an atom at a time (Fig. 9.4). Wilson Ho's research group at Cornell has also been able to create molecules

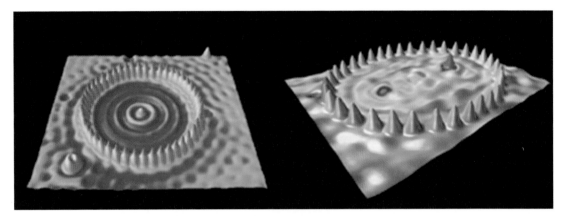

Fig. 9.3 A quantum corral, showing the surface electron waves of the confined electrons.

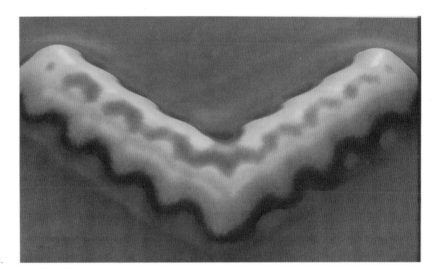

Fig. 9.4 An artificial molecule created by moving the atoms one at a time. The molecule consists of 8 caesium and 8 iodine atoms.

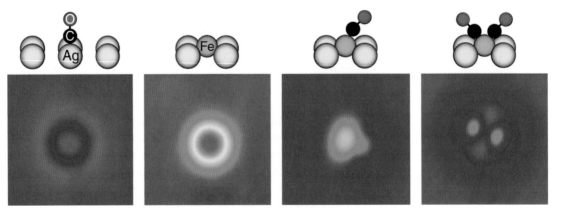

Fig. 9.5 Another hand-made molecule comprising iron and carbon monoxide (Fe(CO)).

in this way. They have combined a single carbon monoxide molecule with an iron atom and studied the vibrational properties of the resultant molecule (Fig. 9.5). Although such molecule building is impressive we will probably need to find different techniques before we can construct organic molecules of any complexity.

In this chapter we will take a look at some other steps towards realizing Feynman's vision. The key feature of such developments is the recognition that life at the bottom is fundamentally quantum mechanical. It is this realization that has fuelled exciting developments in quantum information theory and quantum computing. Before we discuss these applications of quantum mechanics it is worthwhile to look again at the future of the semiconductor industry. It is the efforts of their engineers that have radically transformed our society in the last half of the twentieth century. What is in store for the next 50 years?

From Moore's Law to quantum dots

In chapter 6 we encountered Robert Noyce, a physicist from MIT, who was first to patent a technology that allowed the mass-production of integrated circuits. In 1957, Noyce and Gordon Moore, a physical chemist from Caltech, had been among the first employees of the Shockley Semiconductor Laboratory, Silicon Valley's very first 'high tech' start-up company, long before the region became known as Silicon Valley. Unhappy with the management style and strategic decisions of the company founder, Nobel Prize winner William Shockley, Noyce and Moore left with six other employees – Shockley is reputed to have called them the 'traitorous eight' – to found a new company called Fairchild Semiconductor. In 1961, using Noyce's planar process for manufacturing integrated circuits, Fairchild introduced the first commercial silicon integrated circuits. As Noyce said years later:

> When this [planar process] was accomplished, we had a silicon surface covered with one of the best insulators known to man, so you could etch holes through to make contact with underlying silicon. Obviously, then, you had a whole bunch of transistors embedded in an insulating surface, and the next thing was that, instead of cutting them apart physically, you cut them apart electrically, added other components you needed for circuits, and finally the interconnection wiring.

In 1968, Noyce and Moore left Fairchild to set up Intel with the intention of specializing in memory chips. Their reputation was such that even with only a vague one page business plan they were able to attract investors. In 1965, Moore had written an article for the thirty-fifth anniversary issue of *Electronics* magazine entitled 'Cramming more components onto integrated circuits'. In the paper he noted that the complexity of integrated circuits had been doubling every year since 1962 and he made a bold extrapolation that this would continue for another decade. Moore also speculated that

Gordon Moore, co-founder of Intel, and pioneer from the earliest days of semiconductor technology. Moore, with a degree in chemistry from Caltech, was recruited by Shockley for the first Silicon Valley start-up company in 1956. In an article for the magazine *Electronics* Moore observed that between 1962 and 1965 the number of components on an integrated circuit had been doubling every year and had reached 50 per chip by 1965. He made the audacious suggestion that this annual doubling would continue for another decade until by 1975 chips would contain around 65 000 components. In 1977, his colleague Robert Noyce wrote in Scientific American that there had been no significant deviation from 'Moore's Law'. Although the 12 month doubling is now generally reckoned to be more like 18 months, this trend towards increasing complexity has continued to this day.

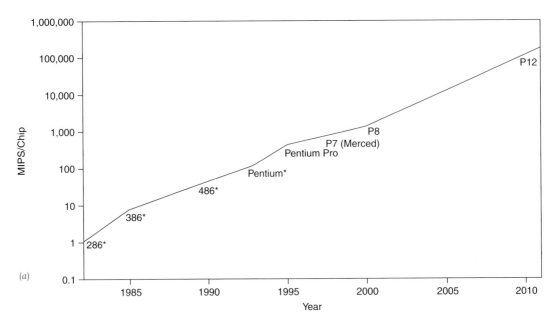

(a)

MIPS – Millions of instructions per second

*Pentium, 286, 386, and 486 are registered trademarks of Intel Corp.

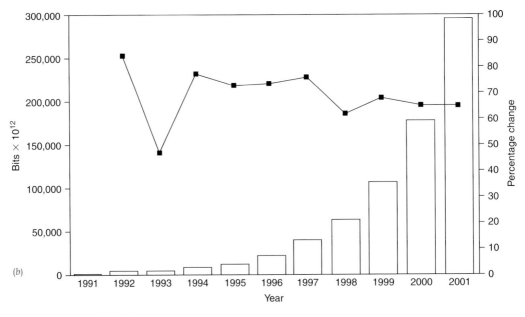

(b)

the eventual impact of such chips would be enormous, not only for industry but for individual consumers:

> Integrated circuits will lead to such wonders as home computers – or at least terminals connected to a central computer – automatic controls for automobiles, or portable communications equipment.

This was over a decade before Steven Jobs and Stephen Wozniak produced the first mass market personal computer and 16 years before the appearance of the IBM PC. Gordon Moore's prediction has become known as *Moore's Law* and this rapid, year-on-year increase in complexity has continued for over 35 years (Fig. 9.6). In 1975 Moore updated his law by suggesting that IC complexity doubling every two years was more realistic. Nowadays, Moore's Law is usually defined as a doubling of the number of transistors on a chip every 18 to 24 months.

The doubling of Moore's Law occurs in part because the transistors can be made smaller and the minimum feature size on the chip decreases. What was not clear in 1965 was whether quantum tunnelling would be a major limitation on how small the transistors could be made. Moore asked Carver Mead at Caltech for advice on this problem. The results of Mead's investigation were stunning. Here is how Mead described his first public presentation of his analysis:

> In 1968, I was invited to give a talk at a workshop on semiconductor devices at Lake of the Ozarks. In those days you could get everyone who was doing cutting-edge work in one room, so that the workshops were where all the action was. I had been thinking about Gordon Moore's question, and decided to make it the subject of my talk. As I prepared for this event, I began to have serious doubts about my sanity. My calculations were telling me that, contrary to all the current lore in the field, we could scale down the technology such that **everything got better**: the circuits got more complex, they ran faster, and they took less power – WOW! That's a violation of Murphy's law that won't quit! But the more I looked at the problem, the more I became convinced that the result was correct, so I went ahead and gave the talk, to hell with Murphy! That talk provoked considerable debate, and at the time most people didn't believe the

Fig. 9.6 A graph of Moore's Law (*a*) for microprocessors (*b*) for memory chips. As can be seen, the performance of microprocessors – 'computers on a chip' – has doubled every 18 months or so for the last 20 or 30 years. A similar increase has taken place for computer memory storage so that, for example, in the years 2001 and 2002 more computer memory will have been installed on this planet than has been installed in the entire previous human history. Along with this relentless increase in performance and memory capacity has come an accompanying drop in the price for a given rate of computing or a given amount of memory. Moore's Law is the reason why our computers need replacing every few years or so and this trend is expected to continue for at least the next 10 years. What happens then is anyone's guess.

result. But by the time the next workshop rolled around, a number of other groups had worked through the problem for themselves, and we were pretty much in agreement. The consequences of this result for modern information technology have, of course, been staggering.

More to the point, as the chips shrunk, not only could more complex chips be designed but also more of them can be produced for the same cost. Amazingly, as the computing power and memory capacity of chips increased exponentially, the cost of computing and memory has also decreased exponentially. Moore's Law is the reason that your PC becomes obsolete every few years – since computers with double the power and memory can be bought for the same price every 18 months! Computer software has been able to become more complex and powerful because both computer memory and computing power are able to cope with greater challenges.

Moore's Law has held true for more than 30 years, fuelling the vast growth in computing and information processing devices. Is there an end to all this? The silicon industry collaborates internationally to produce a 'road map' for future generations of silicon chips. In 1970, Intel produced the first 1024 bit (1kBit) DRAM (dynamic random access memory) chip. A year later, the first microprocessor – the Intel 4004 – was produced, comprising some 2300 transistors etched in circuits 10 microns wide (Fig. 6.20). Twenty-five years later, in 1995, the industry was producing DRAM chips with 64 million bits (64 Megabit) and micoprocessors with 4 million transistors per square centimetre and a minimum feature size of 0.35 microns. At the turn of the millenium, the industry has moved on to 1 thousand million bit (1GigaBit) DRAMs and processors have 13 million transistors per square centimetre with a feature size of 0.18 micron. By 2010 or so, the road map predicts that there will be memory chips with around a hundred thousand million bits – more than the number of stars in our galaxy! The minimum feature size is then predicted to be 0.07 micron or 70 nanometres. It is no wonder that the industry has needed to develop sophisticated CAD (computer-assisted design) software to manage such complexity. More and more powerful computers are needed to design each new generation of chips. The design challenges to fulfil the predictions of the road map are formidable. Although the charts boldly extrapolate Moore's Law to 2010 and beyond, there are many technical problems still to be solved. One obstacle is economic – the sheer cost of building the fabrication facility for each new generation of chips. For 0.25 micron chip plants Moore quotes a cost between $2 billion and $2.5 billion: for 0.18 micron chips the cost rises to between $3 billion and $4 billion. Arthur Rock, the Harvard-trained venture capitalist who helped Moore and Noyce raise funding to start Intel, is credited with Rock's Law: 'A very small addendum to Moore's Law which says that the cost of capital equipment to build semiconductors will double every four years.'

There are also engineering obstacles to be overcome to reach the 2010 Moore's Law target performance. Current generations of chips use a technique called 'photolithography' (Fig. 9.7) to make 'masks' that enable

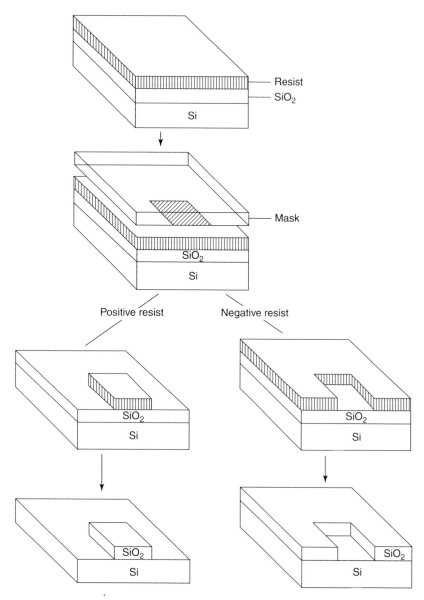

Fig. 9.7 A schematic representation of the sequence of steps required to produce a silicon chip using photolithography. A chip is made up of many different layers of components created from sandwiches of semiconductor, oxide insulator and metal. The designs are so complex that the design is only possible using computer-aided design (CAD) tools to create photomasks to lay out the patterns on the chip.

the patterns of the different layers – silicon, insulator, metal – of the chip to be protected. Between each stage, the unwanted material is etched away in an acid bath. The trouble is that as the feature size gets smaller you need to use light of shorter and shorter wavelengths. It used to be thought that a minimum circuit-line width of one micron was the limit that could be done using optical lithography. Now with deep ultraviolet light we can probably go to 0.13 micron. After that, as Gordon Moore says, 'life gets very interesting'. Two possible technologies are X-ray photolithography and electron beam lithography but both have their problems. Another problem concerns the metallic 'interconnect': the road map states that 'Interconnect has been represented as the technology thrust with the largest potential technology gaps.' A single square centimetre of a modern chip typically contains several tens of metres of interconnect in multiple layers of wiring. The problem is that as the wires get thinner, the resistance increases and they can carry less current. Although the current decreases, the heat generated in the wire increases and can lead to 'electromigration'. In this situation the lattice ions are torn from their sites creating voids in the metal interconnect that cannot be repaired. For reasons such as these, IBM and other manufacturers are now replacing aluminium interconnect by copper, which has a higher conductivity. As the wires get closer there are also capacitance effects between the wires and the silicon dioxide insulator. Such 'crosstalk' can result in logic errors and cause the chip to function incorrectly. Manufacturers are now looking for a replacement for silicon dioxide with a lower dielectric constant so that it will store less charge. Time delays incurred in sending signals across the chip will also cause problems. Chip designers will need considerable ingenuity to overcome such engineering problems and stay on the line of Moore's Law.

Although silicon dominates the mass market for integrated circuits, other semiconductors have advantages for some applications. Gallium arsenide has been used for high-speed transistors and also lends itself rather readily to a process known as 'band gap' engineering. In gallium arsenide, the gallium and arsenic atoms alternate to form a crystal structure rather like diamond. We can also form alloys in which some aluminium atoms are substituted for gallium to form aluminium gallium arsenide or 'algas' for short. Since the atomic spacing in this alloy is almost the same as pure gallium arsenide we can make a crystal with alternating regions of algas and gallium arsenide. Using a technique called chemical vapour deposition, CVD, it is possible to produce a crystal where the changeover between the two regions is only a few atomic layers thick. This turns out to be useful since the energy required to raise an electron from the valence band to the conduction differs in the two materials. A thin layer of gallium arsenide sandwiched between two regions of algas acts like a 'quantum well' in which electrons can be trapped (Fig. 9.8). In fact, the electrons are confined only in one dimension and are free to move parallel to the sides of what is like an artificial 'valley' created in the semiconductor. But, as for waves on a string, changing the width of the well alters the allowed quantum states. This means we have a type of artificial atom – a 'superatom' – whose energy levels and properties we can engineer to suit specific applications. One such application is a quantum well laser where we can tune the wavelength of the light emitted. These devices are already in use in laser printers and compact disc players. Another quantum well device is a resonant tunnelling transistor in which the ease of tunnelling can be controlled by matching the energy levels in the well. If we create many alternate layers of algas and gallium arsenide we create an array of superatoms which act like an artificial crystal or 'superlattice' (Fig. 9.9). Exactly as for the real crystal

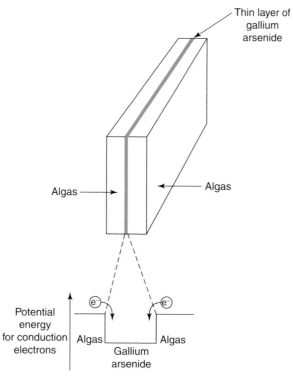

Fig. 9.8 Electrons can be confined to a flat layer of gallium arsenide to form a two-dimensional electron gas. This quantum well is formed because the lowest energy states available to the conduction electrons are in the gallium arsenide layer.

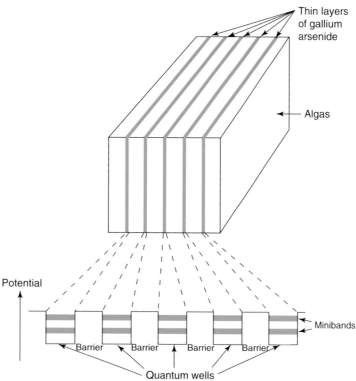

Fig. 9.9 A 'superlattice' structure created by alternate layers of algas and gallium arsenide. Interactions between the quantum wells produce 'minibands' of allowed energy levels as in a natural crystal structure.

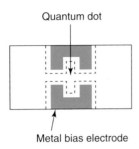

Quantum dot

Metal bias electrode

Fig. 9.10 A 'quantum dot' can be created by forming metal gates on the surface of a quantum well heterostructure such as that shown in Fig. 9.8. The two narrow constrictions form 'quantum point contacts' that isolate the quantum dot when a voltage is applied. Electrons can then only enter the dot by quantum tunnelling.

described in chapter 6, the separate energy levels merge to create narrow bands of allowed energies called 'minibands'. We can also construct superlattices using alternate regions of silicon and germanium with layers only a few atoms thick. Layers with around a hundred or so atoms can be produced by gas deposition. To produce layers only a few atoms wide a technique called molecular beam epitaxy (MBE) must be used. For the MBE process the silicon or germanium is heated in an oven and a beam of atoms from a small hole in the oven is directed onto a surface. The atoms stick to the surface at a controllable rate, typically about one plane of atoms per second. The silicon germanium superlattices produced in this way are 'strained' crystal structures since the silicon and germanium atoms are of different sizes. These strained structures turn out to have better optical properties for integrating optoelectronic devices with conventional integrated circuits. Such combined electronic–optoelectronic devices are likely to be of increasing importance in the next phase of development of communications technology.

In the quantum wells we have constructed, the electrons are confined in the vertical direction, perpendicular to the surface of the semiconductor. However, there is nothing to stop us etching away the superlattice in the other two dimensions to produce a whole array of isolated quantum wells. If the two lateral dimensions of these wells are small enough, quantum effects will again become important. This leads to the creation of a new type of quantum electronic device – the 'quantum dot'. In fact, it is easier to make quantum dots using metal strips on the surface of the layered quantum well structure to confine the conduction electrons (Fig. 9.10). If we apply a negative charge to the metal, the electrons in the quantum well beneath the surface are repelled and cluster underneath the central square defined by the electrodes. Since we can control both the size and the shape of such quantum dots, scientists are experimenting with such 'designer atoms' to create new types of materials with a range of surprising properties.

Another exciting development in semiconductor technology makes use of a phenomenon called 'single-electron tunnelling'. If we fabricate a structure consisting of two small electrodes separated by a thin layer of insulator we expect that electrons will be able to tunnel through the barrier. The electrodes are typically about a tenth of a micron across and the system cooled to a degree or so from absolute zero. If an electron tunnels across the junction the electrostatic energy of the system will be increased. Since the low temperature ensures that the system is in the lowest energy state, such tunnelling will be energetically forbidden. This prohibition is called the 'Coulomb blockade'. By connecting the structure to a current, an equal and opposite charge builds up on the two electrodes and tunnelling can now take place, one electron at a time. This Coulomb blockade mechanism is the basis of 'single-electron turnstile' devices and single electron transistors. Recent developments in technologies have now demonstrated systems that exhibit single or few electron action at room temperature.

There is now the real prospect of combining single-electron structures with conventional semiconductor electronics. Besides leading to the possibility of single-electron memory devices, the extreme sensitivity of these devices to their local electromagnetic environment will make such devices useful for many types of sensor and detector applications.

At the level of integration envisaged by the semiconductor industry in 2010, there will still be many thousands of electrons participating in the storage of a bit or in the action of a transistor. The techniques described above make it possible to envisage devices that work with very small numbers of electrons. This in turn will enable the number of transistors on a chip to continue to increase without excessive power generation. To avoid problems with fluctuations in the number of electrons participating in such devices it will be necessary to use the principle of Coulomb blockade to control individual electrons. There are still many technological problems to be overcome but quantum engineering of new semiconductor devices may be able to keep Moore's Law true for another 35 years.

Quantum information

In his 1959 talk, Feynman calculated that individual letters require the storage of six or seven 'bits' of information – where a bit is a '1' and a '0' as in a computer. Allowing for some redundancy to guard against possible errors, he imagined storing a single bit of information in a little $5 \times 5 \times 5$ cube of 125 atoms. From these rather conservative assumptions, Feynman estimated that:

> ... all of the information that man has carefully accumulated in all the books of the world can be written in a cube of material one two-hundredth of an inch wide – which is the barest piece of dust that can be made out by the human eye.

This was the reason that Feynman called his talk *plenty* of room at the bottom! In fact, in a talk in 1981, Feynman went further and imagined storing a single bit of information using the quantum states of a single atom, electron or photon. For an atom we can use the two lowest energy levels to represent '1' and '0'; for an electron we can use 'spin up' and 'spin down' states; and for a photon we can use the two polarization states, 'V' and 'H', that we discussed in the previous chapter. So far, this would be merely a re-implementation of conventional computer memory. The new feature of 'quantum information' arises from the possibility for quantum systems to be in a quantum superposition of **both** '1' and '0' at the same time. We saw this superposition possibility in our discussion of photons in the previous chapter. After over half a century studying the fundamentals of computation, it came as something of a surprise to computer scientists to find that there was something new to be discovered about information! A bit of information stored in a quantum system requires a new name – a quantum bit or 'qubit'. In later sections we shall see that this insight – the existence of quantum information – coupled with the 'spooky' faster than light signalling implied by the EPR experiment, leads to exciting new possibilities for information theory and computing. We begin with another application that relies on the quantum nature of information – 'quantum cryptography.'

The science of cryptography dates back to ancient times. It consists of techniques for 'encoding' the information in a message so that it can only be 'decoded' by the intended recipient. Julius Caesar used a so-called 'shift cipher' to encode secret government messages. A shift cipher consists of a key number, known only to sender and receiver, which tells you how far to shift a second alphabet that is written under the first one. Nowadays, governments use far more complex schemes to encode their secret messages and other governments employ teams of cryptoanalysts to try to break the codes. There are many famous examples from the second world war. The US broke the Japanese PURPLE code and gained information that enabled them to win the decisive naval battle of Midway. At Bletchley Park in the UK, Alan Turing, with help from Polish intelligence and others, built one of the first primitive computers to break the German Navy's ENIGMA codes. This gave Winston Churchill vital information about the German U-Boat positions and enabled the UK to maintain the North Atlantic conveys that provided essential war supplies. Nowadays, we have need of cryptography for more everyday applications – such as encoding the information about our credit cards and financial transactions before sending them over the Internet. Cryptographic systems will clearly continue to be of great importance for both government and commerce.

There are two main classes of cryptosystems, which are distinguished by whether the 'key' is shared in secret or in public. The secret key system was proposed by Gilbert Vernam of AT&T in 1918. It is the only cryptosystem that provides absolute security. The system requires a key that is as long as the message that is being sent and the same key is never re-used to send another message. Unused keys were given to spies in the form of a tear-off pad: after sending a message the sheet with the used key was torn off. For this reason the system is sometimes known as a 'one-time pad.' When the Bolivian army captured the Marxist revolutionary Che Guevara in 1967 they found a list of random numbers for him to send secret messages to Fidel Castro in Cuba. Guevara could do this securely over an insecure radio link because he and Castro were using Vernam's one-time pad system. Public key cryptosystems, on the other hand, for which part of the key is made public, depend for their security on so-called 'one-way functions'. These are functions that are computationally easy to calculate one way but difficult to work backwards and deduce the input to the function from the answer. A relevant example is the widely used RSA system, developed by Rivest, Shamir and Adelman at MIT. This uses the fact that multiplication of two large prime numbers is easy by computer whereas deducing these prime numbers by factorizing the result is very difficult. As we will see, such public key systems are in principle vulnerable to attack by a quantum computer. For the present we shall concentrate on explaining how quantum information can be of use to Vernam's secret key, one-time pad cryptosystem.

The three participants in any discussion of cryptography are Alice, Bob and Eve. Alice is the sender who wants to encrypt a message and send it securely to Bob. Bob is the receiver, who receives the message and wants to decrypt it and discover its meaning. Eve is a potential eavesdropper who wants to listen in and break the code. The one-time pad is secure because Alice encrypts the message using a random number as a key which is as long as the message itself. Bob has the same key and can easily decrypt the message. The particular random number key is only used once. Although this system is perfectly secure in principle, its weakness in practice lies in the fact that Alice and Bob have to share the same keys and, since these are used only once, they need a lot of them. The keys have to be distributed to Alice and Bob using some secure mechanism – such as a courier or a personal meeting. During the Second World War, the Russians foolishly re-issued some one-time pads. This carelessness allowed the US cryptoanalysts to decrypt a large number of previously undecipherable messages that they had intercepted over the years. This large-scale decoding effort was codenamed the VENONA project. It was transcripts from this project that identified the atom spy CHARLES with the Los Alamos physicist Klaus Fuchs. Where does quantum mechanics come in? In principle, quantum mechanics provides a solution to the problem of key distribution: it allows Alice and Bob to exchange a sequence of random keys in complete security. 'Quantum key distribution' is therefore a better name for the process than quantum cryptography. The use of quantum information permits Alice and Bob to detect the presence of an eavesdropper and to discard any keys that may have been compromised.

The first quantum key distribution scheme was devised by Charles Bennett and Giles Brassard in 1984. It uses just the properties of photon polarization states that we described in the previous chapter. Using photons, let us see how we can transmit a random number key. Alice has equipment that allows her to send photons to Bob in polarization states that she chooses at random. A '0' bit is represented by a V polarized photon and a '1' bit by an H polarised photon. By measuring the polarization of the photons he receives from Alice, Bob can clearly re-construct the random sequence of 1's and 0's that Alice is sending without any difficulty. Unfortunately, so can Eve. She could set up her receiving apparatus to intercept the photons from Alice and then re-transmit an identical sequence of photons to Bob after she has measured their polarization. Neither Bob nor Alice would know that Eve now knew their secret key and the security of the one-time pad system would have been compromised. The scheme that Bennett and Brassard came up with was very ingenious. Alice can now send photons either as V or H, or, using a polarizer rotated at 45 degrees, as DV or DH. She chooses the 'basis ' – V–H or DV–DH – at random. Similarly, Bob can now choose

Quantum engineering

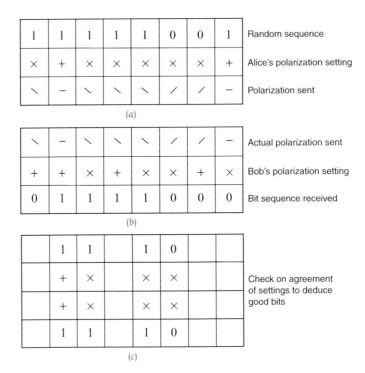

1	1	1	1	1	0	0	1	Random sequence
×	+	×	×	×	×	×	+	Alice's polarization setting
╲	—	╲	╲	╲	╱	╱	—	Polarization sent

(a)

╲	—	╲	╲	╲	╱	╱	—	Actual polarization sent
+	+	×	+	×	×	+	×	Bob's polarization setting
0	1	1	1	1	0	0	0	Bit sequence received

(b)

	1	1		1	0			
	+	×		×	×			Check on agreement of settings to deduce good bits
	+	×		×	×			
	1	1		1	0			

(c)

Fig. 9.11 An example of quantum cryptography at work showing; (a) the random sequence of bits that Alice wants to send, her choice of polarization setting for each bit (denoted '+' for V–H and '×' for DV–DH) and the polarization of the photon she sends; (b) the actual state of the polarized photon received by Bob, his choice of polarization setting to measure the polarization, and the sequence of random bits that he infers from his measurements; (c) Alice calls Bob to tell him which polarization setting she used so that Bob can discard the bits where they have used different settings. If there is an eavesdropper, Alice and Bob will find some disagreements even in cases where their settings were the same.

to measure the polarization either in the V and H directions, or in the DV and DH directions. The important point is that Bob does not know which setting Alice has used when she sent her '1' or '0' photon! So if Alice sent a '1' in the V–H setting, Bob could choose to receive this in his diagonal DV–DH setting. As we discussed in the last chapter, Alice's '1' is now a quantum superposition of Bob's '1' and '0'. Quantum mechanics then says it is impossible to predict which result Bob will find. All that we can say is that if we were to repeat the same experiment many times, Bob will measure the polarization of such a photon 50% of the time as a '1' and 50% of the time as a '0'. Figure 9.11 shows an example of a typical set of results of this process. So sometimes when Bob has chosen the 'wrong' polarization setting he will agree with Alice but other times he will disagree. How does this help? The trick is this. After sending a stream of bits, Alice can call up Bob on a standard phone line and tell him the sequence of settings V–H or DV–DH she used – **not** what bits she sent. Bob then compares these settings with the ones he used and they keep only the bits for which they used the same settings. If all this seems unnecessarily complicated, consider what happens if an eavesdropper has been intercepting the photons sent by Alice. Since Eve does not know which setting Alice used to generate her polarized photons, she has to guess which setting to use to measure the polarization – either V–H or DV–DH. On average she will guess wrong 50% of the time. So if Alice sends a '1' in the V–H basis, but Eve measures a '1' in the DV–DH basis, Eve will re-transmit the wrong type of '1' to Bob. What happens if Bob should choose to measure the polarization of this photon in the same V–H basis originally used by Alice? He now has a 50% chance of recording this as a '0' – even though he was using the 'correct' setting. When Alice telephones to tell Bob which settings she used, they can now check that their photon communications are secure. In addition to comparing settings, Alice and Bob can compare what sequence of 1s and 0s Bob should have received in the cases where their settings agreed. If there has been an eavesdropper they will find that there are some cases where they disagree, even though they used the same settings!

How does all this work in practice when you add in all the imperfections of real polarizers and detectors? In 1989, Bennett and Brassard built a 'toy' system that successfully used this cryptographic protocol with polarized photons to transmit a random key a distance of 30 cm in air. Since then, research groups around the world have successfully demonstrated that the use of this polarization scheme – and other more intricate key distribution schemes using EPR entanglement – can be used to communicate keys over distances of tens of kilometres down ordinary telecommunications optical fibres. There are many practical difficulties arising from the fact that photon sources, detectors and polarizers are not perfect. One needs to be able to distinguish errors caused by experimental imperfections from errors caused by an eavesdropper. Richard Hughes and his group at Los Alamos in the USA are also experimenting with keys sent over long distances in free space. They hope that such a method can eventually be used to send keys to satellites in orbit. This is a case in which there could be real difficulties in renewing the supply of secret keys by any other way!

Quantum computers

> … I'm not happy with all the analyses that go with just the classical theory, because Nature isn't classical, dammit, and if you want to make a simulation of Nature, you'd better make it quantum mechanical, and by golly it's a wonderful problem, because it doesn't look so easy.
> Richard Feynman

The study of the limits imposed by quantum mechanics on computers really became 'respectable' as an academic field after Feynman attended a conference on 'The physics of computation' at MIT in 1981. Feynman had been invited to give the keynote speech at the conference by his friend Ed Fredkin. Feynman, a physicist, and Fredkin, a computer scientist, had a long-standing friendship during the course of which they had 'wonderful, intense and interminable arguments'. Fredkin once spent a year with Feynman at Caltech and they made a deal: Feynman would teach Fredkin about quantum mechanics and Fredkin would teach Feynman about computer science. Fredkin had a hard time: 'It was very hard to teach Feynman something because he didn't want to let anyone teach him anything. What Feynman always wanted was to be told a few hints as to what the problem was and then to figure it out for himself. When you tried to save him time by just telling him what he needed to know, he got angry because you would be depriving him of the satisfaction of learning it for himself.' However, Feynman did not have it all his own way. During one of their arguments Feynman got so exasperated with Fredkin that he broke off the argument and started to quiz Fredkin about quantum mechanics. After a while he stopped the quiz and said: 'The trouble with you is **not** that you don't understand quantum mechanics!' Despite Feynman protesting that he didn't know what was meant by a keynote speech, Fredkin managed to persuade him to fly east to MIT for the meeting. In his talk at the conference Feynman proposed building a computer out of quantum mechanical elements that obey quantum mechanical laws:

> Can you do it [simulate quantum mechanics] with a new kind of computer – a quantum computer? ... It's not a Turing machine, but a machine of a different kind.

Alan Turing (1912–1957) was born in Paddington England and went to Cambridge to read mathematics. In 1936 Turing invented the device that is now called a 'Turing Machine'. This laid the foundations for a new field of computer science and made it possible to define which problems were 'computable'. He spent the Second World War at Bletchley Park in England working to decode messages sent using the German ENIGMA machine. He was also involved in building early computers – the Colossus, the ACE and the Manchester Mk 1, the world's first stored program computer. He was a serious runner and once contemplated training for Olympic trials. He was also interested in Artificial Intelligence and invented the famous 'Turing Test' to determine whether or not computers could think. He died in 1957 after eating an apple containing cyanide.

A	B	A AND B
0	0	0
0	1	0
1	0	0
1	1	1

Fig. 9.12 This shows the symbol for an electronic AND Gate with two inputs and one output. The inputs and output on each wire are binary signals '1' or '0'. The 'truth table' shows the output of the gate for every possible input. As can be seen, if the output is zero the input could have been any one of three pairs of signals.

Turing machines are used by computer scientists as a sort of academic short-hand to encapsulate the underlying principles of all conventional computers. As Feynman correctly appreciated, a computer operating according to the laws of quantum mechanics would be a totally new kind of computer – and one that might be able to do calculations that conventional computers are unable to do. In his talk, Feynman was referring specifically to simulations of quantum systems and quantum probabilities. In fact we will see that quantum computers can do some other types of calculations faster than Turing machines.

The subject of Fredkin's research on his visit to Caltech in 1974 might seem rather strange. Fredkin wanted to devise a *reversible* computer. This is a type of computer that can go backwards 'uncalculating' as well as calculating in the usual way. Conventional computers are built up from elementary 'logic gates' implemented in silicon chips. An example is an 'AND' gate, shown in Fig. 9.12. An AND gate has two inputs and one output. All the possible inputs for an AND gate together with the corresponding outputs are summarized in the 'truth table' shown in the figure. From this table, one can see that an AND gate only outputs a '1' if both inputs are '1': for the other three possible inputs the gate outputs a '0'. It is easy to see that an AND gate is not reversible: from the output, one cannot deduce a unique set of input signals. Fredkin came up with a new set of logic gates that **are** reversible. The simplest example is shown in Fig. 9.13 along with a conventional NOT gate. This new gate is called a 'controlled NOT' or CNOT gate. From the truth table for the CNOT gate, we see that the bottom input either acts as 'do nothing' or acts as a conventional NOT gate, reversing a '1' to a '0' and vice versa. Which action is chosen is determined by the signal on the upper input. If the upper input is a '0', the lower line does nothing: if it is a '1', the lower line acts as a NOT gate. Fredkin showed that it was possible to duplicate everything that conventional logic gates can do using reversible gates (although not with just CNOT gates). This seems all very esoteric! Why do we need to bother about reversible gates? One practical reason is that they may require less energy than conventional gates to perform conventional computing. But the reason such gates are relevant for quantum computing is that the laws of quantum mechanics are reversible in time. Reversibility is also true of conventional waves, not just for probability waves. A wave travelling in one direction along a string, for

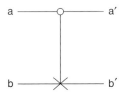

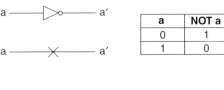

a	NOT a
0	1
1	0

a	b	a′	b′
0	0	0	0
0	1	0	1
1	0	1	1
1	1	1	0

Fig. 9.13 The simplest electronic gate is the NOT gate that has one input and one output wire. As can be seen from the truth table, the NOT gate merely reverses the incoming signal – a '1' to a '0' and vice versa. Since this is clearly a reversible gate in the sense that from the output one can deduce the input it has been represented unconventionally as a 'cross'. The controlled NOT gate or CNOT gate is the most important reversible gate. It has two inputs and two outputs and has the property that the signal on the bottom wire is only reversed if the signal on the upper wire is a '1'. If the upper wire has a '0' the signal on the bottom wire is unchanged. This behaviour is recorded in the truth table.

example, can just as easily travel in the reverse direction. All this means is that if we wish to construct a quantum computer we must use computational elements that are themselves reversible.

The ingredients of a quantum computer are now clear. We must have a physical system for which information can be stored as qubits on individual quantum systems. The information can not only be 1s or 0s but also quantum superpositions of '1' and '0'. To build a quantum computer we have to devise mechanisms by which these qubits can be made to interact and perform the reversible logic operations proposed by Fredkin. Notice that since we can choose to start off our quantum computer in a quantum superposition of all possible initial states, the computer can calculate the results for all these possible logical paths at the same time! David Deutsch, who first proved that quantum computers are, in principle, more powerful than conventional computers has called this 'quantum parallelism'. The trouble is that making a measurement on a quantum superposition only gives one of the possible states, so it remains to be seen whether such quantum parallelism can actually be useful. Is this all there is to quantum computing? In fact there is another key feature of quantum mechanics that we have not yet highlighted – 'quantum entanglement'. The two-particle quantum state involved in the EPR experiment discussed in the last chapter is said to be 'entangled'. The term 'entangled' to describe such states was first used by Schrödinger in the early days of quantum mechanics:

> I would not call that **one** but rather **the** characteristic trait of quantum mechanics, the one that enforces its entire departure from classical lines of thought. By the interaction the two representatives (or ψ-functions) have become entangled.

In the case of the EPR experiment this entanglement led to the 'spooky', faster-than-light correlations so loathed by Einstein. In the case of a quantum computer it is perhaps not surprising that entanglement leads to novel potentialities beyond the power of a classical computer. But how do such entangled states arise in quantum computers? Let us look at a simple example. Consider the action of a quantum CNOT gate on a state of two qubits. When the two qubits contain just 1s and 0s we obtain the exact analogue of the classical result (Fig. 9.13). But in a quantum computer there is nothing to stop us acting with a quantum CNOT gate on a qubit that is in a superposition of '1' and '0'. In this case, the output of the CNOT gate is an entangled two qubit state just like the two photon state in the EPR experiment. It is this new, non-classical feature of quantum mechanics that gives quantum computers their extraordinary properties and leads to quantum teleportation, of which more later.

After Feynman's lecture in 1981, it was David Deutsch, a physicist in Oxford, who took the next step. In 1985, he proved that quantum computers could indeed do calculations that conventional computers were unable to do. However, it was not until 1994 that interest in quantum computing really exploded. The reason was that Peter Shor of Bell Labs had discovered a 'quantum algorithm' that could do something useful! To appreciate his achievement we need to set it in context. Conventional computers are very good at multiplying two numbers together. For example, the time taken to multiply two N digit numbers grows as the square of N. On the other hand, the time needed to factorize an N digit number grows exponentially with N – faster than any power of N. This is an example of a 'one way function' as in our discussion of public key cryptography above. Peter Shor showed that a quantum computer could, in principle, factorize numbers just as easily as multiplication, without requiring exponentially increasing time as the size of the number to be factorized grows. This is an astonishingly powerful result. The whole basis of the RSA cryptosystem is the computational difficulty of factorizing large numbers. For example, in 1994, the 129 digit number known as RSA 129 required 8 months to factorize, using over 1000 computers (Fig. 9.14). If we could build a quantum computer that was roughly the same speed as just one of the computers used in this trial, Shor's algorithm could factorize RSA 129 in less than 10 seconds! For this reason alone, many government agencies around the world are now funding attempts to build a quantum computer.

In 1999, Charles Bennett of IBM Research Laboratories in Yorktown Heights, one of the pioneers of the field of quantum computing and quantum information theory, said:

> I am tempted to say that, as a topic of fundamental scientific inquiry, quantum computing in this sense is so close to finished as not to be interesting any more. Of course there are some practical details to be worked out, like actually building a quantum computer …

So how much progress have we made towards building a quantum computer? This is a fast-moving field and many groups around the world are

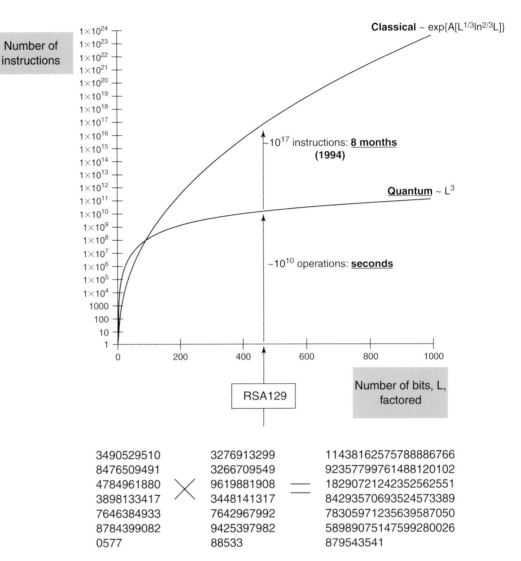

Fig. 9.14 The graph shows the increase in computer power, measured in numbers of computer instructions, required to factorize larger and larger numbers, measured in number of bits. According to classical computing theory this power increases exponentially with the number of bits. Shor's factorization algorithm for a quantum computer predicts that the computer power required will only rise as the cube of the number of bits as shown in the diagram. Also shown is the 129 digit number known as RSA 129 together with its two prime factors. In 1994, using many different computers, this took 8 months to factorize. A quantum computer operating at the same rate would factor this number in seconds!

exploring different ways to store and manipulate qubits. The current front-runner uses stored ions in a laser-cooled ion trap. In 1995, Ignacio Cirac and Peter Zoller from the University of Innsbruck showed how a CNOT gate could be implemented using such trapped ions. Two energy levels of the ion are used as the qubit states. The qubit states can be prepared and measured

Fig. 9.15 A schematic representation of the operations of an ion trap implementation of a quantum computer.

by directing laser beams at specific ions (Fig. 9.15). Coupling between the ions is provided by the vibrational states of the ions in the ion trap. Using these techniques researchers have managed to isolate systems containing a few qubits and implement a quantum gate. Another different physical technology being investigated is the use of conventional nuclear magnetic resonance (NMR) systems to manipulate molecular spins in solution. The problem for both of these technologies is their inability to scale up to systems with hundreds or thousands of gates and qubits. It will be a long time, if ever, before we can create and control systems that are able to factorize a three digit number let alone RSA 129!

There are also other problems for would-be builders of quantum computers. Conventional computer memories suffer from the problem that individual bits can occasionally get 'flipped'. Cosmic rays, for example, are one cause of such errors. To counter this problem, the computer industry has developed a battery of error-detection and correction techniques. A simple example is a 'parity' check. The 1s and 0s are added before and after sending a message. If a '1' has been corrupted to a '0' – or vice versa – this will be revealed by a simple parity check. A similar technique is used for the ISBN numbers of published books such as this one. Computer engineers have devised more complicated techniques to handle situations where more than one error has occurred and also ways of detecting which bit has flipped and correcting it. For qubit memory systems we have all these problems and more. Not only can we have random bit flips but also the crucial phase between the states in a quantum superposition can be affected. Surprisingly, it is possible in principle to detect and correct such quantum errors. Peter Shor of AT&T and Andrew Steane from Oxford independently devised schemes that used quantum entanglement to protect and correct quantum data. It remains to be seen whether such schemes are feasible in practice. As Bennett implies, the problem of actually building quantum computing systems is now becoming a problem of engineering rather than fundamental physics. In view of the difficulties of building large-scale qubit systems with ion traps and NMR, it seems sensible to investigate solid state quantum logic realizations that could be integrated into more conventional computer technology. A group at NEC in Japan have recently demonstrated a qubit system using Josephson Junction superconducting technology (Fig. 9.16). It seems likely that solid state realizations like this may be the best way to deliver interesting novel quantum devices within the next decade or so.

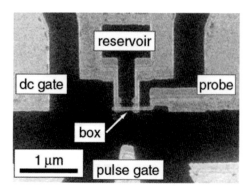

Fig. 9.16 A Josephson Junction qubit device made by researchers at NEC in Japan. A potential disadvantage of this technology is the very low temperatures required to operate such a superconducting device.

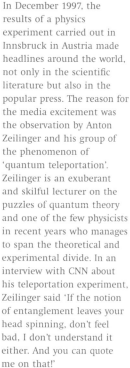

In December 1997, the results of a physics experiment carried out in Innsbruck in Austria made headlines around the world, not only in the scientific literature but also in the popular press. The reason for the media excitement was the observation by Anton Zeilinger and his group of the phenomenon of 'quantum teleportation'. Zeilinger is an exuberant and skilful lecturer on the puzzles of quantum theory and one of the few physicists in recent years who manages to span the theoretical and experimental divide. In an interview with CNN about his teleportation experiment, Zeilinger said 'If the notion of entanglement leaves your head spinning, don't feel bad, I don't understand it either. And you can quote me on that!'

In 1993, these six physicists collaborated to write a paper entitled 'Teleporting an unknown quantum state via dual classical and EPR channels'. The scientists are (top row, left to right) Richard Jozsa, William Wooters, Charles Bennett, and (bottom row, left to right) Gilles Brassard, Claude Crepeau and Asher Peres. Although quantum teleportation has been verified over very short distances in the laboratory we are still a very long way away from the vision of *Star Trek's* 'Beam me up Scotty'.

Quantum teleportation and all that

The February 1996 US edition of *Scientific American* carried an advertisement with the headline: 'Stand by. I'll teleport you some goulash!' Invoking images of *Star Trek*, the ad proclaimed in smaller print 'An IBM scientist and his colleagues have discovered a way to make an object disintegrate in one place and reappear in another'. An elderly lady is pictured talking on the telephone to a friend promising not to give her the recipe but to teleport the actual goulash. Her promise may be 'a little premature', the ad goes on to say, but 'IBM is working on it'. Charles Bennett was embarrassed by these claims for his research and subjected to much teasing from the research community. He later said 'In any organization there's a certain tension between the research end and the advertising end. I struggled hard with them over it, but perhaps I didn't struggle hard enough!'. The advertisement had its origin in a paper published in 1993 by Bennett and others entitled 'Teleporting an unknown quantum state via dual classical and Einstein–Podolsky–Rosen channels'. In fact, the teleporting of the title is a far cry from the teleportation capabilities of the *Starship Enterprise*. Nonetheless, quantum teleportation is a very surprising and exciting application of entangled quantum states.

Suppose Alice has a photon in an unknown quantum state ψ and she wants to send this to Bob. How does she go about doing this? If photons behaved like classical objects she could just send Bob a copy. But in 1992, William Wooters and Wojciech Zurek proved that a quantum state ψ cannot be 'cloned' – so Alice is unable to make a copy of the unknown state. It is easy to understand this result. From our earlier discussion of photon superpositions, we know that if Alice tries to measure the unknown polarization she risks destroying the state. If she measures polarization in the V–H directions and the state ψ is in either a pure V or a pure H state, everything is fine. But if the state ψ starts in a superposition of V and H she will find a photon that is in either the V or H state and lose all knowledge of the initial state. This is the basis of the famous quantum 'no cloning' theorem. How then can Alice send Bob information about the unknown state ψ? This is the problem solved by quantum teleportation. The process works as follows (Fig. 9.17). In preparation for teleportation, Alice and Bob

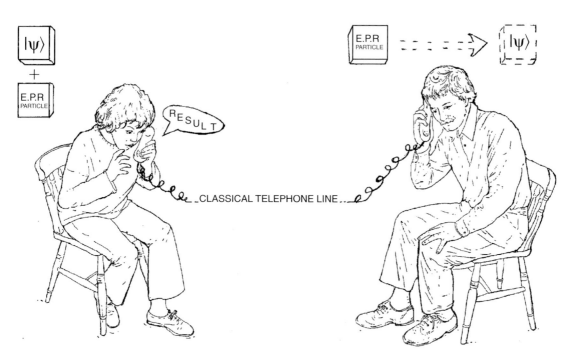

Fig. 9.17 An idealization of a quantum teleportation experiment. Alice and Bob share the particles of an EPR pair state between them without making a measurement of the state. Alice then uses her EPR particle to interact with the unknown quantum state ψ and reports the result of her experiment to Bob on an ordinary classical telephone line. Bob is then able to perform an experiment on his EPR partner particle to re-generate the unknown quantum state ψ.

must first have created a pair of EPR entangled photons. Each has taken one photon of the pair but neither has made any measurement of its polarization state. When Alice receives the photon in the unknown state ψ, she now makes a measurement on the two-photon state arising from combining the unknown photon with her EPR photon. This measurement instantaneously affects Bob's EPR photon in a specific way that depends on the result that Alice has obtained. Alice now telephones Bob to tell him of the result of her measurement. Knowing this result, Bob can perform a simple operation on his EPR photon and generate the unknown state ψ. Even though Bob's EPR photon was affected in the instantaneous, action-at-a-distance manner of the EPR experiment, as Alice has to telephone Bob with news of her measurement before he can re-produce the state ψ, no usable information has been transmitted faster than the speed of light. Several research groups around the world have now demonstrated the feasibility of such quantum teleportation of photons. In addition, a research group at IBM Almaden in California propose using this quantum teleportation principle to create 'quantum software'. Such software would consist of operations 'stored' in an EPR pair and could be used only once since measurement will destroy the fragile quantum state. In this example, the teleportation mechanism acts as a sort of 'quantum internet'!

What of the future? So far, the experiments for quantum key distribution are the most advanced. In the guise described here, such systems involve tried and tested concepts about quantum states. Quantum computers and quantum teleportation on the other hand, depend crucially for their power on EPR entangled states. Can we really have instantaneous action-at-a-distance over large distances? Whether or not quantum computers lead to a new industry it is clear that building such systems will test the ingenuity of a new breed of quantum engineers. We may also discover a limit to the present success of quantum theory. As Feynman says:

> We should always keep in mind the possibility that quantum mechanics may fail, since it has certain difficulties with the philosophical prejudices that we have about measurement and observation.

10 Death of a star

One of the most impressive discoveries was the origin of the energy of the stars, that makes them continue to burn. One of the men who discovered this was out with his girl friend the night after he realized that nuclear reactions must be going on in the stars in order to make them shine. She said, 'Look at how pretty the stars shine!' He said, 'Yes, and right now I am the only man in the world who knows why they shine'. She merely laughed at him. She was not impressed with being out with the only man who, at that moment, knew why the stars shine. Well it is sad to be alone, but that is the way of the world.

Richard Feynman

A failed star

In the previous chapter we have seen how quantum mechanics and the exclusion principle provide the basis for an understanding of all the different types of matter we see around us. What is perhaps more surprising is that quantum mechanics and the exclusion principle also provide the key to understanding stellar evolution and the variety of stars. As a prelude to stars we begin with a planet, Jupiter, which in one sense may be regarded as a star that did not quite make it!

Jupiter is by far the largest planet in our solar system. Some impression of its enormous size can be appreciated from the photo-montage shown in Fig. 10.1. Although Jupiter is huge compared to the Earth, it is still very much smaller than our star, the Sun. In spite of this large difference in size, Jupiter is similar to the Sun in two important respects. Both are mostly made of hydrogen, and both have an average density only slightly greater than water. Given that they consist of more or less the same ingredients, why is Jupiter not a fiery burning ball of gas like the Sun?

Imagine descending through Jupiter's cloud tops (see Fig. 10.2). As we go down towards the centre, the pressure will increase due to the weight of the overlying gas atmosphere above. Soon the pressure becomes so great that molecular hydrogen gas is compressed to liquid molecular hydrogen. If we plunge down into this ocean of hydrogen, just like diving in our oceans, the pressure rises still further. As we go deeper, the density of the

Fig. 10.1 A montage showing the Earth set against the backdrop of Jupiter's cloud tops on the same scale. The giant red spot is visible in the top right of the picture.

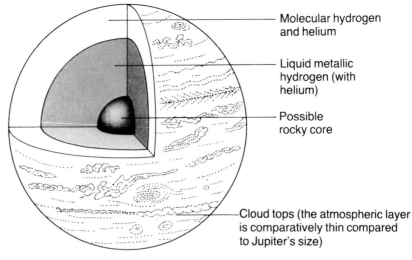

Fig. 10.2 A cut-away model of the interior of Jupiter. At the centre the pressure is 36 million times the ordinary atmospheric pressure on Earth, and the temperature is about 20 000 °C. Nonetheless, these conditions are not severe enough for Jupiter to be a star.

liquid hardly changes since hydrogen molecules have a definite size and the Pauli principle prevents other molecules from coming too close. It is the strength of the covalent hydrogen molecular bonds that resists the enormous pressures deep in Jupiter's hydrogen ocean. But as we continue downwards, the pressure rises well beyond anything experienced on Earth. The hydrogen molecular bonds eventually break so that the ocean is now made up of atomic hydrogen. The atoms of this atomic hydrogen liquid now come close enough for a band structure to develop. Since hydrogen has only one electron in the 1S shell, this atomic hydrogen sea will behave like a liquid metal, similar to liquid mercury familiar on Earth. This metallic sea can therefore sustain large electric currents and is believed to be the source of Jupiter's large magnetic field.

The pressure continues to rise as we press on to the centre of Jupiter, but the hydrogen atoms are strong enough to resist the enormous pressures that are generated. What prevents the hydrogen atom from breaking up? It is the familiar electrical attractive force between an electron and a proton that resists the gravitational pressure of a huge planet like Jupiter. In what sense is Jupiter like a failed star? Stars are very similar to Jupiter except they are more massive – Jupiter is only about one thousandth the mass of our Sun. This means that the pressure at the centre of a star will be even greater than at the centre of Jupiter. In a star, this pressure is so great that the electrons and protons in atoms are squeezed apart. The gravitational forces in a star are strong enough to overcome the electrical attraction between electrons and protons resulting in a soup of electrons and protons known as a plasma.

Planets are held up by atoms. In stars, the atoms have been squashed to form a plasma and gravitational forces tend to make the star collapse further in upon itself. As the plasma is compressed, the electrons and protons move about faster and faster so that the plasma heats up. This thermal motion of the electrons and protons provides a pressure that opposes further gravitational collapse. However, since the star radiates energy in the form of photons, the plasma will eventually begin to cool. To prevent the star undergoing further collapse, there must be a continuous supply of heat from within the star. As the star collapses, conditions at the centre of the star will eventually become dense enough and hot enough for nuclear reactions to take place. To show that famous physicists are human and make mistakes like anyone else, it is amusing to quote Rutherford on the prospects for nuclear power. He once said: 'The energy produced by the breaking down of the atom is a very poor kind of thing. Anyone who expects a source of power from the transformation of these atoms is talking moonshine.' Moonshine or not, nuclear energy is the power that makes the stars shine!

Hydrogen burning

For centuries astronomers and physicists wondered what made the stars shine. Elementary calculations had shown that ordinary chemical 'burning' was hopelessly inadequate to provide energy throughout the

Sir Arthur Eddington (1882–1944) was of Quaker origin and remained a convinced Quaker all his life. His religious beliefs qualified him as a conscientious objector during the First World War. Eddington was one of the first physicists to recognize the importance of general relativity. Indeed, he participated in the 1919 eclipse expedition, which confirmed Einstein's prediction of the bending of light by the Sun. Eddington's major work concerned the theoretical investigation of the interior of stars. He was also a celebrated popularizer of astronomy and was knighted in 1930.

thousands of millions of years of a star's life. The energy had to come from nuclear reactions. It was therefore unfortunate that the famous British astronomer, Sir Arthur Eddington, had shown that the temperatures inside stars were too low for protons to overcome the repulsive energy barrier and get close enough for nuclear reactions to take place! Nonetheless, Eddington remained convinced that nuclear energy was the only possible fuel for stars, and he challenged the doubters with the words: 'We do not argue with the critic who urges that stars are not hot enough for this process; we tell him to go and find a hotter place.' Eddington turned out to be right – but quantum mechanics was required to provide the explanation. Using Gamow's quantum tunnelling, described in chapter 5, a British astronomer, Robert Atkinson, and an Austrian physicist, Fritz Houtermans, solved the problem of energy production in stars. Their paper began with the words: 'Recently Gamow demonstrated that positively charged particles can penetrate the atomic nucleus even if traditional belief holds their energy to be inadequate.' They proposed that light nuclei could act as a 'trap' for protons, and when four protons had been captured an alpha particle could be formed. This could then be ejected from the nucleus thereby liberating the large amount of nuclear binding energy that arises from the fusion process in which the four hydrogen nuclei are converted into helium. Their original paper was entitled 'How can one cook helium nuclei in a potential pot?', but this title was changed to a more conventional one by the editor of the scientific journal! This paper provided the basis for modern theories of stellar thermonuclear reactions, and 10 years later, in 1939, Hans Bethe proposed the so-called carbon cycle in which carbon plays a role similar to Atkinson's and Houtermans' proton-trapping nucleus.

The Sun contains hydrogen and its energy must come from nuclear reactions involving the fusion of hydrogen to make helium and heavier elements. The energy released by a hydrogen bomb also derives from hydrogen fusion reactions. Why does the Sun not explode like a hydrogen bomb? In fact, the rate of energy generation in the Sun is so slow that a human-sized volume of the Sun burns up its nuclear fuel at a much slower rate than a human converts food into energy! The reason for the great difference in the rate of energy generation in a bomb and a star is that different hydrogen fusion reactions are involved. A star consists mostly of ordinary hydrogen nuclei, each containing a single proton, whereas the fusion reaction used in bombs requires the presence of the two rare isotopes of hydrogen, deuterium and tritium, which contain one and two neutrons, respectively, in addition to the proton. These isotopes of hydrogen undergo nuclear reactions with comparative ease. The nuclear reaction involving ordinary hydrogen by which our Sun generates its energy happens so rarely that it has never been observed in the laboratory! This is because the Sun's basic nuclear reaction involves the same mechanism as nuclear beta decay. Such reactions are known as 'weak interactions' and they proceed very slowly compared with the relatively rapid rates of the 'strong' nuclear interactions, such as deuterium–tritium fusion.

Hans Bethe was born in 1906 in Strasbourg, which was then part of Germany. When Hitler came to power Bethe left Germany and, after a brief stay in England, went to Cornell University in the USA. He was an important participant in the development of nuclear weapons at Los Alamos during the war and was also a negotiator for the control of nuclear tests in Geneva, Switzerland. Bethe won the 1967 Nobel Prize for his work on the nuclear processes that make the stars shine.

The weak interaction is the name we give to the force responsible for beta radioactivity. The simplest example of a weak interaction is neutron beta decay. Neutrons are slightly more massive than protons and, left to themselves, will eventually decay to a proton and an electron. These two particles are sufficient for conservation of the electric charge – a neutron with no charge is transformed into two particles of opposite charge – but experiments indicated that momentum and energy could not be conserved unless another, electrically neutral particle was also involved. This bold idea was suggested by Pauli in 1931 – the year before Chadwick discovered the world's first neutral particle, the neutron. To distinguish 'Pauli's neutron' from Chadwick's, Enrico Fermi called this hypothetical particle a neutrino ('little neutral one' in Italian). Since these curious particles have no charge, they do not feel electric forces. Moreover, since all the early attempts to catch a neutrino were unsuccessful, it was evident that they are also unaffected by the nuclear force! Nevertheless, since they are produced by the weak force, neutrinos must be able to interact with nuclear matter via a weak interaction. The difficulty in observing a neutrino lies in the predicted rate for such neutrino reactions – a neutrino has to pass through many 'light-years' of matter to have a fifty–fifty chance of interacting. Since the velocity of light is 300 million metres per second and a light-year is the distance that light would travel in a year (some 30 million seconds) you either need to have a fantastic amount of matter in your detector or have a beam with an enormous number of neutrinos to have a hope of seeing such a neutrino reaction. It is therefore not surprising that it was not until 1956, 25 years after the neutrino was first proposed by Pauli – and long after physicists had accepted that neutrinos must exist – that two US physicists, Frederick Reines and Clyde Cowan, were able to detect weak interactions caused by neutrinos. How did they get enough neutrinos? Since every nuclear fission gives rise to an average of about six beta decay processes, their first idea was to use neutrinos released by a nuclear explosion! Fortunately, they were able to use the neutrinos produced by a nuclear reactor. Out of the huge number of neutrinos that escape from the reactor – more than a million million neutrinos cross a square centimetre per second – about three neutrino events per hour were observed. The basic neutron beta decay reaction is written as follows:

$$n \rightarrow p + e^- + \bar{\nu}$$

where the particle represented by the Greek letter ν with a bar over the top (and pronounced 'new bar') that is created in the decay is, by convention, taken to be an antineutrino – the antiparticle of the neutrino. We will say more about antiparticles in the next chapter. Here, we note that if we move one of the participating particles to the other side of such a nuclear reaction equation, in order to keep the charge and other quantum numbers balanced we must take it to be the antiparticle partner of our original particle. By

Fred Hoyle (1915–2001) developed his theory of stellar evolution with Geoffrey and Margaret Burbidge and William Fowler in 1957. These ideas provide the basis for the theory of the synthesis of heavy elements in the cores of massive stars and supernovas. He was a prominent champion of the Steady State model of the universe that was the main alternative in the 1950s and 1960s to the now generally accepted Big Bang theory. Hoyle has also written several excellent popular books on astronomy as well some very entertaining and perceptive science fiction. Ever controversial, in the last years of his life he pursued the idea that life and diseases come from space. Such ideas are only now gaining some credence.

this process we see that a possible weak interaction is the reaction

$$n + e^+ \rightarrow p + \bar{\nu}$$

which involves the antiparticle of the electron – the positron. In fact, the reaction that Reines and Cowan actually looked for was the reverse of this reaction, namely

$$\bar{\nu} + p \rightarrow n + e^+$$

The news of the discovery of the neutrino reached Pauli not long before he died. It was not until 1995 that Frederick Reines was awarded the Physics Nobel Prize for this discovery. Nowadays, at the large particle accelerator laboratories, we have lost our sense of wonder about these peculiar particles and artificially prepared beams of neutrinos and antineutrinos are commonplace. We can now observe neutrino reactions such as

$$\nu + n \rightarrow p + e^-$$

as well as the antineutrino reaction seen by Reines and Cowan. We will say more about antiparticles later.

In the Sun there are protons, and protons on their own, cannot change into neutrons via the beta decay process

$$p \rightarrow n + e^+ + \nu$$

since neutrons are more massive. Inside a nucleus things are different. This process can, and does, occur if the new nucleus that would be produced by the 'decay' of one of its protons is more tightly bound than the original nucleus. According to the uncertainty principle, the whole system can 'borrow' the extra energy to make this decay possible, since, at the end of the day, the system will have reached a state with lower energy. Thus, although protons by themselves cannot change into neutrons, they can if they are in a suitable nucleus! This is the key to understanding the mechanism by which the Sun produces its energy. Consider the collision of two protons inside the Sun. The electrical repulsion makes it difficult for them to get close enough together to feel the effect of the short-ranged strong force. But, occasionally, due to quantum tunnelling, two protons are able to come together and form an unstable nucleus containing two protons. Usually, after an extremely short time, the two protons will fly apart again. But, because of the weak interaction and the uncertainty principle, there is a very small chance that one of the protons in this unstable nucleus will beta decay to a neutron resulting in the formation of a deuterium nucleus:

$$p + p \rightarrow d + e^+ + \nu$$

On average, it takes any given proton inside the Sun more than a thousand million years of collisions before this happens! This very difficult first step is the secret of the slowly burning Sun. Once deuterium has been made, the remaining nuclear reactions required to make helium proceed much

Subrahmanyan Chandrasekar (1910–1995) was born in Lahore, India (now Pakistan) and educated at Madras University. He obtained a Ph.D. at Cambridge, UK, where he studied under Dirac. He then worked at Chicago University and the Yerkes Observatory. Chandrasekar developed the first consistent model of white dwarf stars and was awarded the Nobel Prize in 1983 together with William Fowler.

more rapidly. These are a strong and electromagnetic interaction between p and d to form ^{3}He:

$$p + d \rightarrow {}^3\mathrm{He} + \gamma$$

followed by a purely strong interaction to produce ^{4}He:

$$^3\mathrm{He} + {}^3\mathrm{He} \rightarrow {}^4\mathrm{He} + p + p$$

This sequence of reactions is called the 'proton–proton cycle', and such reactions are believed to be the main source of our Sun's energy. In many stars, however, the temperatures are hot enough for energy generation to proceed via Bethe's carbon cycle. This does not require a weak interaction to take place during the instant of collision. Bethe's mechanism relies instead on the presence of carbon nuclei to act as a sort of catalyst to 'cook' helium.

Despite this stunning triumph of physicists in explaining the origin of the Sun's energy, there remains a niggling little problem that stubbornly refuses to go away, which we should mention as a cautionary postscript. The puzzle is this. The nuclear reactions in the proton–proton cycle of the Sun are believed to be well understood. As we have seen, some of these reactions produce neutrinos and it is relatively straightforward to predict the rate at which such 'solar neutrinos' should arrive at the Earth. An experiment to detect these solar neutrinos ran from 1968 to 1986 in the Homestake Gold Mine in South Dakota. The experiment was located deep underground to reduce the number of cosmic ray particles from outer space entering the apparatus and causing reactions that could be confused with solar neutrino interactions. Unfortunately, even after much careful checking, only about one-third of the expected number of neutrinos were detected. New experiments using different types of detector were undertaken in the 1980s and 1990s in an attempt to resolve this puzzle. All these new experiments were located underground – a Japanese experiment in the Kamioka mine, a US–Russian experiment in the Caucasus and a European–US–Israeli experiment in the Grand Sasso tunnel near Rome. These experiments confirm the original results of the Homestake Mine experiment of Raymond Davis and only observe about half the expected number of solar neutrinos. There are two possible explanations – either our present understanding of the physics of the interior of the Sun is not completely correct or something happens to the neutrinos on their way to Earth. At the moment physicists are betting on the second alternative. Remarkably, our current understanding of the forces of Nature allows just such a possibility and suggests that neutrinos may indeed change their nature in transit from the Sun to the Earth! In 1998, results from a new experiment at Kamioka support the idea of such 'neutrino oscillations'. These results have recently been confirmed by another experiment carried out in a nickel mine in Sudbury, Ontario. One important consequence of these results is that neutrinos are not massless like photons but have a very small but non-zero mass. We shall return to these mysterious neutrino oscillations in chapter 12.

Fig. 10.3 The neutrino 'telescope' of Raymond Davis in the Homestake Gold Mine in South Dakota, USA.

Red giants and white dwarfs

A star like our Sun has enough fuel to keep burning for several thousands of millions of years. But what happens when the hydrogen begins to run out? Since nuclear reactions occur at the core of the star, the core will eventually consist mainly of helium. Helium needs higher temperatures and pressures than hydrogen before nuclear reactions can take place. As the star generates less and less energy, the gravitational forces start to win and the star begins to collapse. This causes the temperature to rise again and eventually much more rapid hydrogen burning by Bethe's carbon cycle becomes possible. These hydrogen nuclear reactions initially take place in a thin shell about the core. The increased heat production causes the outer layers of the star to expand until its radius is hundreds or thousands of times larger than before. Since the total energy produced by the star is now spread out over a much larger area, the surface temperature is reduced and the now giant star appears redder. Such a star is called a 'red giant' and

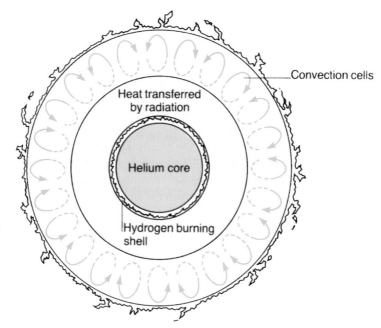

Fig. 10.4 The interior of a red giant star. The diagram is not to scale as the core is small and very dense. The outer layers are extremely tenuous and would occupy a region 100 times the diameter of our Sun.

Fig. 10.5 This painting shows the Sun as it might look about 5000 million years into the future. The Sun has become a red giant and will eventually engulf the Earth.

in this phase of its evolution our Sun will be large enough to engulf both Mercury and Venus!

What happens in the core left behind? This will continue to collapse, becoming more and more dense as more helium is produced by the hydrogen burning shell. As the pressure increases, the electrons are crowded closer and closer together. As ever, the Pauli exclusion principle prevents two electrons with the same quantum numbers occupying the same volume. The size of this minimum volume is determined by the de Broglie wavelength of the electron. Since this wavelength becomes smaller as the momentum of the electron increases, the electrons move faster and faster as the pressure increases. For a star with the mass of our Sun, when the electrons are moving with velocities close to the speed of light, the Pauli principle applied to electrons will prevent any further collapse of the core. There will be a similar Pauli principle effect arising from the protons and neutrons in the core. Protons and neutrons have a much greater mass than electrons and their de Broglie wavelengths are much smaller. The Pauli principle effect from protons and neutrons will therefore not become significant in halting collapse until much higher pressures are reached. Thus it is electrons and the Pauli principle that are responsible for preventing further collapse of the core at this stage in the star's life cycle. Such a core now consists of matter with an incredibly high density – a teaspoonful would weigh several tons!

For a star like the Sun, the temperature and density in the core will eventually become high enough for helium burning to take place. Helium reactions will then occur rapidly until a hot central core of carbon is formed and the outer layers of the star are thrown off into space. A planetary nebula – an expanding shell of gas from such a star – is the result, an example of which is shown in Fig. 10.6. The ring is kept glowing by radiation from the central core of the star. This remnant of the core cools down to become a 'white dwarf': a hot, dense object prevented from further collapse by the 'degeneracy pressure' caused by the electrons and the Pauli principle. A typical white dwarf star would be about the size of the Earth yet contain about the mass of the Sun. The white dwarf is white because it is still very hot and therefore able to radiate light energy. As no more nuclear reactions are possible, the star will gradually cool and grow dim until it enters the final phase of its evolution and becomes a 'black dwarf'. This cooling process is expected to take perhaps a million million years, longer than the present age of the universe, so no black dwarfs have ever been observed.

More complicated evolutions are possible for white dwarf members of a double star system. Figure 10.7 shows Sirius and its white dwarf companion. In this case, it is believed that when the white dwarf star was in its red giant phase, material was transferred to its companion star. Alternatively, if the two stars in the binary system are sufficiently close, matter can be transferred to a white dwarf from a close red giant. The white dwarf accumulates hydrogen from its companion and this can result in a violent nuclear explosion. For a brief period of time, the double star system can

Fig. 10.6 At the end of their red giant phase, stars like the Sun eventually shed their outer layers to form a planetary nebula. The central core that remains will cool to become a white dwarf. This photograph shows the planetary nebula M27, which was ejected from its star some 50 000 years ago. Intense ultraviolet light from the central star keeps the nebula glowing.

Fig. 10.7 Sirius A, the brightest star in the night sky, together with its white dwarf companion Sirius B. The dwarf star is hotter than Sirius A, but because it is only about the size of the Earth it is about 10, 000 times less brilliant.

become several tens of thousands of times brighter. In the days before telescopes it seemed as if a new star had been born, which then faded and died in a matter of weeks. Such stars are called 'novas', from the Latin for new.

A white dwarf in a double star system may also be involved in the most violent of all star explosions – a 'supernova'. In this case, the new star may appear as bright as a whole galaxy. No supernovas have been observed in our own Galaxy after the invention of the telescope. In 1054, however, the Chinese recorded the appearance of a 'guest star' that was bright enough to be visible in the daytime for many days. At the position

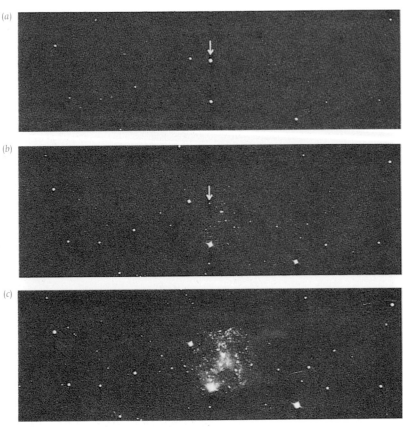

Fig. 10.9 Three photographs of a supernova in the galaxy IC4182. (*a*) In this short 20-minute exposure, taken on August 23rd, 1937, the supernova shows up clearly while the galaxy is too faint to be seen at this exposure. (*b*) On this photograph, taken just over a year later on November 24th, 1938, the supernova is much fainter. With this 45-minute exposure it can just be seen and the galaxy is starting to become visible. (*c*) On January 19th, 1942, the supernova is too faint to be seen. This 85-minute exposure shows the galaxy clearly, and this series of photographs emphasizes the extreme brilliance of supernova explosions. Supernovae are now being found at a rate of about 100 per year but the last one in our Galaxy is still that observed by Kepler in 1604.

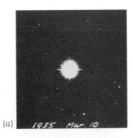

(*a*)   *1935 Mar. 10*

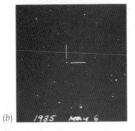

(*b*)   *1935 May 6*

Fig. 10.8 (*a*) The nova that exploded in 1934 in the constellation of Hercules as it appeared some months later, in March, 1935. (*b*) The nova as it appeared 8 weeks later.

of this explosion, we now see the spectacular Crab Nebula, which certainly looks like the remains of some enormous explosion. In fact, if the star is sufficiently massive, supernovas can occur without the aid of a companion star. The Crab supernova observed by the Chinese is believed to have been of this type. In the next section, we shall see that quantum mechanics and the Pauli principle have still more to say about the evolution of such very massive stars into even more exotic objects than white dwarfs. The range of applicability of quantum mechanics is truly breathtaking!

As a footnote to this section, we ought to complete our family of dwarfs with a mention of 'brown dwarfs'. These are objects that are intermediate between Jupiter-like planets and true stars. The interior of such systems collapses as we have described but the plasma does not become

(a) (b)

Fig. 10.10 In 1987, the first supernova visible to the naked eye since Kepler's day was observed in the Large Magellanic Cloud – a companion galaxy to the Milky Way. This was only visible from the southern hemisphere ((a) before, (b) after).

hot enough for the main hydrogen burning nuclear reaction to take place. The first such brown dwarf, Kelu 1, was discovered in 1997.

Neutron stars and black holes

In very massive stars, nuclear reactions are able to continue beyond the stage of helium burning to carbon. When the core becomes sufficiently hot, new nuclear processes can occur. A very complicated series of reactions can make successively heavier elements up to iron, ^{56}Fe. After iron, it is not possible to gain more energy from fusion reactions, since iron has the highest binding energy of all the elements. The iron accumulates in the core until the nuclear fuel begins to run out. As this happens, the core begins to contract again until it is prevented from further collapse by the exclusion principle.

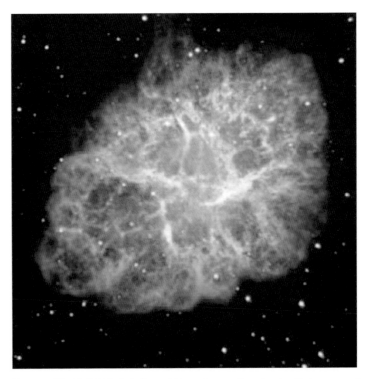

Fig. 10.11 The Crab Nebula is formed from the remains of a star that was seen to explode by the Chinese two hours after midnight in the summer of 1054. Chinese astronomers noted the appearance of a 'guest star' between the horns of the constellation we call Taurus – the Bull. It outshone Venus and Jupiter and was even visible during the day for about 3 weeks. At the centre of the Crab Nebula there is a pulsar – a rapidly rotating neutron star formed in the explosion.

Can the electron Pauli principle prevent the collapse of any star, no matter how massive? The answer is no. There is a critical mass – called the Chandrasekar limit – beyond which the Pauli principle applied to electrons cannot prevent further gravitational collapse. How does this come about? As the iron core of a very massive star collapses, the electrons eventually get squeezed together so much that a significant number of them have enough energy to initiate a weak interaction process. This changes protons into neutrons via the reaction

$$e^- + p \rightarrow n + \nu$$

This has the effect of removing electrons and protons from the core as well as allowing energy – in the form of neutrinos – to escape from the star. Once this process starts to reduce the Pauli pressure due to the electrons, an incredibly rapid and violent collapse of the core is initiated. The precise details of this collapse and how the spectacular supernova explosion is generated are very complicated and still debated by astrophysicists. What seems clear is that the supernova will leave behind a compressed ball of hot neutrons – a 'neutron star'. As the hot neutron star cools, further collapse is prevented by the Pauli principle applied to the neutrons – unless the mass is so great that the star can become a 'black hole', as we discuss later. For a star remnant about twice as massive as our Sun, the resulting neutron star will be about ten miles in diameter. The density is over a million million times that of water and is roughly the same as that inside an atomic nucleus. In a sense, therefore, a neutron star is just like a gigantic nucleus!

The application of quantum mechanics to neutron stars may sound like a very speculative application of quantum mechanics, but the idea was suggested more than 50 years ago by J. Robert Oppenheimer. Oppenheimer is such an interesting character who played such a pivotal role in the development of the post Second World War nuclear 'MAD' world – MAD stood for mutually assured destruction – that we cannot resist a brief historical digression. The austere, academic scientist who predicted neutron stars was the same man who later directed the physicists in the Manhattan project to make the first atomic bomb. At the height of anti-communist paranoia in the USA, in 1954, Oppenheimer was declared a security risk and 'unfit to serve his country'. Edward Teller, a colleague of Oppenheimer at Los Alamos during the war and later popularly known as the 'father of the hydrogen bomb', split the scientific community by giving evidence against Oppenheimer. The background to Oppenheimer's indictment was turbulent and confused with Senator Joseph McCarthy's infamous witch-hunts against communists reaching their peak. In 1950, Klaus Fuchs, who had worked with Oppenheimer at Los Alamos and who was co-author with John von Neumann of a top-secret document called a 'Disclosure of invention' that contained a summary of every significant advance made towards a hydrogen fusion thermonuclear bomb, had been identified as a Russian spy. This disaster had been followed, in August of 1953, by the successful Russian test of the world's first true, 'droppable' H-bomb. It would not be until 1956 that the USA would have a usable fusion bomb. It was not until the morning of November 22nd, 1963, that the White House announced that President Kennedy would personally present Oppenheimer with the prestigious Fermi award. This was intended as a first step towards a public apology for the anti-communist hysteria that had led to his indictment 10 years earlier. Alas, on the afternoon of the same day, John Kennedy was assassinated, and it was left to President Johnson, in the face of opposition from his political advisors, to make a personal presentation of the award to Oppenheimer. One US senator who had been prominent in the campaign against Oppenheimer called the award ceremony 'shocking and revolting'.

After this historical diversion, we return to neutron stars and explain why astronomers believe they exist. The observational evidence for neutron stars is associated with the discovery of 'pulsars' by Jocelyn Bell, a research student of Anthony Hewish, in Cambridge, UK, in 1967. Pulsars are rapid and remarkably regular radio pulses of extra-terrestrial origin. Soon after the first pulsar was discovered, another was found at the centre of the Crab Nebula, at the site of the Chinese supernova explosion. The Crab pulsar

Jocelyn Bell was a graduate student supervised by Anthony Hewish in Cambridge, UK, when she first noticed the regular signal of pulsars. Her thesis contains results on the angular diameter of some 200 scintillating radio sources and only mentions pulsars in an appendix! Anthony Hewish, who developed the scintillation technique and supervised the project, won the Nobel Prize in 1974.

Fig. 10.12 More than 2000 dipole antennae made up the radio telescope with which pulsars were discovered. The sloping bars support a reflecting screen that increases the telescope's sensitivity. The original purpose of the instrument was to study the twinkling of radio sources in interstellar space.

flashes on and off about 30 times a second (Fig. 10.13), emitting energy across most of the electromagnetic spectrum (see appendix 1). Pulsars were originally designated by the acronym LGM – standing for Little Green Men – since their extraordinary regularity was at first suspected to be a signal from an extra-terrestrial civilization. The truth is now believed to be somewhat less romantic – they are almost certainly rapidly rotating neutron stars!

Tommy Gold of Cornell, USA, first realized that pulsars could be understood as rotating neutron stars. The rate of rotation required was very much greater than that of normal stars. But just as a skater draws in her arms to make a slow spin become a rapid spin – an elegant demonstration of angular momentum conservation – so does the rotation of a star increase as it collapses to form a neutron star. The magnetic fields of the star are also increased to much higher values by the collapse. As illustrated in Fig. 10.14, the magnetic poles of the neutron star will not usually coincide with the poles of the axis of rotation. By a rather complicated mechanism involving

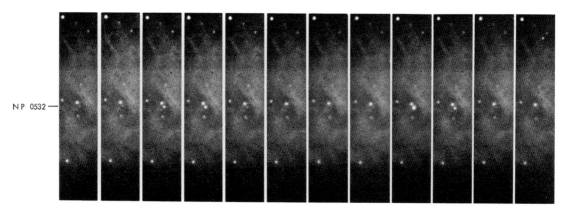

N P 0532 —

Fig. 10.13 This sequence of photographs shows the Crab pulsar NP0532 flashing on and off for a complete cycle. The whole cycle takes only 1/30th of a second, which is the period of rotation of the neutron star.

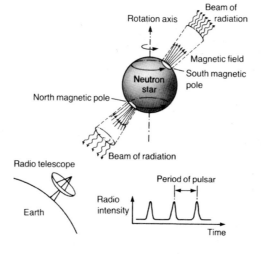

Fig. 10.14 A pulsar is a rapidly rotating neutron star with an enormous magnetic field. As the star rotates, it emits a narrow beam of radiation from the polar regions. If the beam crosses the Earth, the pulsar can be detected by the regular series of pulses of radio energy received.

both the magnetic and electric fields of the neutron star, it is believed that an intense narrow beam of radiation is produced in the direction of the magnetic axis. It is this beam of radiation sweeping regularly across the Earth as the neutron star rotates that causes the observed pulsing of the pulsar.

A neutron star is an amazingly compact and dense object. Nevertheless, the immense gravitational forces generated within such an object are countered by the neutron Pauli principle. But if the star is sufficiently massive (more than about three times the mass of our Sun) even the Pauli principle for the quarks inside neutrons (see chapter 12) cannot prevent the star collapsing to form an even more bizarre object – a 'black hole'. Such curious objects are permitted by Einstein's general theory of relativity – actually a theory of gravity – which is not the subject of this book (see our companion volume *Einstein's Mirror*). A black hole corresponds to a special type of

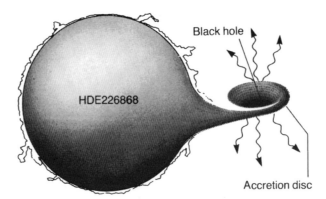

Fig. 10.15 A sketch of a black hole model for the X-ray source Cygnus X-1. Measurements of the period of rotation of this binary system suggest that the mass of the unseen X-ray source is larger than the mass of a neutron star. It is suggested that the X-rays are produced as material streaming from the companion star falls onto the 'accretion disc' of material rotating around the black hole, before eventually falling beyond the region of no return.

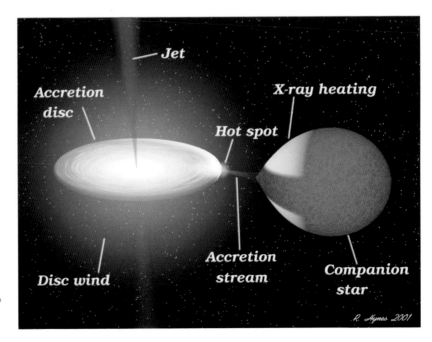

Fig. 10.16 The best candidates for black holes are now believed to be so-called 'X-ray transients'. This simulation of such a system was produced by Rob Bynes and Phil Charles in Southampton.

solution to the Einstein equations. To form a black hole enormously high densities are required. For example, for our Sun to become a black hole, it would need to be compressed to a ball about four miles in diameter. Once a star has been compressed smaller than a critical radius – the Schwarzschild radius – the effects of gravity are so strong that nothing, not even light, can escape. It is truly a black hole!

At present, we have no complete theory that combines both quantum mechanics and general relativity in a totally satisfactory way. Thus, we cannot be sure of the details of a star's collapse to a black hole, or even be absolutely sure that such objects must exist. Our doubts would be removed if we could observe a black hole experimentally. But since no radiation can ever leave the black hole, how can this be done? One way that has been suggested is to look for binary star systems. Such systems consist of two stars rotating around each other, like a couple on a dance floor. If one of the objects in such a system is a black hole, its mass can be estimated from the behaviour of the visible companion. The black hole sucks in matter from the partner star and this material will radiate X-rays – high energy photons – as it falls towards the black hole. The first candidate black hole was suggested in the Cygnus constellation (Fig. 10.15) but astronomers have now found about 15 such black hole candidates (Fig. 10.16). We shall have more to say about quantum mechanics and black holes in chapter 11.

11 Feynman rules

It is as though a bombardier flying low over a
road suddenly sees three roads and it is only
when two of them come together and disappear
again that he realizes he has simply passed over
a long switchback in a single road.

Richard Feynman

Albert Einstein worked at
the Patent Office in Berne,
Switzerland. After a crucial
conversation with his friend
and colleague at the Patent
Office, Michele Angelo Besso,
Einstein realized that a
radical rethinking of the
nature of time was needed
and this led to his Special
Theory of Relativity. Paul
Dirac has said that if
Einstein had not published
his theory in 1905, someone
else would have done so,
soon after. But Dirac also
went on to say that without
Einstein we would probably
still be waiting for the
General Theory of Relativity.
About Dirac, Einstein once
said: 'I have trouble with
Dirac. This balancing on the
dizzying path between
genius and madness is
awful'.

Dirac and antiparticles

We have seen in the earlier chapters that quantum mechanics, despite its inherent probabilistic nature, is capable of making successful predictions for an enormous range of phenomena. In the microscopic domain,
there is no doubt that Newton's laws of classical mechanics must give way
to quantum theory. There is another area where Newton's laws have been
shown to be in need of modification, namely, when the velocities of objects are close to the speed of light. Since light travels at approximately
300 000 km/s, the effects of these modifications, like those of quantum mechanics, are not usually apparent in everyday life. According to Einstein's
theory of special relativity the energy, E, and momentum, p, of a particle
are related by the equation

$$E^2 = p^2c^2 + m^2c^4$$

where c is the velocity of light and m is the mass of the particle when it is
at rest. The more familiar relation between energy and momentum

$$E = p^2/2m$$

may be derived from the relativistic equation as an approximation, valid
when the velocity of the particle is much less than the velocity of light.
It is also customary not to include the rest mass energy $- mc^2 -$ in this
non-relativistic expression for the energy. Einstein, Technical Expert Third
Class at the Patent Office in Berne, Switzerland, developed the special theory of relativity in 1905. The theory was well understood and accepted by
the physics community by the 1920s, when quantum mechanics was being
invented. It was natural for Schrödinger to attempt to develop quantum mechanics starting from the relativistic energy–momentum equation above.
After trying unsuccessfully to find a relativistic wave equation that agreed

Paul Dirac and Werner Heisenberg in 1933. Dirac's father was Swiss but he emigrated to England and became a language teacher in Bristol. Dirac was brought up to be bilingual in French and English, but remained extremely reserved in both languages. He was married to the sister of physicist Eugene Wigner and in 1933 he shared the Nobel Prize with Schrödinger.

with experiment, Schrödinger resorted to the approximate non-relativistic relation. His famous equation was published in January, 1926. In spite of the great success of the Schrödinger equation, this version of quantum mechanics cannot be valid for high speed electrons. Furthermore, the spin angular momentum of the electron that we discussed in chapter 6 had to be added to the theory in a very *ad hoc* way. Clearly, a relativistic equation was needed.

Paul Dirac was born in Bristol, UK, in 1902 and obtained a BSc degree in electrical engineering from the University of Bristol in 1921. Twelve years later he shared the Nobel Prize with Schrödinger 'for their discovery of new and productive forms of atomic theory', and he had also predicted the existence of antimatter. Dirac was an extremely original thinker but notoriously reserved and sparing with his conversation. Heisenberg told an amusing story about Dirac which illustrates both these qualities. The two of them were travelling to Japan from the USA by boat and Heisenberg liked to join in the social activities that went on in the evenings. At a dance one night, Heisenberg was enjoying himself dancing and Dirac, as usual, was sitting watching. As Heisenberg came back to his chair after a dance Dirac asked him 'Why do you dance?' Heisenberg replied 'Well, when there are some nice girls it is a pleasure to dance'. Dirac thought about this for a while. After about five minutes, he said 'Heisenberg, how do you know beforehand that the girls are nice?'

$$E\,\psi = (-i\vec{\alpha} \cdot \vec{\nabla} + \beta m)\psi$$

Dirac equation for a relativistic electron

It is rather curious that Dirac is still relatively unknown to the general public. He was certainly one of the twentieth century's greatest physicists and his achievements are on a par with such great figures as Newton, Maxwell and Einstein. What did Dirac actually do? As Feynman says, 'Dirac got his answers by ... guessing an equation'. The Dirac equation looks deceptively simple when written in the usual, very compact mathematical notation (see the box). For a relativistic version of quantum mechanics, the solutions of this equation must give rise to the correct relativistic relation between E and p. But, for a given momentum p, there are two possible solutions for the energy, namely

$$E = \pm\sqrt{(p^2c^2 + m^2c^4)}$$

One solution has positive energy as we would expect, but the second solution appears to have negative energy! How can negative energy solutions have any physical sense? Dirac's great achievement was to take them seriously and turn these apparently unwelcome solutions into a triumph of theoretical physics. His ingenious suggestion was that the negative energy levels did exist but were normally already occupied by electrons. Then,

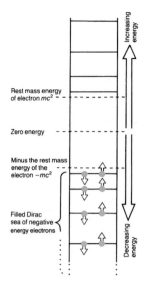

Fig. 11.1 Dirac's picture of the vacuum. Solving the Dirac equation for relativistic electrons in a box leads to both positive and negative energy levels. Dirac made sense of this by supposing that in an empty box – the vacuum state – all the negative energy levels were filled. Then, according to the Pauli exclusion principle, if a positive energy electron is put into the box it cannot lose energy by falling into the negative energy levels.

because of Pauli's exclusion principle, no ordinary positive energy electron can make a transition to any of these levels. According to Dirac, a quantum box, which is apparently empty – which contains no positive energy electrons – in fact has a fully occupied 'sea' of negative energy electron levels (Fig. 11.1)! This is not as ridiculous as it first sounds. If we put some positive energy electrons into this box, the charge and energy of the resulting system are measured relative to the charge and energy of the empty box state. The infinite negative charge and negative energy of Dirac's empty box sea is therefore unobservable. Although this sounds like some wild theoretical fantasy, like any good theory, Dirac's picture of the 'vacuum' – as the empty box state is called – makes some dramatic predictions. We know that if we shine light on an atom, electrons can absorb energy from a light photon and jump to an excited state. Consider what happens if we shine light on Dirac's empty box. According to Dirac, it should be possible to excite one of the negative energy electrons to a positive energy level. Instead of an empty box, we would then have a positive energy electron, together with a 'hole' in the Dirac sea. Relative to the normal empty box state, a box with a hole in the sea is lacking some negative energy and an electron's negative charge. Compared with the empty vacuum state, a hole in the sea has positive energy and positive charge! The physical process we have described

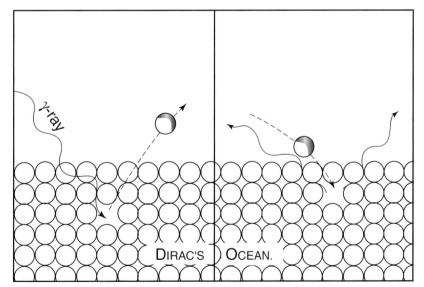

Fig. 11.2 Dirac's sea of negative energy electrons has observable consequences. In this picture from Mr Tompkins' adventures we see that a high-energy photon or gamma ray can excite an electron from a negative energy level in the sea so that it appears as an ordinary positive energy electron. The 'hole' in the sea acts like a particle of positive charge and energy relative to the normal vacuum state. The photon has therefore created a particle–antiparticle pair. The second picture shows an electron jumping into a hole in the sea. This corresponds to the reverse process of annihilation of an electron with its positron antiparticle to produce high-energy photons.

On the right of this photograph is Carl Anderson, who won the Nobel Prize for the discovery of the positron, while on the left is his student, Donald Glaser, who won the Nobel Prize for the invention of the bubble chamber. There is a bar in Ann Arbor, Michigan, where Glaser is said to have got the idea for the bubble chamber by watching the bubbles in a glass of beer.

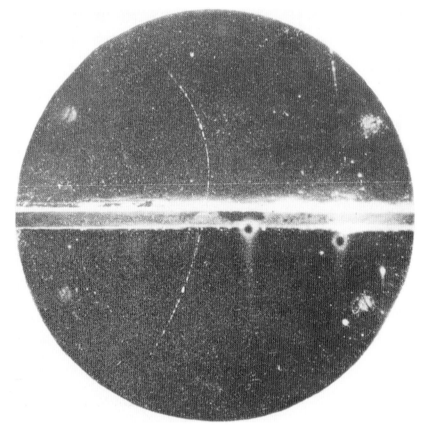

Fig. 11.3 A curved line of droplets shows the passage of a positron through a cloud chamber. The tracks are curved because there is a magnetic field throughout the chamber and the direction of curvature gives the sign of the electric charge. This photograph was taken by Carl Anderson and conclusively verified Dirac's prediction of antimatter. The lead plate in the middle of the chamber slowed down the positron as it passed through and resulted in the track above the plate being more tightly curved. This meant that Anderson knew that the track represented a positive charge going upwards rather than an ordinary electron going downwards. Anderson had to take elaborate precautions to be sure that none of the Caltech undergraduates had played a joke on him by reversing the magnetic field.

using Dirac's negative energy sea picture is the creation of an electron–positron pair by a photon. The positron is the antiparticle of the electron – a particle with the same mass but opposite charge.

As always in physics, a theory is only as good as its predictions. The positron was found in cosmic ray experiments by Carl Anderson in 1932, four years after Dirac wrote down his equation. The antiproton – the

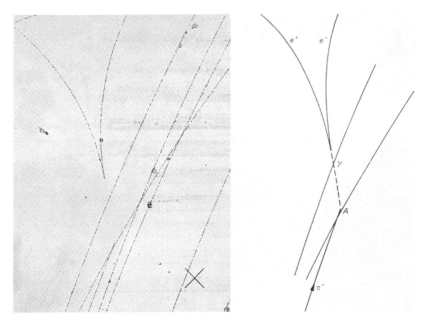

Fig. 11.4 A bubble chamber photograph of electron–positron pair creation. Only charged particles leave tracks in the chamber and the photon leaves no direct trace.

antiparticle of the proton – was discovered at Berkeley, California, in 1955. Its discovery had to wait for the construction of an accelerator that could provide enough energy for the creation of proton–antiproton pairs via the reaction

$$p + p \rightarrow p + p + p + \bar{p}$$

The converse process to pair creation was also predicted by Dirac. What happens if we have a positive energy electron plus a positron hole in our quantum box? The electron can now jump back into the sea and fill up the hole, leaving an empty box plus two photons to take away the electron–positron annihilation energy

$$e^+ + e^- \rightarrow \gamma + \gamma$$

Although the notion of Dirac's negative energy sea was the way antimatter was predicted, it is a very awkward and asymmetrical way of looking at antiparticles. The new feature of relativistic quantum mechanics – illustrated by both the pair creation and the pair annihilation processes – is the possibility of transforming energy into matter. Unlike the non-relativistic quantum mechanics of Schrödinger and Heisenberg, the number of quantum particles can change. Dirac's invocation of a filled negative energy sea of electrons is just a trick that enabled him to continue to use a single particle wave equation in a regime where a genuinely many particle theory is required. It is also unhelpful since it relies on the Pauli principle to prevent positive energy particles jumping to a lower energy state which implies that bosons – particles that do not obey Pauli's exclusion principle – will behave differently. In fact, as we shall see in the next chapter, there is

Fig. 11.5 The Bevatron accelerator in Berkeley, California. This machine was the first accelerator to have enough energy to produce antiprotons.

an elementary particle called a pion which is a spin zero boson and whose positively charged and negatively charged varieties – called π^+ and π^- – take part in pair creation and annihilation processes in the same way as for electrons and positrons. The Dirac sea, with its apparently infinite negative charge and mass, disappears from a proper many-body formulation of quantum theory that allows particle creation and annihilation processes right from the outset. Instead of single particle wave mechanics, this approach is called 'quantum field theory'.

The relativistic quantum field theory describing the interactions of electrons and photons is known as quantum electrodynamics (QED). QED combines Maxwell's equations of electromagnetism, quantum mechanics and relativity. It is the most successful theory physicists have yet constructed and it has been tested to an astonishing accuracy. To demonstrate that this is no idle boast, consider the spin of the electron that caused Schrödinger so much trouble. The spinning electron acts like a little magnet and the strength of the 'magnetic moment' of the electron can be calculated in

QED. The result may be expressed in terms of the 'g-factor' of the electron. The classical value predicted for g is

$$g_{classical} = 1.0$$

which is less than half the value predicted by QED

$$g_{quantum} = 2.002\,319\,304$$

When this is compared with the measured value

$$g_{experimental} = 2.002\,319\,304$$

it is evident that the agreement between QED and experiment is impressive. Attempts to extend this agreement beyond nine places of decimals are hindered both by some numerical limitations in the theoretical predictions and by some experimental uncertainties. The major experimental uncertainty arises from the measurement of the so-called 'fine structure constant' from which the charge on the electron may be determined to high accuracy. A detailed description of quantum field theory such as QED would involve the use of advanced mathematical techniques that are far beyond the scope of this book. Fortunately for us, Feynman has provided us with a beautiful intuitive and pictorial approach to quantum field theory. This is the subject of the next section.

Feynman diagrams and virtual particles

Feynman's way of looking at negative energy states takes a little getting used to but in the end is very helpful. To see what is involved let us imagine a 'scattering' experiment with electrons. Picture this experiment rather as if you made two attempts to flip a ball in a pin-ball machine so that the ball makes two collisions on its way to the top of the table (see Fig. 11.6). In the first attempt, both the collisions are glancing collisions that merely deflect the ball a little but do not stop it moving up the table. In our second attempt the ball undergoes a much more energetic collision in which the ball is first sent back down the table before it undergoes another violent collision and heads up the table again. In the case of a scattering experiment with electrons Feynman showed that we can draw two similar trajectories   for the electron – with the key difference that we must now interpret the electron paths on our pin-table for the electrons as a graph of the motion of the electrons in space and time! The motion of the electron in a spatial direction corresponds to going across the table; motion up the table corresponds to the time evolution of the electron's trajectory. The electron trajectory for our first attempt looks entirely normal – the electron is slightly deflected by both the collisions but continues on its way in the same general direction in both space and time. The second case now looks very peculiar – the electron appears to have been scattered backwards in time! Feynman proposed that in relativistic quantum mechanics this bizarre

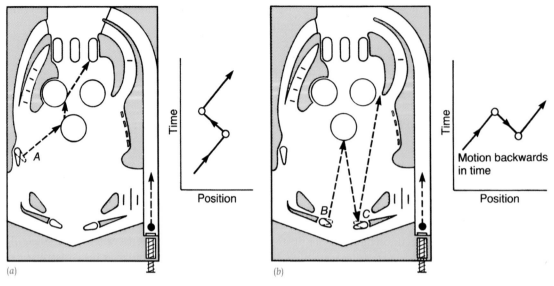

Fig. 11.6 Two possible paths of a ball in a pin-ball machine. (*a*) The flipper has hit the ball so that it makes a glancing collision on its way up the table. Alongside this we show a similar path for the scattering of an electron. The axes of the graph are labelled with time going upwards and space position across. (*b*) In this case, the flipper has hit the ball so that it suffers a very hard collision and bounces back down the table before being hit back up again. In relativistic quantum mechanics there is a similar path for an electron on a 'space–time' graph. However, things now look very peculiar since the first scattering appears to cause the electron to be scattered 'backwards in time'!

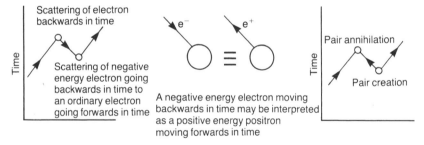

Fig. 11.7 Feynman realized that diagrams with 'backwards in time' electron paths could be understood as the physical process of pair creation followed by pair annihilation. Negative energy electrons travelling backwards in time are equivalent to positive energy positrons travelling forwards in time.

possibility was allowed if the electron going 'backwards in time' had negative energy. The physical interpretation of the negative energy electron travelling backwards in time again corresponds to electron–positron pair creation (Fig. 11.7). Absorbing negative energy and charge 'from the future' decreases the total energy and charge of a scattering centre. This has the same effect on the energy and charge of the centre as if positive energy and charge had been emitted into the future. The apparently absurd trajectory for the electron can be interpreted physically as the creation of an electron–positron pair at the first scattering centre. The positron then travels forwards in time to the second scattering centre where it is annihilated by the original incoming electron.

We can now see the relevance of Feynman's quotation at the beginning of this chapter, which is taken from his original paper entitled 'Space–time approach to quantum electrodynamics', published in the *Physical Review* in 1949. We have been able to avoid dealing explicitly with the many-body nature of the pair creation process by treating the trajectory of the electron and positron as a single electron 'world line' in the same way that the road referred to in the quotation is actually a single road, although it sometimes appears differently to the pilot. There is one great advantage of Feynman's interpretation of the relativistic negative energy states: using negative energy states moving backwards in time to correspond to positive energy antiparticles moving forwards in time, works just as well for bosons as it does for fermions. Needless to say, we should stress that all this is only a device to get the right answer without having to use the complicated machinery of quantum field theory. Nothing, as far as we know, actually travels backwards in time!

There is one other key idea that we must introduce in our account of relativistic quantum mechanics. This is the idea of a virtual particle. In our discussion of tunnelling, back in chapter 5, we pointed out that a helpful way of looking at tunnelling was in terms of the energy–time uncertainty relation

$$(\Delta E)(\Delta t) \approx h$$

In this context, this means that we can borrow an energy ΔE and this loan will be undetected so long as we put it back within a time $\Delta t \approx h/\Delta E$. With the possibility of particle creation in relativistic quantum mechanics this means that a particle need not always remain the same particle. Enough energy could be borrowed to create another particle or pair of particles for a very short time. A photon, for example, may borrow enough energy to turn into a virtual electron–positron pair. These particles can only exist fleetingly before they have to recombine back into a photon. Such transient processes are called 'virtual processes' and the particles created on the borrowed energy are referred to as 'virtual particles'. The probability amplitude for such virtual interactions may be calculated in QED, and Feynman invented a system of diagrams to evaluate these amplitudes (Fig. 11.8).

The importance of Feynman diagrams lies not only in their visual power but also in the fact that Feynman gave precise rules for calculating the quantum amplitude for any diagram, no matter how complicated. Figure 11.9 shows some of the diagrams needed for the calculation of the electron magnetic moment that we mentioned earlier. Feynman diagrams also have the property that the same diagram can be used to represent processes involving antiparticles. Figure 11.10 shows the simplest diagram for electron–quark scattering by exchange of one virtual photon. In the diagram, the time axis is assumed to be pointing towards the top of the

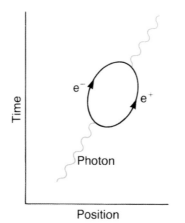

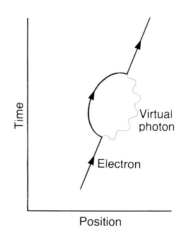

Fig. 11.8 Feynman diagrams for virtual processes. According to the uncertainty principle, energy can be borrowed to allow particle creation provided it is repaid in a short enough time.

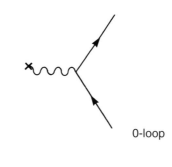

0-loop

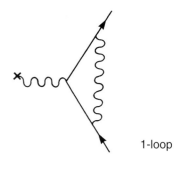

1-loop

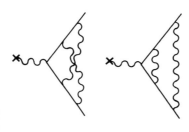

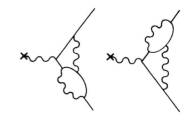

Fig. 11.9 Feynman diagrams needed for the calculation of the magnetic moment of the electron. The straight lines represent electrons and the wiggly lines photons. The 'x' represents interaction with an electromagnetic field. Diagrams with more internal photon lines are numerically smaller than diagrams with fewer photons, but each such 'loop' diagram involves doing very complicated integrals.

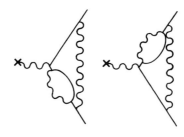

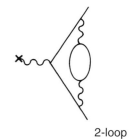

2-loop

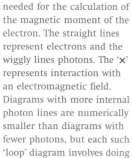

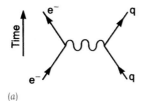

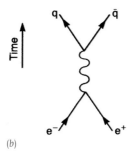

(a)

(b)

Fig. 11.10 The process of electron–quark scattering is represented by the Feynman diagram shown in (a). The same diagram, rotated on its side, (b), is able to predict electron–positron annihilation to create a quark–antiquark pair.

page. We now rotate the diagram and remember that lines with arrows pointing backwards in time must be interpreted as antiparticles moving forwards in time. We see that the diagram now represents the physical process of electron–positron annihilation into one virtual photon which then creates a quark–antiquark pair.

Zero-point motion and vacuum fluctuations

Dirac's picture of the vacuum appeared complicated enough but we have now replaced it with one that looks even more complex. Instead of a place where nothing happens, our empty box should be regarded as a bubbling soup of virtual particle–antiparticle pairs! For example, in our discussion of an electron scattering backwards in time, it may have seemed as if the positron that was created by the first scattering centre needed to know in advance that it was going to be annihilated by the incoming electron in the later scattering. We now realize that the vacuum is full of such virtual electron–positron pairs, all of them only existing for fleeting moments. This was just one positron that got caught in the act by the incoming electron!

The vacuum of relativistic quantum mechanics – or, strictly speaking, the ground state of the relativistic quantum field theory – has other interesting observable effects. When we apply quantum mechanics to the vibrations of atoms in a crystal, the vibrational waves set up in the crystal turn out to have particle-like aspects, just like photons. These quantum lattice vibrations are called phonons. If the crystal lattice is cooled down so that no vibrational phonons are excited, there must still be some zero-point motion of the atoms to satisfy Heisenberg's uncertainty principle. It is this zero-point motion that prevents liquid helium from solidifying, as we discussed in chapter 7. What is the relevance of this crystal lattice to our discussion of the real physical vacuum? Well, in the same way that phonons are quantum objects associated with vibrations of the crystal positions, so we may regard photons as associated with vibrations of the electromagnetic field. Now we see that, as in the case of phonons, there must be a zero-point motion for the electromagnetic fields. Remarkably, these vacuum fluctuations of the electromagnetic field have some experimentally testable consequences.

The most famous application of these ideas is the so-called Lamb shift in the hydrogen spectrum. If one looks very carefully at the spectral lines of hydrogen one finds that there is a tiny splitting between the $n = 2$, orbital angular momentum $L = 1$ and $L = 0$ energy levels that cannot be accounted for, even after including the effects of relativistic electron spin. There is one quantum effect that we have left out. The vacuum fluctuations of the electromagnetic field will cause the electron in the hydrogen atom to jiggle about a little. The effect of this motion can be calculated and the result agrees remarkably well with Lamb's experimental

Hendrik Casimir was born in 1909 in the Netherlands and was a student of Niels Bohr. He later became the research director of the Philips laboratories in Eindhoven.

measurement. The existence of these vacuum fluctuations also enabled Dirac to explain how excited electrons could undergo 'spontaneous emission' of a photon and jump to a lower state. It is the zero-point motion of the electromagnetic field that stimulates these apparently spontaneous transitions.

There is another curious observable effect that originates with these vacuum fluctuations – this is a 'vacuum force'. In the physical vacuum, all possible fluctuation wavelengths are allowed for the electromagnetic field. Now consider two large metal plates set up face to face. For an isolated metal plate the electromagnetic field must vanish since no current is flowing. Of all the vibrations of the vacuum electromagnetic field, only those which vanish at the plates are allowed between the plates. Since some of the normal vacuum vibrations are now absent, the zero-point energy of the electromagnetic field is altered. Detailed calculations show that this results in a tiny attractive force between the plates. This phenomenon is known as the Casimir effect, after Hendrik Casimir, the famous Dutch physicist who first proposed the existence of such a force.

The existence of Casimir's vacuum force was verified experimentally in 1958. In the 1980s, experiments were carried out to show how these vacuum vibrations influence the so-called spontaneous emission rate of excited atoms. The experiments used excited atoms travelling between two metal Casimir plates. A magnetic field oriented the atoms so that the spontaneously emitted radiation was most likely to take place in a direction perpendicular to the plates. For widely separated plates no observable effect is expected. If we bring the plates close together so that the separation becomes close to half a wavelength, the gap can only accommodate a single crest or trough (see Figs. 4.7 and 4.8). In this situation we predict that spontaneous emission will be inhibited. In 1995, this effect was confirmed experimentally by Daniel Kleppner and his research group. They were able to verify that spontaneous emission from caesium atoms was reduced by such Casimir plates. Another prediction of Casimir has recently been confirmed. In 1948, Casimir and his colleague Polder had predicted that the altered vacuum state produced by Casimir plates should exert a force on neutral atoms. In 1993, Charles Sukenik and colleagues at Yale succeeded in detecting such a force acting on neutral sodium atoms.

Hawking radiation and black holes

Particle–antiparticle pair creation is now thought to be of relevance in the theory of black holes. In chapter 10 we encountered black holes as the final stage in the evolution of a very massive star. Once a black hole is born from a dying star it is easy to see how it might grow. Indeed, many galaxies and the mysterious objects called 'quasars' – enormously powerful quasi-stellar radio sources – are now thought to contain gigantic black holes. If black holes absorb all the radiation that arrives, how can such objects be seen? As stars and other interstellar matter are attracted towards the black

Stephen Hawking pictured in Cambridge. Despite his appalling physical handicap Hawking has made many remarkable contributions to astrophysics and cosmology. He also wrote one of the best-selling popular science books ever with his book *A Brief History of Time*.

George Gamow (1904–1968) is probably best remembered for his work on the Big Bang theory of cosmology. The picture shows Gamow emerging from a bottle of 'YLEM' – Gamow's word for the primordial neutron-dominated material out of which he believed the universe was born. Nowadays, physicists no longer believe that the matter in the early universe was mostly neutrons, but Gamow's idea of a hot and dense origin to the universe is now a cornerstone of modern cosmology.

hole, charged particles are accelerated and radiate electromagnetic energy that can be observed. However, once matter has been sucked inside the so-called Schwarzchild radius of the black hole, nothing, not even radiation, can overcome the gravitational forces and escape from the black hole.

The discussion above relates to black holes with masses ranging from three or four times the mass of our Sun to masses hundreds of millions of times greater than this in a quasar. It is much harder to see how very-low-mass black holes might be formed. Stephen Hawking, a cosmologist in Cambridge, UK, has suggested that black holes with a whole range of masses could be formed in the very early stages of the universe. To understand something of the conditions that most cosmologists believe prevailed soon after the creation of the universe, we must make a short detour and discuss some of the evidence for the expanding universe and the 'Big Bang' of creation.

The Big Bang theory of cosmology was put forward by George Gamow and others to explain the observed expansion of the universe as we see it today. Figure 11.11 shows the core of a large cluster of galaxies. Each galaxy contains a vast number of stars, as does our own Galaxy, the Milky Way. In the universe, we observe that clusters of galaxies are roughly evenly distributed throughout space. Figure 11.12 shows galaxies in increasingly distant clusters, together with the measured spectra of light emitted from each galaxy. These spectra can be used to tell us how fast the galaxy is moving away from or towards us. Like the Sun, these galaxies emit light of all wavelengths – a continuous spectrum of light. As light from the Sun travels through the outer layers of the Sun's atmosphere, some of the photons have just the right energy to excite electrons in the gas atoms making up these outer layers. Photons with wavelengths characteristic of elements in this solar atmosphere will be missing from the light spectrum that we observe. This type of line spectrum is called an absorption spectrum and such spectra can be used to identify elements present in the Sun. Figure 11.12 shows similar absorption spectra from the various galaxies. What is remarkable

in these absorption spectra is that the characteristic absorption lines of the various elements appear at longer wavelengths – they are shifted towards the red end of the spectrum. This effect is similar to the familiar Doppler shift in the pitch of a train's whistle as it first comes towards and then travels away from you. These red shifts are interpreted as evidence that the galaxies are all moving away from us, and from each other.

The universe appears to be expanding and the further away from us a galaxy is, the faster it appears to be moving. This famous observation is known as Hubble's law:

$$v = H \times d$$

speed of recession = Hubble constant times distance away from us

This painting by Wimmer shows Joseph Fraunhofer and his spectroscope. Fraunhofer discovered that the spectrum of the Sun was crossed by many dark lines. Fifty years later Kirchhoff was able to interpret these lines as absorption line spectra characteristic of the atoms in the Sun's atmosphere. On Fraunhofer's tombstone is the epitaph 'Approximavit sidera' ('He approached the stars').

after the US astronomer Edwin Hubble. Incidentally, this observation does not put us at the centre of the universe. Imagine baking a currant loaf; as the dough rises all the currants see all the other currants moving further away from each other, and the further away a currant is, the faster it will be moving (Fig. 11.14).

This picture of an expanding universe carries with it the implication that at earlier times all the galaxies and matter must have been much closer

Fig. 11.11 The core of the cluster of galaxies in the constellation of Coma Berenices. The very bright object is a star in our own galaxy, but nearly all the other objects in the picture are galaxies about 300 million light years away.

together. This is the motivation for the Big Bang model of cosmology that extrapolates this expansion back in time to a time when all the matter in the universe was compressed together. The huge energy densities present in the early stages of this Big Bang could have compressed matter so much that black holes with very small masses were created. These mini black holes could have masses from as small as a few grams up to the mass of a small planet. Hawking called these objects primordial black holes. All observational attempts to find such objects have so far been unsuccessful.

Edwin Hubble (1889–1953) shown here with his cat, Nikolas Copernicus. He was born in Missouri and was a good enough athlete to consider a career in heavyweight boxing. After initially taking up law, Hubble returned to astronomy and said 'Even if I were second rate or third rate, it was astronomy that mattered'.

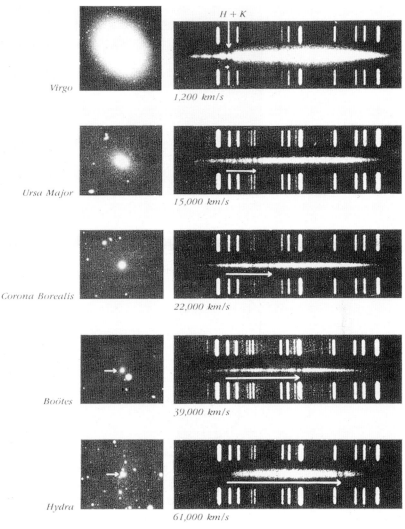

Fig. 11.12 Five galaxies at increasing distances (top to bottom) together with their absorption line spectra. Compared with the reference spectra above and below, the absorption lines are seen to be further to the right for the more distant galaxies. Since this Doppler shift is believed to be due to the speed of the galaxy, this shows that the further away a galaxy is from us, the more rapidly it is moving away. This is Hubble's law.

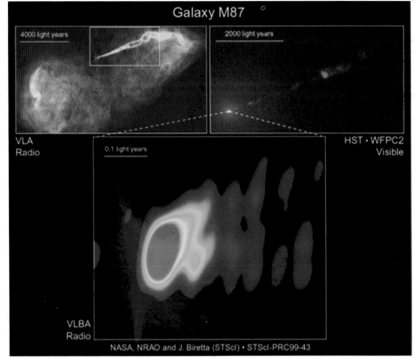

Fig. 11.13 Using radio telescopes in Europe and the USA, including the Very Long Baseline Array consisting of ten radio telescopes located from New Hampshire to Hawaii, astronomers have photographed the formation of a giant cosmic jet near the black hole believed to be at the centre of the galaxy M87.

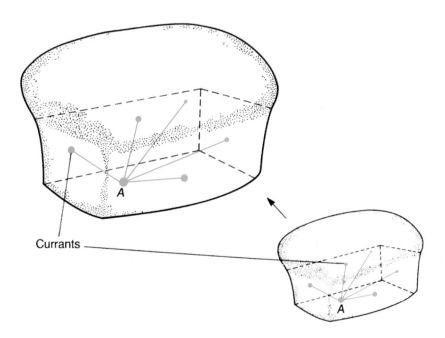

Fig. 11.14 A currant loaf model for the expansion of the universe. As the loaf is baked the whole loaf expands and the currants move apart. Each currant sees all the others moving away.

Hawking also developed the theory of particle creation near such black holes. From our previous discussion of virtual particles, we know that the vacuum can be regarded as a sort of bubbling soup of virtual particle–antiparticle pairs. Hawking proposed that one member of such a pair could be captured by the black hole while the other member escapes into the surrounding space. How could this come about? Surprisingly, the key ingredient in the mechanism of such 'Hawking radiation' is an understanding of ordinary gravitational tidal forces.

There is a short story called *Neutron Star*, written by Larry Niven, a science fiction writer popular amongst science students at Massachusetts

Fig. 11.15 The cover picture of Larry Niven's book *Neutron Star*. The story first appeared in 1966 – one year before pulsars were discovered.

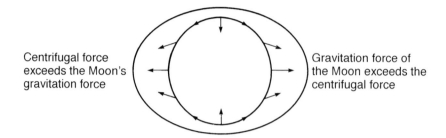

Centrifugal force
exceeds the Moon's
gravitation force

Gravitation force of
the Moon exceeds the
centrifugal force

Fig. 11.16 A simplified
diagram to illustrate the
origin of the Earth's tides.

and California Institutes of Technology. The background to the story is that
the galaxy's main manufacturer of space-ship hulls – an alien race known as
the puppeteers – are worried. Some unknown force has been able to penetrate their supposedly invulnerable 'No. 2 General Products hull' and kill the
occupants who were on an exploratory voyage to a neutron star. The puppeteers, being inveterate cowards, blackmail the hero, Beowulf Schaeffer,
into repeating the trip. Needless to say, the hero survives, and in the process realizes that the mysterious force is nothing but the familiar tidal force
due to gravity. Since the puppeteers do not know what a tide is, Schaeffer
is able to deduce that their secret home planet does not have a moon, and
blackmails them into paying him a fortune.

What are these mysterious tidal forces and how could they kill or
cause Hawking radiation? Figure. 11.16 shows an idealized picture of the
Earth surrounded by a continuous ocean of water. On the side nearest
the Moon, gravity attracts the water most strongly and is greater than the
centrifugal force caused by the rotation of the Earth–Moon system. This
causes the water surface to bulge towards the Moon. On the other side
of the Earth, further away from the Moon, the gravitational force on the
water is smaller than the centrifugal force and the water surface bulges
away from the Moon. The existence of these two bulges is the reason why
we have two high tides for every daily rotation of the Earth. This is the effect
of the Moon's gravitational attraction on the Earth. What is the effect of
that of the Earth on the Moon? Since the Moon does not rotate relative to
the Earth, a rock at the nearest point of the Moon and a rock at the point
furthest away from the Earth are moving in two concentric orbits with the
same orbital speed. If they were not part of the same Moon, the two rocks
would naturally move in different orbits – since they experience different
gravitational forces. Thus, the tidal force of the Earth on the Moon is trying
to pull the Moon apart. Similarly, a spaceman in orbit round a neutron star
or a black hole would experience huge gravitational tidal forces trying to
tear him apart. It is these tidal forces that cause a particle–antiparticle pair
to become separated if they are created in the enormous gravitational field
of a primordial black hole. In this way it is possible for one of the pair to
fall into the hole while the other escapes into the surrounding space. The
black hole will thus appear to radiate particles.

12 Weak photons and strong glue

> Now we are in a position in physics that is different from any other time in history (it's always different!). We have a theory, ... so why can't we test the theory right away to see if it's right or wrong? Because what we have to do is calculate the consequences of the theory to test it. This time, the difficulty is this first step.
>
> Richard Feynman

James Clerk Maxwell (1831–1879) made original contributions to many areas of physics and was the first to suggest that Saturn's rings were composed of myriads of tiny particles. His most important work was putting Faraday's 'field ideas' into a precise mathematical form and unifying electricity and magnetism in one theory of electromagnetism. Maxwell's equations for electromagnetism were first published in 1865 and remain unchanged today, despite the development of both quantum mechanics and relativity. He died of cancer at a relatively young age and did not live to see Hertz verify his prediction of electromagnetic waves.

The double-slit experiment revisited

In this chapter we turn to recent advances in our understanding of the fundamental forces of Nature. As we have said in earlier chapters, the combination of classical electromagnetism, quantum mechanics and relativity provides an astonishingly successful description of electromagnetic forces. The resulting theory is called Quantum ElectroDynamics, or QED for short. For over 50 years physicists searched for similarly successful theories to describe not only the weak forces responsible for natural radioactivity but also the strong forces that hold the nucleus together. It was not until the mid 1970s that real progress was made and these remarkable developments are the subject of this chapter.

Particle physicists now have a unified theory that combines both the electromagnetic and weak forces. The major predictions of this theory have been spectacularly verified by experiments at the CERN high-energy particle physics laboratory in Geneva. We shall describe the theory and these experiments in more detail in this chapter. But particle physicists also believe that they have at last discovered the correct theory of the strong nuclear force. We now have a theory of the proton and neutron in terms of their quark constituents (see chapter 3). This theory is called Quantum ChromoDynamics, or QCD for short. As Feynman suggests in the introductory quotation to this chapter, there are unique difficulties for calculating the predictions of this theory. QCD describes the strong forces in terms of forces between quarks – but no free quarks have ever been seen, nor do physicists expect to see them! Instead, physicists believe that the interactions between quarks are cunningly arranged so that we can only ever observe states that are combinations of three quarks – like the proton and neutron – or combinations of a quark and an antiquark – like the mesons we shall meet in the next section. This property of non-observability of free quarks is called quark confinement. It is impressive that in spite of this seemingly show-stopping difficulty, physicists have been able to make predictions that can be validated by experiment and have built up a convincing circumstantial case for quarks and QCD. The unified theory of the weak and electromagnetic forces together with QCD to describe the strong force constitutes what physicists call the 'Standard Model'. As we shall see, the Standard Model has been remarkably successful and has survived twenty years of detailed examination by experiment. Now at last, there are some hints that we need to go beyond the Standard Model. One way to go further is to seek to combine all three of the weak, electromagnetic and strong forces in a 'Grand Unified Theory' or GUT. Although there is no direct experimental evidence in support of such a unification, theoretical physicists are continuing to do their job and are already diligently looking beyond GUTs to new ideas of 'supersymmetry' and 'strings' in the hope of finding a theory that includes gravity in a way that is consistent with quantum mechanics. A detailed exploration of these topics is beyond the scope of this book and we shall only scratch the surface of such developments. Let us begin by returning to the double-slit experiment with which we started our discussion of quantum mechanics.

How is the double-slit experiment relevant to these developments? It appears that Nature has been surprisingly kind to us. All the theories in the Standard Model have at their heart same basic principle. We will be able to gain some understanding of this principle by looking again at our double-slit experiment with electrons that we discussed in chapters 1 and 2. Although this key principle is usually referred to by the rather intimidating name of gauge invariance, we will try to convince you that the fundamental idea is really quite simple and appealing. Figure 12.1 shows the double-slit experiment with electrons once again. As we discussed in chapter 1, the probability of arrival of electrons at the screen can be predicted by assuming that electron waves spreading out from each slit overlap and interfere. Whether many electrons arrive at any given point or none at all, depends on whether the waves arriving from the two slits are both crests (in phase) or a crest arriving with a trough (out of phase). Suppose we now insert, between the slits and the screen, a thin sheet of material, as shown. In a similar fashion to the interference experiment with neutrons (see chapter 3), the matter in the sheet will interact with the electrons and alter the phase of the electron waves that pass through. The phases of the electron waves arriving at the screen will therefore be altered – where there was originally a crest there may now be a trough, and so on. Now this is the important point. If the phases of the waves from both slits are both affected by the same amount, the interference pattern will not change. Since the interference pattern does not change when we insert the thin sheet of matter, physicists call this an *invariance* of the double-slit experiment. More precisely, since the only relevant effect of the sheet of matter on the electrons is to cause a shift in the phase of the electron waves, this property is called a *phase invariance*.

There is another feature of this invariance property that we want to stress. This type of phase invariance requires that we insert a phase-changing sheet of material that spans the entire area of the screen. If we insert just a small piece of the material behind only one of the slits we find that the interference pattern will change. This is because the material now only affects the phase of waves from one of the two slits. At the screen, at points where we used to have two crests arriving, we may now have a trough arriving with a crest. We can summarize this as follows. If we only affect the phase of the electron waves behind one slit – a *local* phase change – the interference pattern is changed by the insertion of the piece of material and there is no invariance. Only if we alter the phase everywhere – a *global* phase change – does the interference pattern of the electrons remain unchanged so that we have an invariance. We see that the double-slit electron interference experiment exhibits only *global phase invariance*. The fact that interference pattern changes if we make a local phase change means that the double-slit experiment does *not* possess *local phase invariance*.

To illustrate more graphically the difference between global and local effects, Feynman once gave this example. Suppose we were interested in the total number of cats in the world at any given time. If we look at the population of cats for a short enough time so that no cats are born or die during this time, the total number of cats remains constant. We can then say that the number of cats is conserved. But we know more than this. We know from experience that cats are conserved in a local way. For example, if five cats disappear in Pasadena and reappear at the same instant in Southampton, that would be an example of global conservation of cats. But we know that cats don't work like that! The number of cats in every small local region is conserved and it is this local conservation of cats that leads to the global conservation of the total number of cats.

This story has a serious point to make. Physicists are always intrigued whenever they find an invariance principle and immediately try to see if they can find a better one. In particular, in the case of our double-slit experiment with electrons, we have found a global phase invariance. How can we do better? Well, the necessity for the phase of the electron wave to be changed everywhere, at the same time, in order to have an invariance seems a somewhat irksome and unnatural restriction. Would it not be more natural to have a theory that allowed us to make a phase change in some small local region without having to worry about what was happening elsewhere? In other words, is there some way of fixing things up so that we can allow a local alteration of the phase and yet still have an invariance? There is – and the resulting theory is quantum electrodynamics (QED)!

To see how the connection with QED comes about, we must first explain what happens if we perform the double-slit experiment in the presence of a magnetic field. In Fig. 12.1*d* we show the experimental set-up, with a magnet in position behind the slits. In classical electromagnetism, magnetic fields cause the trajectory of a charged particle to curve so it should be no surprise that the interference pattern changes. In terms of the quantum

Fig. 12.1 The double-slit experiment with electrons – reprise: (a) A reminder of the interference pattern observed for the usual double-slit experiment with electrons. The electrons arrive individually at the detector and they are represented as black-and-white circles to remind us that we cannot tell which slit they passed through. Notice that the pattern is highest at the central detector marked A. (b) The interference pattern is unchanged if a thin sheet of matter is inserted between the slits and the detectors. The electron waves coming from both of the slits undergo the same change in phase. Thus, at the detector, the two waves will still add up to give a peak or still cancel to give a dip. We say that there is a 'global' phase invariance since the pattern is unchanged – 'invariant' – provided the sheet crosses the entire region behind the slits. (c) If a thin sheet of matter is inserted behind only one of the slits the pattern changes. Instead of the detector at A registering a maximum number of electron counts, it now records a dip in the interference pattern. The pattern has changed because the matter has altered the phase of only one of the electron waves. This shows that a 'local' change in the phase does not leave the interference pattern 'invariant'.

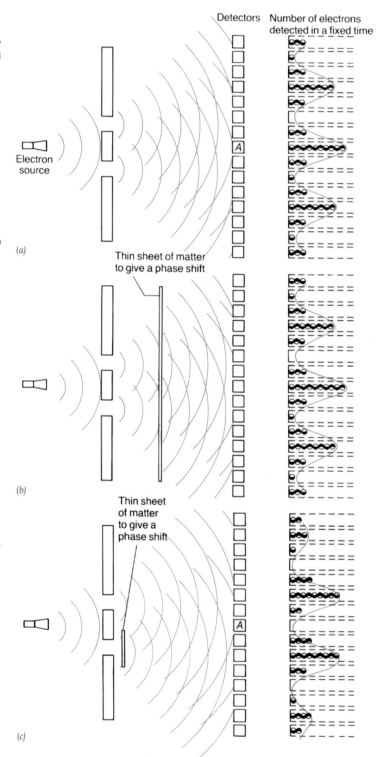

(d) The presence of a magnetic field will also cause the interference pattern to change. Apart from the fact that we cannot say which slit each electron passed through, this is more or less what we would expect, since in classical physics electrons are expected to be deflected by magnetic fields. (e) The famous Bohm–Aharanov experiment showed that there was a shift in the interference pattern even if the magnetic field was shielded so that the electron paths from each slit to the detector did not pass through any magnetic field! The shielded magnetic field has been achieved in practice by using a long, thin electromagnetic coil, thinner in diameter than a human hair. (f) The phase shift caused by the insertion of a sheet of matter behind one of the slits can be exactly compensated for by adjusting the magnetic field. This shows that 'local invariance' can be achieved provided the magnetic field interacts with the electron in just the right way. This is the basic principle behind all 'gauge' theories.

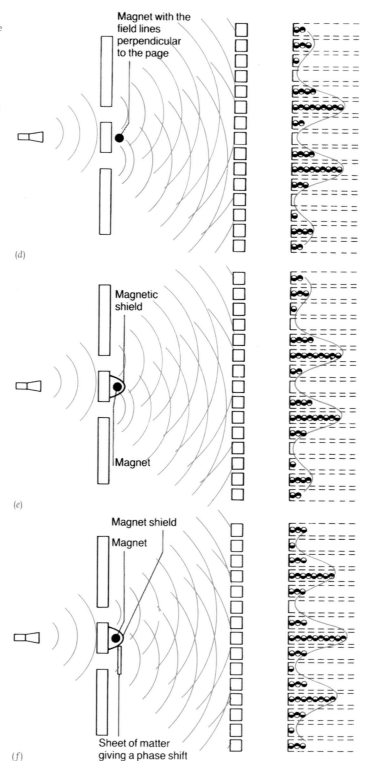

(d)

(e)

(f)

Magnet with the field lines perpendicular to the page

Magnetic shield

Magnet

Magnet shield

Magnet

Sheet of matter giving a phase shift

mechanical wave picture of electrons the reason for this change is not so obvious. Since the interference pattern does change, the effect of the magnetic field must be to alter the phase of the electron waves. This is all roughly in accord with our classical expectation, but quantum mechanics has another surprise in store for us. If we arrange to shield the magnetic field so that no magnetic field can penetrate the region where the two electron waves travel and overlap, the interference pattern still changes! This remarkable phenomenon goes by the name of the Bohm–Aharanov effect, after the two physicists who suggested it. Their prediction aroused much controversy amongst physicists until the effect was conclusively confirmed by experiment in the early 1960s.

It is this effect of magnetic fields on the phase of electron waves that leads to the possibility of achieving a local phase invariance. At this point we must confess that the precise details of how this comes about are too complicated to explain here. Instead, we will try to give some idea of how a local phase invariance can be achieved. Suppose we insert a thin sheet of material behind only one of the slits. As we have argued, the interference pattern will be changed. But what happens if at the same time as this sheet is put into position behind one slit, we insert a magnet behind the slits? It is certainly plausible that we could arrange for the effect of the magnetic field to compensate the phase change caused by the small sheet of matter. We would then observe no change in the original interference pattern and have achieved a curious sort of local phase invariance. To summarize: we can make a local phase change in one of the electron waves yet maintain the invariance of the interference pattern by introducing a magnetic field at the same time. The full story of the connection between local phase invariance and magnetic fields is more subtle than it appears from this argument but the conclusion suggested by this example is correct. It is only because electrons interact in a very specific way with a magnetic field that a local phase invariance is possible.

This is the vital clue from QED that enables us to construct theories of the weak and strong forces. QED is a theory in which the interactions of electrons and electromagnetic photons are carefully arranged so that theory is unchanged by local changes in the phase of the electron wavefunction. For rather obscure historical reasons, physicists usually use the term **gauge invariance** to describe this state of affairs, instead of a more informative term such as **local phase invariance**. Hence, QED is known as a **gauge theory** rather than what it actually is, namely a **phase theory**. How does all this help us to find a theory of the weak or the strong interactions? The trick is to reverse the argument. In other words, suppose we did not know how electrons and photons interacted. If we now demand that any theory with electrons must have a local phase invariance, we would be forced to introduce magnetic fields that interact with electrons in a specific way and thereby invent QED! This reversal of our invariance argument is called the **gauge principle**: demanding that the theory has a local phase invariance determines the interactions of its constituents. We will use this beautifully

Hermann Weyl (1885–1955) was an outstanding mathematician who also made important contributions to physics. At the height of his career, in 1933, he resigned his post at the University of Gottingen in protest at the dismissal of his Jewish colleagues. Like so many other German scientists, Weyl went to the USA and became a member of the Institute for Advanced Study in Princeton, New Jersey. In the 1920s, when trying unsuccessfully to unify gravity and electromagnetism, he introduced some of the ideas of modern gauge theories. The term 'gauge theory' is a relic of these attempts: it would be far more appropriate to use the term 'phase theories' in the modern context.

simple idea to show how the theories that are believed to describe the weak and strong forces are constructed.

The birth of particle physics

Before we can describe the application of the gauge principle to weak and strong interactions, we must first provide a rapid overview of some of the most significant discoveries in elementary particle physics and introduce some terminology. In 1932, when Chadwick discovered the neutron, all was simple: there seemed only to be three elementary building blocks of matter – the proton, the neutron and the electron. The proton and neutron are much more massive than the electron and are called *baryons*, from the Greek word *barys* ($\beta\alpha\rho\nu\zeta$), meaning heavy. The electron, on the other hand, is now known to be one of a family of particles known as *leptons*, from the Greek word *leptos* ($\lambda\varepsilon\pi\tau o\zeta$), meaning light. We have already met another type of lepton – Pauli's neutrino, the mysterious particle involved in the radioactive decay of a neutron into a proton, which we discussed in chapter 10. This classification into baryons and leptons seems rather elaborate to describe these four particles. Its usefulness only becomes apparent when we appreciate that, over the past 50 years or so, hundreds more such 'elementary' particles have been discovered. After decades of confusion, the emergence of the Standard Model has restored a large measure of order to elementary particle physics. As we shall see, our new understanding is due to the leading roles played by quarks and by local phase invariance.

In the preceding chapters, we encountered Feynman diagrams as a pictorial way of representing interactions of particles. For example, a diagram that occurs in the scattering of electrons by a proton is shown in Fig. 12.2. The diagram shows a virtual photon being exchanged between the

Chen Ning Yang won the Nobel Prize with T. D. Lee in 1957 for predicting the violation of left–right symmetry by weak interactions. Earlier, in 1954, together with Robert Mills, Yang had written down a generalization of the gauge theory of ordinary electromagnetism. This was also proposed independently by Robert Shaw, a Ph.D. student of Abdus Salam in Cambridge, UK, at about the same time. These 'Yang–Mills' theories were the precursors of modern gauge theories.

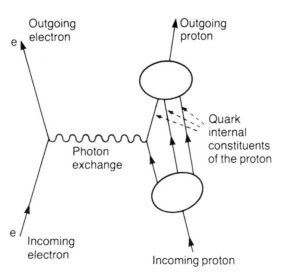

Fig. 12.2 Feynman diagram for electron–proton scattering. The scattering process is represented by the 'exchange' of a virtual photon between the electron and one of the quarks in the proton.

Hideki Yukawa (1907–1981) won the 1949 Nobel Prize for his prediction of mesons as carriers of the strong force. Yukawa was the first Japanese scientist to win the Nobel Prize.

electron and one of the quarks that make up the proton. The concept of a force arising from the exchange of virtual particles gives us an idea of the distance over which this force can act. From the uncertainty principle we argued that we could borrow energy ΔE for a time $\Delta t \approx h/\Delta E$ without spoiling energy conservation. If we multiply this time, Δt, by the velocity v of the particle, we obtain an estimate of the typical distance such a particle can travel

$$R = v \times (h/\Delta E)$$
$$\text{range} = \text{velocity} \times \text{time}$$

It was this argument applied to the known range of nuclear forces that led the Japanese physicist Hideki Yukawa to predict the existence of a particle with a mass in between that of the electron and the proton.

Yukawa predicted the mass of this particle to be about 200 or 300 times heavier than the electron. Remember that the proton is about 2000 times heavier than an electron. This prediction was made in 1935 and no such particle had ever been observed. Not surprisingly, when particles with about the right mass were found in cosmic ray experiments two years later, this seemed to be dramatic confirmation of Yukawa's prediction. Although the Second World War intervened to slow down research on these new particles, it did not stop research completely. Three young Italian physicists, Marcello Conversi, Ettore Pancini and Oreste Piccioni, were in hiding from the Germans to avoid being deported from Italy to forced labour in Germany. Working in a cellar in Rome, they discovered some very puzzling properties of the new particles. The particle did not behave at all like the carrier of the strong force. Instead of interacting strongly with a nucleus, it seemed to interact more like an electron. The mystery was not cleared up until 1947 when it was suggested that there might be two new particles with about the same mass. One was the particle that had already been observed – which behaved like a heavy electron – while the other, yet to be observed particle, would be Yukawa's strong force carrier. This guess was shown to be correct when Cecil Frank Powell and Guiseppe Occhialini working in Bristol, UK, obtained cosmic ray tracks in photographic emulsion that conclusively established the existence of Yukawa's elusive particles. After some debate as to whether these new particles should be called *yukons* in honour of Yukawa, these intermediate mass particles are now called **mesons**, from the Greek word *mesos* ($\mu\varepsilon\sigma o\zeta$), meaning in the middle. The lightest of these mesons is called a pi-meson or **pion** for short. The heavy electrons discovered first are now called **muons**. Yukawa's reasons for choosing to become a theoretical physicist rather than an experimental physicist provide an amusing footnote to all this. He has said that his decision was due, in part, to his 'inability to master the art of making simple glass laboratory equipment'!

The discovery of Yukawa's meson marks the birth of modern particle physics. It was the result of the development of new and better methods

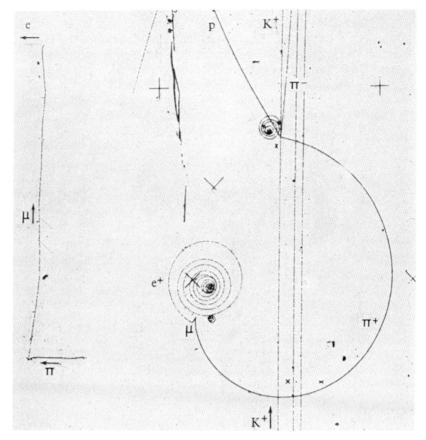

Fig. 12.3 Pions decay to a muon and an unseen neutrino. The muon in turn disintegrates into an electron and two more neutrinos. The left-hand picture shows the tracks left by this decay chain in a photographic emulsion. On the right the same process is shown in a bubble chamber. Because of the magnetic field present in the chamber, the tracks are curved and the slow moving electron winds up like a watch spring.

of observing the collisions of high-energy particles. This search for new techniques continues to this day. Powell and Occhialini had been collaborating with the photographic laboratories of Ilford Ltd to manufacture better emulsions for showing particle tracks. Occhialini took some of these new plates to the top of a mountain in the French Pyrenees and exposed them to high-energy cosmic rays. What then happened to these plates is best described in Powell's own words:

> When they were recovered and developed in Bristol it was immediately apparent that a whole new world had been revealed. The track of a slow proton was so packed with developed grains that it appeared almost like a solid rod of silver, and the tiny volume of emulsion appeared under the microscope to be crowded with disintegrations produced by fast cosmic ray particles with much greater energies than any which could be generated artificially at the time. It was as if, suddenly, an entry had been gained into a walled orchard, where protected trees had flourished and all kinds of exotic fruits had ripened undisturbed in great profusion.

Fig. 12.4 One of the neutrino experiments at CERN laboratory in Geneva. The detector weighs 1400 tons and consists of thick iron plates interleaved with scintillation counters and drift chambers to detect the charged particles coming from the neutrino interactions.

Even with the great theoretical advances made in particle physics in recent years, a deep reason for the existence of the muon remains a mystery. The Nobel Prize winner, Isidor Rabi, is reported to have said 'Who ordered that?' when told of the muon's discovery, and his question remains unanswered. We have now uncovered a further clue that may yet prove vital to solve the puzzle of the leptons. Curiously, this discovery involves a repetition of the confusion surrounding the discovery of the muon and Yukawa's meson. In the mid-1970s, physicists were looking for a new meson to confirm their theories about a new type of quark. Instead, at about the same mass expected for this new meson, another heavy electron was identified. This discovery was largely due to the efforts of the US physicist, Martin Perl, who named the new particle the *tau* lepton. The expected new meson was found shortly afterwards, thus completing a very strange and still unexplained re-run of history.

Other strange cosmic ray events were discovered in the same year by George Rochester and Clifford Butler in Manchester, UK. (see Fig. 12.5). The characteristic feature of these peculiar new events was the existence of two 'vees' pointing back to the initial interaction point. Since only charged particles leave tracks in the detector, we deduce that the two vees are the charged decay fragments of two neutral particles created at the point of the original collision. These neutral particles, soon to become known as

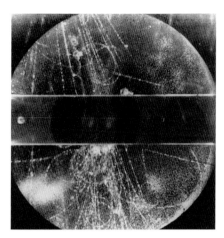

Fig. 12.5 The discovery of strange particles. A neutral K meson is produced by a cosmic ray interaction in the lead plate that passes through the cloud chamber. Its decay products are charged pions and can be seen as a 'vee' in the lower right of the picture.

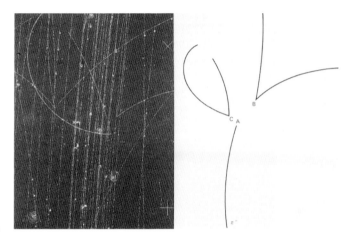

Fig. 12.6 Double 'vee' events are common in bubble chambers. A negatively charged pion collides at A with a proton of the hydrogen that fills the chamber. In the reaction two strange particles are produced, a neutral K meson and a neutral Λ baryon. The Λ decays at B to a proton and a π^-, and the kaon decays at C to a π^+ and a π^-.

strange particles, then travel some distance before decaying. By taking photographs of these events, in the presence of a magnetic field, and making careful measurements of the curvature of these tracks, we can apply the laws of energy and momentum conservation to deduce the masses of all the particles participating in the event. In this way, new strange baryons and mesons were identified. Why were these new particles called strange particles, apart from the obvious strangeness of their production in pairs of vees? Consider a typical 'double vee' event (see Fig. 12.6). This corresponds to the reaction

$$\pi^- + p \rightarrow \Lambda^0 + K^0$$

where the lambda (Λ) is a strange baryon and the kaon (K) a strange meson. The most puzzling thing about these strange particle events was that, while it was easy to create pairs of strange particles from collisions of pions and protons, these strange particles, left to themselves, showed a marked reluctance to turn back into protons and pions. In other words, we must deduce that the production of pairs of strange particles takes place via the strong interactions but the decay of individual strange particles is governed by weak interactions

$$\Lambda^0 \to p + \pi^-$$
$$K^0 \to \pi^+ + \pi^-$$

This interpretation has been confirmed by observation of other weak decay modes of the lambda and the kaon such as

$$\Lambda^0 \to p + e^- + \bar{\nu}$$
$$K^0 \to \pi^- + e^+ + \nu$$

We now know that strange particles possess a new type of charge that distinguishes them from ordinary matter such as protons, neutrons and pions. In the strong interaction reactions, the final state must have the same strangeness as the initial state. Thus, in the example above, if we assign the kaon to have strangeness $+1$ the lambda particle must have strangeness -1: the final state now has zero net strangeness, the same as the initial pion–proton state. In the decays of strange particles, strangeness is not the same on each side of the reaction. This means that the process is not allowed to proceed via the fast strong interactions, but can only take place, reluctantly, by the much slower weak interactions of beta radioactivity.

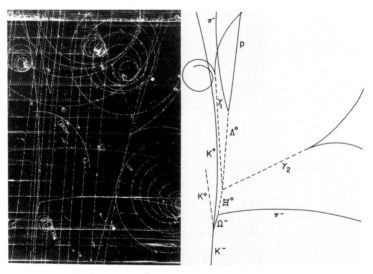

Fig. 12.7 The world's first Ω^- event. The Ω^- contains three strange quarks and decays via a three stage process to a Ξ^0 and then to Λ^0 before finally shedding the last of its strangeness in the Λ decay. A neutral pion is also created in the decay chain which itself decays to two photons. The photons are neutral and leave no track, so reconstruction of what happened would normally be very difficult. This event is remarkable because both of the photons from the pion decay have been converted to electron–positron pairs in the bubble chamber. This was the happy chance that enabled Brookhaven to win the race to find the Ω^- particle. Lightning seems to have struck twice in the same place, for Brookhaven were similarly fortunate in their finding of a charmed baryon many years later!

Looking back, it is clear that the discovery of strange particles was a turning point in our understanding of the strong and weak interactions. At the time of Rochester and Butler's discovery of the 'Manchester V-particles', there was much controversy and confusion. Rochester himself has written that: 'The two years following 1947 were tantalizing and embarrassing to the Manchester group because no more V-particles were found.' It was not until 1950 that a group from Caltech confirmed their discovery using a cloud chamber based on the top of nearby Mount Wilson. By April 1953, V-particles were observed in bubble chamber photographs at accelerators and the modern particle physics era had begun. Given the significance of their discovery and the battle they had to convince the particle physics community that they were right, it seems all the more remarkable that Rochester and Butler were never awarded the Nobel Prize. The discoverers of every new quark flavour found since then have received Nobel Prizes for their achievements.

During the 1950s and 1960s experimentalists found many short-lived excited states – 'resonances' – of the proton, neutron and pion, as well as of the strange particles. How could all these particles be fundamental? Gell-Mann and Zweig brought some sense to all this with their introduction of

Fig. 12.8 Experimental particle physics now involves collaborations of large numbers of physicists from many countries. In this photograph are some of the 114 people who shared in the hunt for the Ω^-.

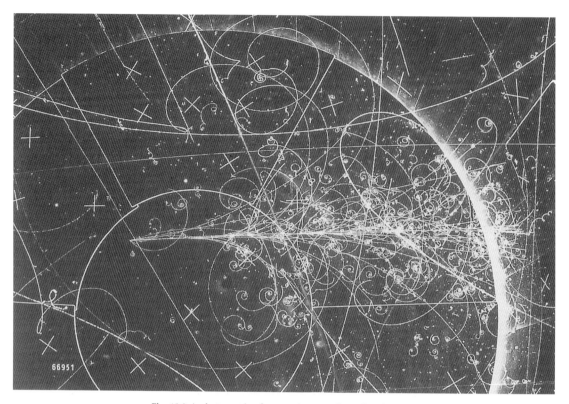

Fig. 12.9 A photograph of a neutrino reaction taken in BEBC – the Big European Bubble Chamber at CERN. The neutrino beam enters from the left and collides with a quark in the proton to produce a very complicated spray of particles.

quarks. Baryons are made up of three quarks, and mesons of a quark and an antiquark. Dramatic confirmation that Gell-Mann was on the right track had come a year or so earlier with his prediction of a particle called the Ω^-. The proton and neutron are made of the two types of non-strange quarks: the Ω^- was predicted to be composed entirely of strange quarks. The short-lived resonances could now be understood as excited states of these quark systems, much in the way that we understand excited states of atoms and nuclei. Although such a constituent quark approach to this problem now seems fairly obvious, at the beginning of the 1960s it was fashionable to suppose that all particles were equally elementary and 'nuclear democracy' was the slogan. It took physicists some time to come round to the idea that there really were fundamental constituents of matter. During the late 1960s only a few perceptive physicists, like Dick Dalitz in Oxford, UK, persevered with the idea of explaining the excited states in terms of quarks, often in the face of ridicule and disbelief from some of their colleagues. It was not until the end of the 1960s that the quark picture of elementary particles received dramatic confirmation from the electron–proton scattering experiments performed at Stanford, California (see chapter 3). These experiments had a natural explanation in terms of the electrons being scattered by collisions with quarks inside the proton. Since these early experiments, further experiments with electrons and neutrinos at CERN in Geneva, at DESY in Hamburg and Fermilab near Chicago have confirmed this picture. It is now generally accepted that both baryons and mesons contain quarks – despite the fact that no-one has ever seen a free, isolated quark.

We conclude this section by introducing yet another new word. Baryons and mesons interact via the strong nuclear force. Leptons, on the other hand, feel only the weak and electromagnetic forces. Particles that

Fig. 12.10 The Big European Bubble Chamber (BEBC) at CERN. This was filled with either liquid hydrogen or a mixture of neon and hydrogen. The chamber is surrounded by superconducting niobium–titanium coils which produce a very large magnetic field inside the chamber.

```
T
261503  IMPCOL G
22931   GNTC G
D970  LA989 UOP411
GXXX CO SWSM 094
STOCKHOLM 94/89 15 1145- PAGE 1/50


PROFESSOR ABDUS SALAM
IMPERIAL COLLEGE OF SCIENCE
AND TECHNOLOGY
PRINCE CONSORT ROAD
LONDON(SW7 2AZ)


DEAR PROFESSOR SALAM,
I HAVE THE PLEASURE TO INFORM YOU THAT THE ROYAL SWEDISH ACADEMY
OF SCIENCES TODAY HAS DECIDED TO AWARD THE 1979 NOBEL PRIZE
IN PHYSICS TO BE SHARED EQUALLY BETWEEN YOU, PROFESSOR SHELDON L.
GLASHOW AND PROFESSOR STEVEN WEINBERG, BOTH AT HARVARD UNIVERSITY,
FOR YOUR CONTRIBUTIONS TO THE THEORY OF THE UNIFIED WEAK AND
ELECTROMAGNETIC INTERACTION BETWEEN ELEMENTARY PARTICLES,
INCLUDING INTER ALIA THE PREDICTION OF THE WEAK NEUTRAL CURRENT.
        C.G. BERNHARD
        SECRETARY GENERAL


IV SENT 1200 JC
22931   GNTC G
261503  IMPCOL G


T
```

Fig. 12.11 The telex sent by the Nobel Prize Committee to Abdus Salam informing him that he had won the Nobel Prize.

interact via the strong force are called **hadrons**. This word was first coined by the Russian physicist, Okun, from a Greek word that at first sight does not seem very appropriate – since *hadros* ($\alpha\delta\rho o\zeta$) is Greek for bulky. However, *leptos*, from which lepton is derived, has another meaning besides light, namely fine-grained. It is in this sense that *hadros* is the opposite of *leptos* and therefore an appropriate term to use to distinguish between particles that feel or do not feel the strong nuclear force.

Weak photons and the Higgs vacuum

The 1979 Nobel Prize for physics was awarded to three physicists, Sheldon Glashow, Abdus Salam and Steven Weinberg, for their 'contributions to theory of the unified weak interaction between elementary particles'. This was a bold move by the Nobel Prize committee since the unified theory of Glashow, Salam and Weinberg predicted the existence of two new particles, the W and Z, with masses 80 or 90 times heavier than the proton. At the time, no such particles had been observed. The committee must have heaved a collective sigh of relief when these predictions were spectacularly verified at the proton–antiproton collider at CERN in Geneva (see Fig. 12.23). Carlo Rubbia and Simon van de Meer were awarded the 1984 Nobel Prize for their part in making these experiments possible. How does the unification of weak and electromagnetic interactions come about? And what has all this got to do with the gauge principle? To understand the answers to these questions we must look again at Yukawa's argument about the range of forces and see how this is related to the mass of the virtual particle being exchanged.

Yukawa was able to deduce the mass of the pion from the observed range of nuclear forces. The heavier a particle is, the more energy that has to be borrowed to create it, and the shorter the distance it can travel on

Sheldon Glashow (left) and Steven Weinberg shown together at a press conference in Harvard on the day they won the Nobel Prize. Together with Abdus Salam, they were awarded the prize for contributing to the unification of the weak and electromagnetic forces in a single theory.

Abdus Salam (1926–1996) was born in what is now Pakistan, and he studied mathematics at Lahore University. He originally intended to become a civil servant but eventually ended up on a scholarship to Cambridge, UK, studying physics. Until his death Salam was one of the most prominent scientists of Islamic faith. He donated his share of the Nobel Prize to his institute in Trieste, Italy, which encourages scientists from the developing countries.

borrowed time. The energy E, momentum p, and mass m, of a particle moving at very high, 'relativistic' speeds are related by the equation

$$E^2 = p^2c^2 + m^2c^4$$

where c is the velocity of light. At low, 'non-relativistic' velocities this equation reduces to a more familiar result

$$E = (p^2/2m) + mc^2$$

This equation says that the total energy of a non-relativistic particle is the usual kinetic energy plus the mass energy from Einstein's famous mass–energy equivalence. For photons we must use the relativistic formula since photons always travel with the velocity of light. Moreover, photons are found to have zero mass – which just means that the energy and momentum of a photon are related according to the first equation above, with m set to zero. If we now go back to Yukawa's borrowed energy argument, this means that virtual photons with very low momentum will have almost zero total energy. Such virtual photons can travel almost as far as they like without getting into trouble with the energy–time uncertainty relation. This argument then suggests that electromagnetic interactions can be effective over very large distances – and this expectation is confirmed by experiment.

At first sight, the requirement of local phase invariance seems to require that the gauge particles exchanged must have zero mass like the photon. This is because we need to be able to compensate for the effects of a local phase change over all positions on the screen and this can involve very large distances. In fact, this requirement of zero mass turns out not to be the case, but massive gauge particles are only possible in a rather peculiar way. We can illustrate this by looking again at magnetic fields and superconductors. In chapter 7 we saw that magnetic fields do not penetrate far inside a superconductor. On entering the superconductor the magnetic field falls off very rapidly over a very short distance. The effect is caused by currents set up within the superconductor when it is placed in a magnetic field. These currents produce magnetic fields that tend to screen out or cancel the applied magnetic field inside the metal. Such a *diamagnetic* effect takes place in all metals but in a superconductor, because there is no electrical resistance, these induced currents produce a magnetic field that almost completely cancels the applied magnetic field inside the metal, apart from in a very thin surface layer. Consider this situation in terms of the range of the magnetic field inside the superconductor. Since the field only penetrates a very short distance, the effect is as if, inside the superconductor, the photon has acquired a very large mass.

In this case of course, we know that this effective photon mass is caused by the superconducting screening currents induced by the applied magnetic field and that outside the metal, photons are massless. But now try and imagine what the world would look like from the point of view of someone small enough to live permanently inside such a superconductor. Such tiny people may not be clever enough to realize that they are living

Fig. 12.12 A view of the enormous UA2 detector installed in Underground Area 2 at CERN. The apparatus is placed in the SPS tunnel and protons and antiprotons circulating round the SPS are brought together to collide inside the detector.

in the presence of currents that screen out the external magnetic field. Instead, they would deduce that photons have a mass related to the distance that magnetic fields can travel in the metal. It is in this sense that gauge particles can acquire mass and still preserve the local phase invariance.

What has all this got to do with weak interactions? In the previous chapter we drew Feynman diagrams for electron–quark scattering with the electromagnetic interaction mediated by the exchange of a virtual photon. For weak interactions we can draw similar diagrams. In the beta decay of the neutron, for example, a **down** quark changes into an **up** quark emitting a virtual W particle that decays into an electron and an antineutrino. In this case, unlike the electromagnetic interaction, the range of the weak force is found experimentally to be very small. Using Yukawa's argument we deduce that the mass of the W particle must be rather large. The W particle must also be charged – unlike the photon, which is electrically neutral. At first glance, there appears to be little similarity between the local phase

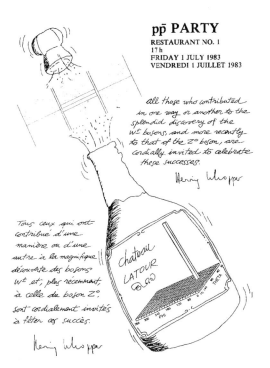

pp̄ PARTY
RESTAURANT NO. 1
17 h
FRIDAY 1 JULY 1983
VENDREDI 1 JUILLET 1983

all those who contributed in one way or another to the splendid discovery of the W± bosons, and more recently to that of the Z⁰ boson, are cordially invited to celebrate these successes.

Henry Whopper

Tous ceux qui ont contribué d'une manière ou d'une autre à la magnifique découverte des bosons W± et, plus récemment, à celle du boson Z⁰, sont cordialement invités à fêter ces succès.

Henry Whopper

Fig. 12.13 A poster for the party held at CERN to celebrate the discovery of the W and Z bosons.

theory of QED, with its massless neutral photon, and any theory of weak interactions involving massive charged W particles.

It is here that the relevance of our discussion of the superconductor becomes apparent. Imagine that we are like the tiny people living inside a superconductor. Because their normal background – the 'vacuum' – has screening currents that make it look like the photon can only travel a short range, they think the photon has a mass. So, if the 'vacuum' in which we live is analogous to a 'weak superconductor', then similar 'vacuum screening currents' can make it appear that the W particles have mass. This is the key idea behind the 'Higgs mechanism'. Perhaps not surprisingly, given the close relation to superconductivity, this mechanism for giving mass to gauge particles was first suggested by Philip Anderson, the distinguished solid-state physicist whom we encountered briefly in chapter 7. In a superconductor, the screening currents are due to circulating Cooper pairs of electrons. In the case of a gauge theory of weak interactions, these currents are believed to be due to particles called Higgs bosons. Peter Higgs is a British theoretical physicist, based in Edinburgh, who was one of the first to work out Anderson's ideas in a relativistic context.

Why did the Nobel Prize committee have enough confidence to award the prize to Glashow, Salam and Weinberg before the W particle was discovered? One of the reasons was the successful prediction of a new type of quark – the so-called *charmed* quark. This came about as follows. The theory that is now known as the standard model, or the GSW model, of electroweak interactions predicts that, in addition to the charged W particles,

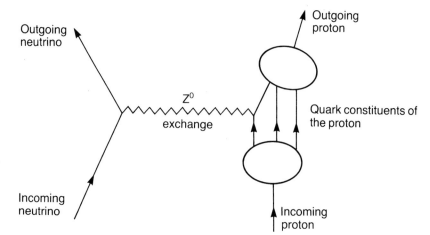

Fig. 12.14 Feynman diagram for neutrino–proton scattering. The neutrino exchanges a virtual Z boson with one of the quarks in the proton.

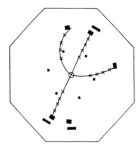

Fig. 12.15 The J/ψ particle was discovered more or less simultaneously at Brookhaven, New York, and at SLAC in California. Here is a reconstruction of a 'ψ looking' event from the electron–positron collider at SLAC. It is caused by a heavier version of the ψ decaying to the usual ψ and positively and negatively charged pions. The ψ, is identified by its electron–positron pair decay products.

there should exist a neutral heavy Z particle – a genuine weak photon. If this particle exists, then it must contribute to neutrino scattering via Feynman diagrams like that shown in Fig. 12.14. Unlike interactions involving W exchange, the quark charge is unchanged in such Z exchange interactions. These reactions correspond to the 'neutral currents' referred to in the telex from the Nobel committee to Salam. After many rumours and false alarms, the discovery of such events was finally announced to the world at an international conference in London, UK, in 1974. The invited review speaker on weak interactions at the conference was the Greek physicist John Iliopoulos. In his talk he threw down a famous challenge. Just as the discovery of neutral currents was the sensation at the London conference, so Iliopoulos offered to bet anybody a case of wine that the sensation of the next conference would be the discovery of the charmed quark. Iliopoulos won his bet.

Exactly how the existence of neutral currents implies the existence of a new type of quark is quite a complicated story. If one believes in a gauge theory of weak interactions, a fourth quark is necessary in order to avoid a conflict with well-established experimental data. This new quark must have a new quantum number which Glashow called **charm**. Just as the electromagnetic force is different for particles with different electrical charges, so the strength of the weak force depends on both the strangeness and charm of the quarks. This was the situation in the summer of 1974, and it is probably fair to say that not many physicists thought it likely that Iliopoulos would win his bet. But in the autumn of that year there was great excitement in the physics community when a spectacular new meson was discovered simultaneously at both Stanford and Brookhaven in the USA. Burton Richter and the experimentalist team at Stanford named the new particle the ψ meson: Sam Ting and his group at Brookhaven chose the name J. Today, nobody seriously doubts that the J/ψ meson is made up from a charmed quark bound to a charmed antiquark. At the time, of course, it was much less clear and there were many other competing

Samuel Ting with other members of his group that discovered the J/ψ particle in Brookhaven, New York. This particle is believed to be a bound state of a charmed quark and its antiparticle. Ting shared the 1976 Nobel Prize with Burton Richter, who led the team that discovered the same particle in SLAC, California.

The leaders of the collaboration of physicists that discovered the J/ψ particle in November, 1974, at the electron–positron collider at SLAC, California. From left to right they are Gerson Goldhaber, Marty Perl and Burton Richter. Marty Perl was later awarded the 1995 Nobel Prize for physics for his discovery of the tau lepton.

ingenious 'explanations' for the J/ψ and its properties. These other theories have now all disappeared as a whole new and complex spectroscopy of **charmonium** states has now been uncovered along with a new family of mesons containing charmed quarks bound to uncharmed antiquarks.

A postscript to this success story is in order. At the time that Glashow, Salam and Weinberg made their contributions to the standard model of the electroweak interactions, there was a serious problem. Although their theory looked like a promising candidate to describe the experimental data,

Gerard 't Hooft was born in 1947 and is now professor of physics at the University of Utrecht in the Netherlands. While studying for his Ph.D. under Tini Veltman in Utrecht, 't Hooft made a vital breakthrough in discovering how to make consistent Feynman diagram calculations for gauge theories. Veltman and 't Hooft were awarded the 1999 Nobel Prize for physics.

nobody knew how to do calculations beyond the *tree* diagrams – diagrams with no closed loops. *Loop* graphs usually involve more powers of the charge, e, the quantity that governs the strength of the coupling of Ws and Zs to quarks and leptons. Since e^2 is found to be very small, e^4 will be much smaller so these such 'higher order' loop diagrams should be relatively unimportant. Unfortunately, all attempts at calculating the effects of these loop diagrams had ended in failure and untameable infinities so that no-one knew what to make of these theories. It was not until a young Dutchman named Gerard 't Hooft came along that all became clear. In the words of the physicist Sidney Coleman, "t Hooft's work changed the Weinberg–Salam frog into an enchanted prince'. Some years earlier, Coleman had reproached Tini Veltman, 't Hooft's thesis advisor, for persisting in his research in 'sweeping out a forgotten corner of theoretical physics'. It is fortunate that Veltman held firm against the prevailing fashions of the time and was one of the first to recognize the importance of gauge theories. It was fitting that both Veltman and 't Hooft were rewarded for their pioneering work with the award of the 1999 Nobel Prize for Physics.

Quarks and gluons

Since the early days of nuclear physics, physicists had hoped that the theory of the strong force would be simple and elegant. With the discovery of the pion and the menagerie of all the other hadrons, together with their excited states, it rapidly became apparent that the force between neutrons and protons was very complicated. During the time that physicists were discovering all these new particles, they also learnt that hadrons are built out of quarks. If there was to be a simple theory of hadronic forces, it was natural to look for an explanation in terms of quarks. Perhaps the so-called strong interactions are merely a feeble shadow of enormously powerful inter-quark forces that can be described by a simple and elegant law?

We have seen that quarks come in several different varieties: non-strange, strange, charmed, and so on. It is the electroweak force that distinguishes between these different *flavours* of quarks: the strong force is the same whether it acts on a strange or a charmed quark. We must apologize for the light-hearted names that particle physicists give to these new quantum numbers. A quantum number like strangeness refers to a well-defined physical property. Early on, some physicists used to prefer the name *hypercharge* instead of strangeness. Hypercharge certainly invokes a more formal and imposing image for particle physics, but most physicists prefer to use strangeness. Similarly, the non-strange quarks, as the name implies, have no strangeness but have different electric charges. Instead of referring to these quarks by the 'eigenvalues of the third component of isotopic spin', physicists prefer to use the shorthand *up* and *down*. Given that the first three quarks are called *up*, *down* and *strange*, it not such a surprise that the next three quarks are called *charm*, *top* and *bottom*. This is not an elaborate joke at the tax-payers' expense – it merely shows that physicists are human!

In 1977, Leon Lederman announced the discovery of the Upsilon particle at Fermilab. The Upsilon is analogous to the J/ψ and is believed to contain a **bottom** quark and its antiparticle. In the same way as for charmed quarks, physicists have now discovered a whole new spectroscopy of mesons containing bottom quarks. Just as Iliopoulos predicted the discovery of charm, it was now clear that another quark was needed to complete the quark pairings **up** and **down**, **charm** and **strange**, and **top** and **bottom**. These three quark pairs mirror the three lepton pairs consisting of the electron, muon and tau together with their respective neutrinos. It was not until 1995 that this long expected **top** quark was discovered, again at Fermilab. What was surprising about its discovery was its very large mass – about 180 times heavier than the mass of the proton.

Strong forces do not notice the different flavours of quarks. Instead, they are sensitive to yet another type of charge carried by all the quarks. Particle physicists refer to this new quantum number as **colour**, but again it should be remembered that this is only a shorthand for a very specific mathematical property. We could say, with pedantic accuracy, that 'quarks transform according to the fundamental representation of the special unitary group SU(3)'; it is surely preferable to say they carry a colour charge. We can make the necessity for this colour quantum number somewhat more physical by the following argument. Consider the Ω^- particle predicted by Gell-Mann. This is a baryon and therefore contains three quarks. In order to make up its electric charge of −1 unit and have strangeness −3, the three quarks must all be strange quarks (Fig. 12.16). The spin angular momentum of the Ω^- is 3/2 units. Since the quarks are all in the lowest energy level with zero orbital angular momentum, the spin of the Ω^- must be made up from the spin of the quarks. Roughly speaking, each of the quarks has spin 1/2 and all of these spins must point in the same direction to add up to a total of 3/2. This all looks very satisfactory – where is the problem? The problem lies with the Pauli exclusion principle that we discussed in chapter 6. The quarks are fermions and must obey the Pauli principle. As

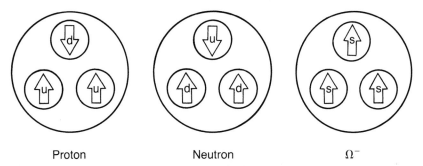

Proton   Neutron Ω^-

Fig. 12.16 The quark content of the proton, neutron and Ω^-. For the proton and neutron we have indicated possible quark spin orientations adding up to a net spin of 1/2. The Ω^- has spin 3/2 so that all three strange quarks must have their spins pointing in the same direction. This would be forbidden by the Pauli exclusion principle unless the quarks had some extra hidden quantum number.

things stand, all of the quarks in the Ω^- have the same quantum numbers and Pauli does not allow this. The introduction of a colour quantum number for the quarks allows us to solve this problem. Colour has something to do with a mathematical construct known as a group, and, in particular, the so-called 'special unitary group SU(3)'. The threeness of this group means that there are three different possible states for the quark. Again, we usually refer to this situation more informally by talking about quarks coming in three different colours. It is important to remember that this is just a shorthand for the mathematics: quarks do not have real physical colours that control the strong forces! We can now see how colour solves our problem with the Ω^-. There are three different possible colours for the quarks, so each quark in the Ω^- must have a different colour – **red**, **green** and **blue**, say – to satisfy the exclusion principle.

We are now in a position to describe the ingredients that make up the theory called Quantum ChromoDynamics, QCD – the long-sought theory of the strong force. QCD is a gauge theory based on the local phase invariance of the colour properties of the quantum amplitudes of the quarks. Although this may sound intimidating, it is difficult to imagine that any theory of the strong interactions could be simpler. Just as the electromagnetic forces are mediated by zero mass gauge particles – the photons we have met so often – so we expect that the quark–quark interactions are described in terms of the exchange of similar 'strong photons'. Physicists have given these particles the name **gluons**, because, in a very real sense, they are the glue that holds everything together. Photons couple to the ordinary electric charge of the quarks: gluons couple to the colour charge of the quarks. Moreover, the gluons themselves carry a colour charge and the gauge principle dictates that, unlike our photon example, gluons must interact with themselves. Physicists believe that it is this key feature that makes quantum chromodynamics (QCD) – 'chromo' is for colour – so different from quantum electrodynamics (QED). Why do we say that QCD is so different from QED? This is because it is easy for us to observe electrons in the laboratory. No-one has ever been able to observe a quark all by itself. Quarks have only been 'seen' in combination with other quarks and antiquarks inside hadrons. Physicists believe that this is not an accident and that the interactions between quarks and gluons arrange themselves to make it impossible for us to isolate a single quark. This property is called quark confinement and we shall examine some of the ideas in which this may come about in the next section.

Superconductors, magnetic monopoles and quark confinement

In experiments involving the collision of high-energy particles, only ordinary hadrons have ever been observed. Despite occasional flurries of

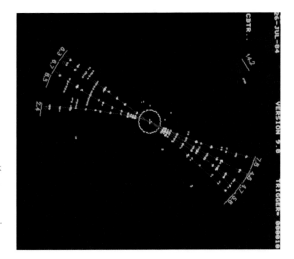

Fig. 12.17 An example of a quark–antiquark back-to-back 'jet' event seen at the PETRA electron–positron collider in Hamburg, Germany. Most of the tracks are made by pions. This event was observed in the TASSO detector.

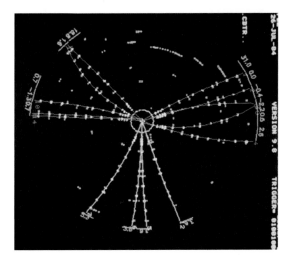

Fig. 12.18 Electron–positron annihilation at PETRA sometimes gives rise to 'three-jet' events as in this event seen in the TASSO detector. Such events are believed to be due to the fragmentation into ordinary hadrons of a quark and an antiquark, together with a gluon.

excitement, no fractionally charged quark-like objects have been conclusively identified. In the collision of two very energetic protons we do not observe the protons breaking up into quarks. Instead, the collision energy is used to create a whole host of mesons, baryons and antibaryons. Even in a reaction in which we believe an electron is annihilated by a positron to produce a quark and an antiquark going off in opposite directions, we still do not see any quarks. Instead, all that remains is a relic of the initial quark and antiquark motion in the form of two jets of normal hadrons. Three-jet events, corresponding to a Feynman diagram in which one of the quarks emits a high energy gluon, have also been seen, but in none of these jets do we see quarks or gluons by themselves.

We have now accumulated much circumstantial evidence that suggests hadrons contain quarks and gluons, yet it seems that their interactions arrange things so that we can never isolate an individual quark or gluon.

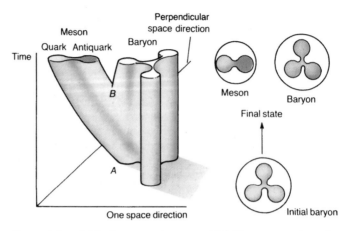

Fig. 12.19 A model for baryons and mesons which illustrates quark confinement. In order to draw the picture, only two space dimensions are shown. A quark is represented as a pale blue open circle and an antiquark by a dark blue circle. The gluon forces are shown as an elastic rubber-like sheet that keeps the quarks inside the hadron. As quark A is pulled away from the other quarks, so much energy has been put into the system that eventually a quark–antiquark pair are created at point B and two ordinary hadrons emerge.

If we try to pull a quark out of a baryon, we have to put in so much energy that we create a quark–antiquark pair (Fig. 12.19). Instead of breaking up the baryon, we end up with a baryon and a meson. According to this quark picture, Yukawa's meson exchange model of strong interactions is clearly not at all fundamental. Measurements of the contribution of pion exchange to nuclear forces can only tell us very indirectly about the basic quark and gluon force. But the basic question remains: how does confinement come about? Nobody knows for sure, but there are several clues and speculations. One of the most interesting ideas involves another solid-state analogy with superconductors – as for the weak force – but this time with a new twist to the argument.

There are two new strands of thought that we must introduce. The first concerns classical electromagnetism. Most people are aware that although electric charges can exist separately, magnetic charges apparently only exist as north and south pole pairs, as in a bar magnet. Cutting a magnet in two does not isolate a *magnetic monopole* but, instead, produces two smaller magnets. This magnet analogy is sometimes used to illustrate a type of confinement – of monopoles in this case – but the mechanism proposed for the confinement of quarks is more subtle. The electric and magnetic fields generated by a system of charges and currents are described by a set of equations known as the Maxwell equations. Because free magnetic monopoles are not found in Nature, these equations are not symmetrical with respect to interchanging electric and magnetic fields. If individual magnetic charges and currents did exist, there would be a curious dual symmetry of the resulting equations under the interchange of electric and magnetic fields. It is typical of Dirac's quirky type of originality that he

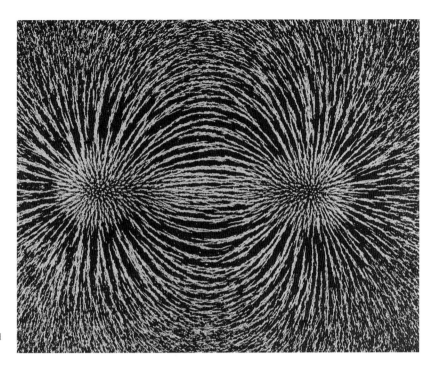

Fig. 12.20 Iron filings scattered on a card above a bar magnet reveal the pattern of the magnetic field around the magnet.

was the first to consider seriously the implications of the existence of magnetic monopoles for quantum mechanics. By an argument too complicated to attempt here, Dirac showed that the existence of just one quantum mechanical magnetic monopole would imply that all electric charges must be exact multiples of the charge of the electron! If the above discussion of hypothetical magnetic monopoles seems rather remote from reality, the other strand in the argument is firmly rooted in experiment. Earlier in this chapter we explained how screening currents act to cancel out any applied magnetic field in a superconductor. Actually, it is found experimentally that there are two types of 'classic' superconductor – as well as the new high temperature superconductors. Type-I superconductors are those in which the magnetic field is screened out as we have described. In type-II superconductors, on the other hand, the magnetic field is not entirely expelled from the metal, but is allowed to thread its way through in thin filaments (see Fig. 7.22). Here we encounter another unexpected result of quantum mechanics – the magnetic field threading each filament is quantized and can only have certain values.

We can now explain how quark confinement could come about. We suppose that the vacuum state of QCD is like a type-II superconductor. Since QCD is very similar to QED, it should come as no surprise that QCD has both colour electric fields and colour magnetic fields. A vacuum that behaved as a type-II superconductor would only allow the colour magnetic field to exist in thin filaments as shown in Fig. 12.21. The quantized magnetic field allowed through the filament turns out to be just the right amount for the field to

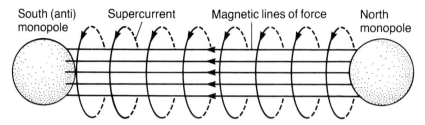

Fig. 12.21 The magnetic field lines joining a magnetic monopole–antimonopole pair in a superconductor. The magnetic field is squeezed into a narrow tube by supercurrents of Cooper pairs circulating around the tube. Quark confinement is thought to be due to a similar mechanism – the ordinary vacuum acts like a 'dual superconductor' in which circulating magnetic monopole currents squeeze the electric field between a quark and an antiquark into a thin tube.

begin and end on a colour magnetic monopole. What has this to do with confinement? Figure 12.20 shows the magnetic field lines of a bar magnet, displayed by the pattern of iron filings. Compared with this pattern, the magnetic field of the monopole–antimonopole pair that threads its way through a type-II superconductor is very different. The electric screening currents circulating around the filament have caused the colour magnetic field lines to be squeezed into a narrow tube. If we could arrange for this to happen for the colour electric field with quarks and antiquarks at the ends instead of monopoles we would have just the sort of field pattern that we need for quark confinement. This is because if the colour electric field lines are squeezed into a thin tube, the energy required to separate a quark–antiquark pair would grow in direct proportion to their separation. In this situation it would require an infinite amount of energy to separate the quark and antiquark by an infinite distance. This is the hallmark of quark confinement.

So far, we have confined monopoles not quarks. But, as promised, there is a final twist to this story. A quark and an antiquark system is bound together by a colour electric field, not by a colour magnetic field. Moreover, quarks do not have a magnetic charge and are not magnetic monopoles. Now we remember the electric–magnetic interchange symmetry of Maxwell's equations with magnetic monopoles. Using this dual symmetry, we see that electric fields will be squeezed into the thin tubes required for confinement if the physical vacuum behaves as the dual version of a type-II superconductor. Instead of Cooper pairs encircling filaments containing magnetic fields, the physical vacuum would now have magnetic monopole currents trapping tubes of electric fields. This is a beautiful idea and would provide us with a model for confinement but how do we know if the vacuum state of QCD is like a dual type-II superconductor?

Is there any way in which we can test these ideas of confinement and investigate the long distance aspects of QCD? In 1973, David Politzer, and independently David Gross and Frank Wilczek, showed that QCD has the remarkable property that the effective coupling becomes smaller at shorter and shorter distances. This means that although QCD describes the ultra-strong forces between quarks that lead to confinement at long distances, at short distances it is allowable to use an expansion in terms of the coupling 'constant' and familiar Feynman diagrams to make predictions that can be compared with experiment. This property – known as 'Asymptotic Freedom' – has meant that QCD can be applied very successfully to describe certain features of high-energy electron and neutrino scattering experiments. The downside of asymptotic freedom is that it implies the effective coupling strength increases with distance. In this regime we are clearly

Fig. 12.22 Mr Tompkins inside the proton. In this picture we add an extra chapter to Mr Tompkins' adventures. Inside the proton he meets three mischievous quark dwarfs sentenced to eternal confinement in Gell-Mann's quark prison. Mr Tompkins was appalled at the prison conditions. The dwarfs were all chained together by their ankles but seemed almost oblivious of their cramped confinement. Indeed, they happily maintained they were free, huddled together in the middle of the prison. 'It is positively cosy', they said, but Mr Tompkins was not so sure. 'Now let me show you a trick', said one of the quarks, who was standing on his hands. Mr Tompkins and the Professor agreed to hold on to the quark and pull. The quark became more and more excited and kept urging further effort. Suddenly the chain broke and they all tumbled out of prison. When Mr Tompkins had collected his wits he saw an extraordinary sight. Outside the prison were two quarks chained together! Mr Tompkins rubbed his eyes and turned to the Professor who explained that the second quark was actually an antiquark. Mr Tompkins was still bemused and was further surprised to see that back in the prison, there were three happy-looking quarks still chained together. The Professor started to explain that a quark and an antiquark always appear when a gluon chain is broken but Mr Tompkins was no longer listening. It was all too much for him! (Dwarfs courtesy of Frank Close.)

not allowed to make a 'perturbation' expansion in terms of powers of the coupling constant since the terms with more powers of the coupling are at least as important as the lowest-order terms. If we are not allowed to make an expansion in terms of Feynman diagrams how can we hope to extract any information on these 'non-perturbative' aspects of QCD such as confinement? One answer to this question that many physicists are attempting is to solve the equations of QCD numerically on a very powerful computer! In order to put the QCD equations in a form that can be programmed for a computer, we must first make some rather drastic approximations. The most significant approximation is to approximate continuous space-time by a discrete set of points on a four-dimensional 'lattice'! Furthermore, in order to handle rapidly oscillating quantum amplitudes, physicists boldly rotate the time direction into an 'imaginary' time direction. The computer techniques required to solve the resulting lattice gauge theory uses Feynman's quantum path formulation of quantum mechanics. This lattice QCD programme was pioneered by US physicists Ken Wilson and Michael Creutz in the 1970s. For our approximation of space-time by a fixed set of discrete points to give us results that can be related to the continuous version of QCD, a large number of lattice points in each of the four space-time directions are required. This in turn requires a lot of computation. For this reason, physicists all over the world have been putting together large assemblies of computer chips to form a parallel supercomputer powerful enough to give realistic results. So far, the results of such computer simulations are very encouraging and provide us with more evidence that QCD is the correct underlying theory.

Beyond the Standard Model

What remains to be done? The Large Electron Positron accelerator, LEP, at CERN has confirmed the existence of the Ws and the Z and tested other predictions of the GSW theory (Fig. 12.23). Despite these stunning experimental triumphs for the Standard Model, there are several fundamental unanswered questions. Although the model can accommodate both the muon and the tau leptons along with their neutrinos, in addition to the electron–neutrino doublet, there is no compelling reason for their existence nor any predictions of their masses. Nor do we have a real understanding of why there are three doublets of quarks – (up, down), (strange, charm) and (top, bottom) – to accompany these lepton doublets or any real understanding of the large range of masses of the different quarks.

There is also one crucial piece of the jigsaw that has not yet been confirmed. This concerns the mysterious Higgs particle. Any experimental search for this particle is hampered by the fact that the model makes no prediction for the mass of the Higgs particle. Moreover, the solid-state analogue of the Higgs particle in a real superconductor is a Cooper pair of electrons. Thus, the Higgs particle may not be a genuine elementary

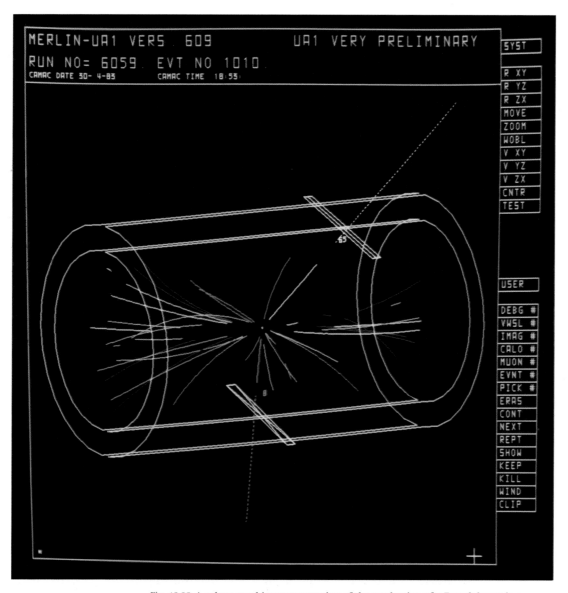

Fig. 12.23 A colour graphics reconstruction of the production of a Z weak boson in a proton–antiproton collision. The Z is identified by the blue and white tracks emerging from the sides of the cylindrical detector. These are left by the electron and positron coming from its decay. This event was seen in the UA1 experiment led by Nobel Prize winner Carlo Rubbia.

particle but may turn out to be composite. In any event, it is believed that some phenomena characteristic of the Higgs must show up at energies of around a million million electron volts (TeV). For this reason, the Large Hadron Collider (LHC) is under construction at CERN using the same tunnel as for the LEP accelerator. Physicists from all over the world are now

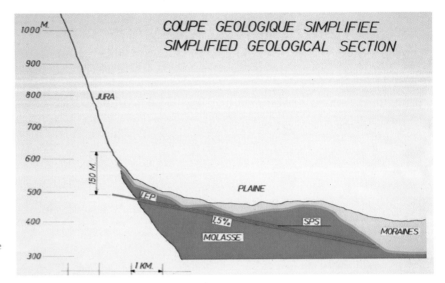

Fig. 12.24 A cross-section through the mountains near Geneva, showing the location of the LEP/LHC tunnel.

Fig. 12.25 A mechanical mole was used to bore the tunnel for the LEP accelerator, now to be used for the LHC, deep underneath the French–Swiss border near Geneva. The ring of superconducting magnets that forms the LHC accelerator is located in this tunnel.

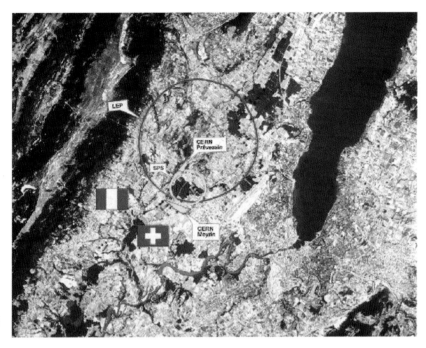

Fig. 12.26 An aerial view of the CERN laboratory located on the France-Switzerland border just outside Geneva, Switzerland. The large circle marks the position of the underground LEP tunnel that is now used for the LHC machine.

busy designing and building gigantic new detectors for the LHC that will be capable of finding the Higgs boson.

Are there any clues that point to new physics beyond the Standard Model? The answer is yes! There are now some intriguing indications that suggest that our Standard Model may need to be extended. The first clue is concerned with the question of whether neutrinos have mass, or are massless like the photon. The simplest version of the Standard Model assumes all three neutrinos are massless. As we discussed in chapter 10, experiments to detect solar neutrinos generated by nuclear reactions in the Sun see fewer electron neutrinos than are expected on the basis of careful calculations. One of the possible explanations for this observed deficit of electron neutrinos is that the neutrinos have a non-zero mass that allows electron neutrinos to 'oscillate' into some combination of muon and tau neutrinos. Until recently, this solar neutrino result remained a tantalizing curiosity. What was needed was more experimental evidence about possible neutrino oscillations. All this has recently changed with exciting results from a huge new neutrino experiment called Super-Kamiokande. This experiment consists of 50 000 tons of water and 11 200 photomultiplier tubes located at a depth of 2700 metres below the surface of the Earth, in the Kamioka Mozumi mine in Japan. Photomultipliers are very sensitive light detectors that can detect single photons. Neutrinos interact inside the detector producing either electrons or muons that generate Cherenkov light. Cherenkov light is generated when a charged particle travels with a speed greater than the speed of light in water. It is similar to the sonic shock wave caused by

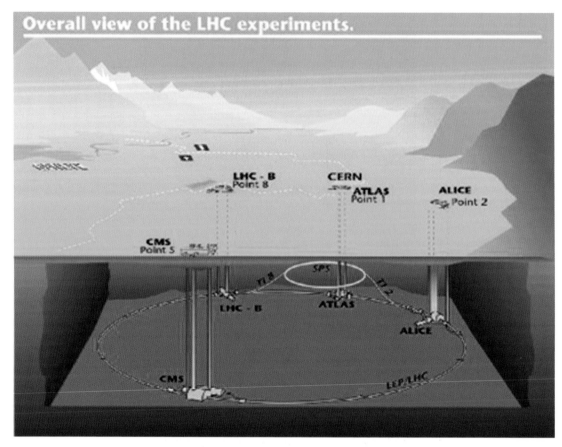

Fig. 12.27 A cross section of the Large Hadron Collider (LHC) at CERN showing the four experimental pits (only three will be occupied initially). The experimental groups for the LHC experiments each contain well over 1000 physicists from over 100 institutions from Europe, the USA and Japan. The data and computing demands are so large that the particle physicists have helped develop a new distributed computing infrastructure known as the 'Grid'. Just as Tim Berners-Lee and the particle physics community at CERN developed the World Wide Web for sharing information, so the Grid enables them to share data and computing resources.

a plane travelling faster than the speed of sound in air. The detector is located deep underground to shield it from cosmic ray muons which could swamp the signal from muons created in a neutrino interaction in the water. Besides confirming the solar neutrino result, the experiment is able to measure the number of 'atmospheric neutrinos' – electron and muon neutrinos that arrive at the detector as a result of cosmic ray interactions in the upper atmosphere. The number of muon neutrinos is found to be significantly smaller than expected and the results provide compelling evidence that muon neutrinos must also undergo neutrino oscillations. The Sudbury Neutrino Observatory (SNO), located underground in a nickel mine in Ontario, has also recently announced results confirming the need for solar neutrino oscillations. The SNO project is unique in using heavy water

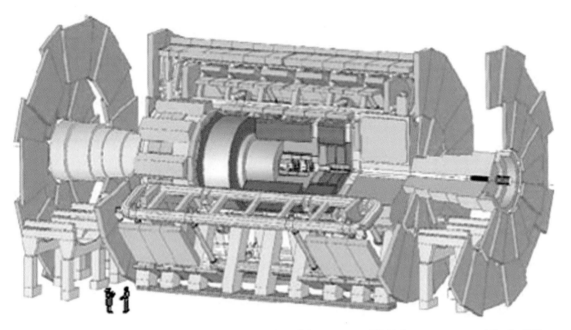

Fig. 12.28 A schematic drawing of the enormous ATLAS detector constructed for the LHC at CERN. Note the scale of the people.

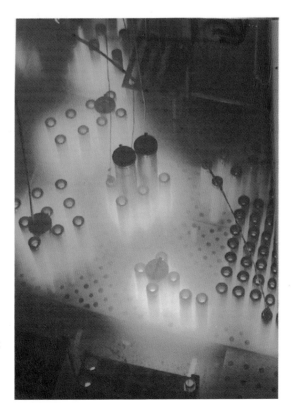

Fig. 12.29 The eerie blue glow from this nuclear fuel storage pond is due to Cherenkov radiation. This radiation is caused by radioactive decay particles travelling at speeds faster than that of light in water. It is similar to the sonic shock wave caused by a plane travelling faster than the speed of sound in air.

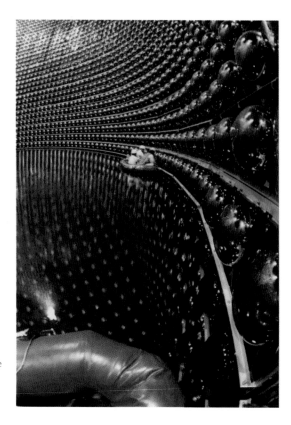

Fig. 12.30 Technicians in a rubber boat examine the interior of the Super-Kamiokande detector. When operational the tank is filled with 50 000 tons of ultra-pure water. Some of the 13 000 photomultiplier detector tubes are seen in this photograph.

containing deuterium. The Sun's nuclear reactions produce electron neutrinos and SNO's heavy water detector singles out these neutrinos. Only this flavour of neutrino can be absorbed by the neutron in the deuterium nucleus, transforming the nucleus into two protons that fly apart rapidly. All these new results confirm the need to extend the Standard Model.

The next two clues are less direct but again suggest that the Standard Model needs to be extended. The first of these two clues concerns the observed matter–antimatter asymmetry in the universe. We are made of particles and not antiparticles. As far as we know there is no evidence for galaxies with stars and planets (and people) of antimatter elsewhere in the galaxy. The Standard Model cannot explain how a universe that was initially symmetric between particles and antiparticles could evolve to our present asymmetric situation. Of course, it is possible that the universe started off in an asymmetric state but most physicists find this an unappealing solution. The second clue from the universe at large concerns the question of 'dark matter'. The existence of dark matter is deduced as follows. Astronomers have observed many spiral galaxies like our own galaxy. These are disk-shaped collections of stars and gas that are rotating about their centre. The puzzle is that the rate of rotation is too high for the galaxy to stay together if it only contained the mass of the stars and gas that we can observe. A similar observation holds true for clusters of galaxies. We

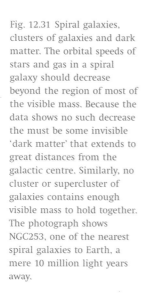

Fig. 12.31 Spiral galaxies, clusters of galaxies and dark matter. The orbital speeds of stars and gas in a spiral galaxy should decrease beyond the region of most of the visible mass. Because the data shows no such decrease the must be some invisible 'dark matter' that extends to great distances from the galactic centre. Similarly, no cluster or supercluster of galaxies contains enough visible mass to hold together. The photograph shows NGC253, one of the nearest spiral galaxies to Earth, a mere 10 million light years away.

deduce that there must exist some as-yet-unobserved dark matter that provides the gravitational attraction that holds these galaxies and clusters of galaxies together. What can this dark matter be? One candidate for at least a substantial fraction of this missing dark matter arises in a novel extension to the Standard Model based on a peculiar type of symmetry known as 'Supersymmetry'.

Up to now, quantum mechanics has distinguished between particles that are 'fermions' and obey the Pauli Exclusion Principle – like leptons and quarks – and particles that are 'bosons' and mediate the forces – like photons, W and Z bosons and gluons. Supersymmetry is a new kind of symmetry that requires that the equations of the fundamental theory should be unchanged when fermions are replaced by bosons and bosons by fermions! This statement implies that there are many new 'sparticles' – supersymmetric partners of the particles with which we are familiar – waiting to be discovered. Besides the leptons and quarks that we know about, the theory predicts new bosonic partners, playfully called 'sleptons' and 'squarks'. Similarly, besides photons, Ws, Zs and gluons, we should expect to find photinos, Winos, Zinos and gluinos. The photino, for example, could contribute to the mysterious dark matter that must pervade the universe. Alas, at present there is no direct experimental evidence for any of these curious sparticles. Hopes are high that the new generation of accelerators at Fermilab in the USA and the LHC at CERN in Europe will reveal some more clues that will tell us whether supersymmetry has a role to play outside the speculations of theorists.

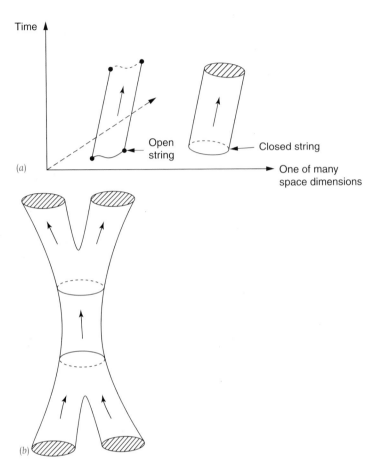

Fig. 12.32 (a) Open and closed strings. The closed string mode is characteristic of the spin 2 massless graviton. (b) Strings interact by splitting and joining.

There is one final challenge for quantum mechanics. This is the unification of quantum mechanics with gravity to produce a consistent theory of quantum gravity. Attempts to define a gauge field theory of gravity along the lines of QED and QCD have proved unsuccessful. Instead, some theoretical physicists are looking at a new way of constructing a theory that could encompass both the Standard Model and gravity in a consistent and calculable way. Such theories are called 'String Theories' – since instead of describing a particle such as an electron by a point in space-time, such theories describe fundamental particles as one–dimensional strings. The characteristic mass scale for quantum gravity is given by the so-called Planck Mass, defined in terms of a combination of Planck's constant, the speed of light and Newton's gravitational constant. Numerically, the Planck Mass is about 10^{19} (1 followed by 19 zeros) times larger than the mass of the proton. This mass sets the scale of the length of the string. This turns out to be around 10^{-33} (a decimal point followed by 32 zeros and a 1) centimetres – so that for our ordinary length scales string particles will be indistinguishable from point particles. The strings can either be open or closed in a loop, and have certain vibrational modes that can be related to properties like mass and spin. Supersymmetry also arises naturally in string theories and the string theorists were excited to find that only a small number of consistent 'superstring' theories were possible, and only in 10 space-time dimensions! Presumably, since we live in four dimensional space-time, for these theories to describe our world correctly, six of these dimensions must curl up into a compact 'ball' too tiny to measure. An interesting question is whether we could ever see any observable effects caused by these 'extra' dimensions. There has also been much excitement amongst string theorists by the recent discovery of 'M-theory': the realization that all the different superstring models in 10 dimensions can be derived from a single unique 11–dimensional supergravity theory. At present physicists are still exploring the fascinating possibilities raised by such theories – and they are still a long way from answering questions such as how the supersymmetry is broken and how the observed particle masses are derived. But from all the above it can be seen that particle physics, gravity and quantum mechanics are still likely to hold some surprises for the next generation of physicists.

13 Afterword – quantum physics and science fiction

> You read too many novels!
> Richard Feynman

Prelude: the atom and the nucleus

Now that the atomic basis of matter is taught routinely in schools, it is difficult to imagine the suspicion and hostility towards atoms that existed at the end of the nineteenth century. This seems especially strange since the idea of atoms has been around since the fifth century BC in the writings of the Greek philosophers Leucippus and Democritus. Such distrust of the 'atomic hypothesis' is all the more surprising given that Daniel Bernoulli, James Clerk Maxwell and Ludwig Boltzmann had all successfully used an atomic model of gases – with atoms as tiny hard spheres that could move and collide like billiard balls – to explain many thermodynamic properties of gases. Nonetheless, it was only with Einstein's famous 1905 paper on 'Brownian' motion – which explained the observed random jiggling motion of grains of pollen floating in water in terms of collisions with water molecules – that almost all of the doubters were silenced and the atomic hypothesis became generally accepted.

As we have seen, the idea of atoms as tiny, hard, indestructible spheres only survived until 1911. It was then that Ernest Rutherford came up with the startling discovery that most of the atom was empty space! From his calculations of the scattering of alpha particles by atoms, Rutherford deduced that almost all the mass of the atom, and all the positive charge, must be concentrated in a tiny sphere much smaller than the apparent size of the atom. He called this the atomic nucleus and calculated that its radius was roughly 10 000 times smaller than the atomic radius. Two years later, the young Niels Bohr, working in Manchester with Rutherford, suggested his famous 'electronic orbital' model of the atom. As we have seen in chapter 4, the problem was that this model – with negatively charged electrons orbiting a positively charged nucleus – was unstable according to Maxwell's laws of electromagnetism. Bohr knew this was a problem but had found he could explain the frequencies of the spectral lines of hydrogen in terms of energy differences caused by electrons 'jumping' between different orbits. He therefore proposed that electrons do circulate around the nucleus – but only in certain stable 'quantum' orbits that defy the laws of classical physics. This picture of the atom caught the popular imagination. Instead of being solid hard spheres, atoms were mostly empty space. Atoms looked like miniature solar systems with the nucleus playing the part of the Sun and the orbiting electrons playing the part of the planets.

Recognition of these developments soon found its way into the newly emerging form of literature known as science fiction or SF. Many of the early attempts to incorporate the new discoveries about the atom in an SF novel were wildly inaccurate. Around the turn of the century, 'lost world' stories – in which bold explorers grappled with unfamiliar dangers in novel environments – were a popular theme of this new type of fiction. *Voyage to the Centre of the Earth* by Jules Verne and *The Lost World* by Sir Arthur Conan Doyle are two classic examples. By the 1920s, science fiction writers were finding it difficult to find plausible places in which to locate new lost worlds. As news of the discoveries about atomic physics entered the public consciousness, some SF writers turned to the now wide-open spaces of the atom for inspiration. The idea that the atom itself was a miniature solar system, populated by miniature, intelligent beings, was the setting for many early SF stories. The SF writer Brian Aldis has characterised these as a new form of *Gulliver's Travels*, 'Gulliver down the microscope'. A typical example is the story *Submicroscopic* by Captain S.P. Meek. Meek's hero is inspired to be a scientist after listening to a lecture on 'modern' theories of the atom. In the lecture, the 'doctor' describes the size of an atom by reference to a cubic millimetre of hydrogen gas:

> It contains roughly ninety quadrillions of atoms, an almost inconceivable number. Consider this enormous number of particles packed into a cube with an edge less than one-twentieth of an inch long; yet so small are the individual atoms compared with the space between them that the solar system is crowded by comparison.

Meek then describes the 'violent motion' of the electrons revolving around the protons and states that 'this combination of centrifugal force and electrical attraction holds the atom in a state of dynamic equilibrium'. This introduction serves to lend plausibility to the construction of an 'Electronic Vibration Adjustor'. As the hero explains:

> The work of Bohr and Langmuir particularly attracted me, and I bent my energies to investigating the supposed motion of electrons about the nuclear protons. This line of investigation led me to the suspicion that the motion was not circular and steady, but was periodic and simple harmonic except as the harmonic periods were interfered with by the frequent collisions.

Herbert George Wells
(1866–1946), better known as
H.G. Wells, was inspired by
the lectures of T.H. Huxley
on Darwin's theory of
evolution by natural
selection. The theme of
humanity evolving according
to these inexorable forces is
evident in his first and
probably most famous book,
The Time Machine, published
in 1895, as well as in many
of his later works. He is
often credited with forseeing
the development of the tank,
of aircraft and air warfare,
of the atomic bomb and of
the nuclear stalemate, and
even of a sort of genetic
engineering in his novel *The
Island of Doctor Moreau*. Hugo
Gernsback reprinted almost
all of Wells' novels in the
1920 editions of his *Amazing
Stories* magazine. As a result
Wells' work has been and
remains enormously
influential in the USA to this
day. His book *The War of the
Worlds* was the first 'alien
invasion' story and Orson
Wells' dramatization of it
caused a riot in New York in
1938.

Scientific 'explanation' has now descended into jargon-sprinkled nonsense!
More follows with the discovery of new civilisations living at a microscopic
scale. The 'formula' of such novels was simple: fantastic extrapolations from
the recent discoveries in atomic physics were used to stimulate the reader
and to act as a backdrop to a swash-buckling adventure story. Apart from
our greatly increased scientific sophistication, this formula is not so very
different from that used by two, hugely popular, modern day successors –
Star Trek and *Star Wars*!

What is the crucial ingredient of science fiction? According to the
SF writer Fred Pohl, 'A good science fiction story should be able to predict not only the automobile but also the traffic jam'. Probably no science
fiction writer has been more successful in looking two steps ahead than
Herbert George Wells, better known as H.G. Wells. Wells was born in 1866
in the south of England in a period of great upheaval in the scientific
world. Darwin, Maxwell, Mendeleev, Joule and Kelvin were laying the theoretical foundations of evolution, electricity and magnetism, chemistry,
statistical physics and thermodynamics. Following on from these great scientific triumphs, the first decade of the twentieth century saw Planck,
Einstein, Rutherford and Bohr over-turning classical physics with their discovery of quantum physics. At the same time, Einstein was forcing physicists to re-think their ideas about the fundamental nature of space and
time. Curiously then, it was in 1895, fully 10 years before Einstein published his special theory of relativity, that Wells wrote his famous book *The
Time Machine*. With this story, Wells introduced time travel and gave science
fiction a whole new dimension to explore. It is less well known that Wells
was equally imaginative in the context of atomic and nuclear physics. His
novel *The World Set Free*, written shortly before the First World War, is worth
a detailed look.

In the early years of the twentieth century, Ernest Rutherford was
in Canada making a painstakingly thorough study of the radioactive decays
of radium and other heavy nuclei. By 1903, Rutherford and his collaborator
Frederick Soddy were able to quantify the huge amount of energy released
in such decay processes. Their calculations were performed over a decade
before Einstein discovered his famous relation between mass and energy
and physicists had little idea about the nature of the atomic nucleus. It
would be nearly 30 years before James Chadwick discovered the neutron and
over 40 years before Francis Aston proposed the existence of a strong nuclear
force. It is therefore not surprising that the origin of these radioactive decay
energies was a mystery – but both men were uncomfortably aware that
the very large energies released by radioactive processes were potentially
dangerous. In 1904 Soddy wrote:

> It is probable that all heavy matter possesses – latent and bound up
> within the structure of the atom – a similar quantity of energy to
> that possessed by radium. If it could be tapped and controlled what

Leo Szilard was a Hungarian theoretical physicist and was born in Budapest in 1898. In 1928, when working in Berlin, he read H.G. Wells' manifesto called *The Open Conspiracy* in which Wells appealed for a public association of scientifically minded industrialists and financiers to establish a world republic to save the world. Typically ambitious, Szilard then travelled to London to meet Wells and to bid for the European rights to his books. He was also inventive: he had a patent on a new type of refrigerator with a co-inventor, one Albert Einstein. As a Jew he was wise enough to leave Germany one day before the Nazis starting checking the trains. He wrote: 'This just goes to show that if you want to succeed in this world you don't have to be much cleverer than other people, you just have to be one day earlier.' Szilard read *The World Set Free* in 1932: 30 years later he could still summarize the book in detail. On September 12th 1933, Szilard was angered by a report in the London *Times* that Lord Rutherford had declared that anyone thinking of producing power from nuclear energy 'was talking moonshine'. As he walked back to his London hotel, at a traffic light at Southampton Row, Szilard conceived of the idea of a nuclear chain reaction. In 1934, he patented the key idea and in a later

an agent it would be in shaping the world's destiny. The man who put his hand on the lever by which a parsimonius Nature regulates so jealously the output of this store of energy would possess a weapon by which he could destroy the Earth if he chose.

In 1909 Soddy's wrote a book called *The Interpretation of Radium* and this stimulated Wells to write a novel about this new nuclear energy. *The World Set Free* was the result. In the novel, Wells speculates that an explosive chain reaction could be used to make an atomic bomb. He imagined these bombs being constructed from a new, man-made element called 'carolinium' – supposedly 'most heavily stored with energy and the most dangerous to make and handle'. Although Wells was understandably wrong in details, his fictional element carolinium has an eerie parallel with the element plutonium, discovered many years later by Glenn Seaborg. Wells not only accurately predicted the horrifying impact of nuclear weapons but he also foresaw the nuclear stalemate that widespread possession of such weapons would inevitably bring about. In his fictional world, peace and regulation of nuclear weapons were only achieved after the atomic annihilation of all the major cities of Europe. We should be thankful that this is one area where fact has been a little better than fiction. Although Wells' novel was not a commercial success, it can claim credit for affecting the course of the Second World War. After reading *The World Set Free*, Hungarian physicist Leo Szilard became seriously alarmed about the danger of a Nazi atomic bomb. He was afraid that Heisenberg and the other great German physicists of the day were well capable of producing such a weapon if it was technically feasible. Szilard persuaded his friend and fellow refugee, Albert Einstein, to write to President Roosevelt to alert him of the danger. In their letter Einstein and Szilard explained the new threat posed by nuclear weapons in very graphic terms: 'A single bomb of this type … exploded in a port might very well destroy the whole port together with some of the surrounding territory.' This is an interesting case study of how science fiction can influence the course of history. By convincingly describing the possible application of the new scientific discoveries in a fictional context, Wells was able to alert Szilard to the dangers of nuclear weapons. Perhaps the SF writers of today will play a similar role in the case of genetic engineering and the coming bioinformatics revolution.

Nuclear energy and the 'Golden Age' of science fiction

The origins of modern SF remain a controversial topic of debate in the science fiction community. Brian Aldis suggests that modern SF can be traced back to Mary Shelley's *Frankenstein* and has links to other early Gothic novels. Others point to the works of Jules Verne and H.G. Wells as the starting point. A third faction maintains that SF only came of age with

amendment to his patent, spelt out the necessity for obtaining a critical mass and the possibility of creating a nuclear explosion. Rutherford refused to give him funds and facilities to do the necessary experiments.

Hugo Gernsback (1884–1967) was born in Luxembourg and emigrated to America when he was 20. He published his novel *Ralph 124C 41+*, subtitled *A Romance of the year 2660*, in the first of the many issues of magazine *Modern Electrics* in 1911. By 1926, Gernsback was ready to launch the first magazine to be exclusively devoted to science fiction *Amazing Stories*. The slogan on the title page proclaimed its mission 'Extravagant Fiction Today ... Cold Fact Tomorrow'. Gernsback invented the term 'science fiction' and the annual SF Hugo awards are named in his honour.

the emergence of the 'pulp' magazines – so-called because of their cheap paper format – devoted to science fiction. A European immigrant to the USA, Hugo Gernsback, launched the SF magazine *Amazing Stories* in 1926. Although *Amazing Stories* was probably not the world's first SF magazine, it rapidly became the most influential. It was Gernsback who also introduced the term 'science fiction'. From these beginnings, an SF magazine emerged that has had a profound influence on the development of modern science fiction. The magazine was *Astounding Science Fiction* and its editor was John W. Campbell, Jr. In his *Illustrated History of Science Fiction*, James Gunn goes so far as to say:

> The dozen years between 1938 and 1950 were 'Astounding' years. During these years the first major science fiction editor began developing the first modern science fiction magazine, the first modern science fiction writers, and, indeed, modern science fiction itself.

Campbell was a scientist by training but also wrote science fiction. One of the most famous SF writers of all-time, Isaac Asimov, designates the August 1938 issue of *Astounding* as the beginning of the 'Golden Age' of science fiction. This issue contains *Who Goes There?* – a story by Campbell written under the pseudonym Don A. Stuart. Asimov rates this story as 'one of the best science fiction stories ever written' and claims that it served as a 'training manual' for a whole generation of aspiring science fiction writers. Of Campbell and *Astounding* Asimov has said that:

> During the Golden Age, he [Campbell] and the magazine he edited so dominated science fiction that to read *Astounding* was to know the field entire.

In 1944, Campbell's office at *Astounding* was visited by agents from Military Intelligence. They were interested in an article that Campbell had published in the March issue of the magazine. The story that had attracted the interest of the military was *Deadline* by Cleve Cartmill. Reading the story now, it is easy to see what triggered their alarm. It was over a year before the Manhattan project would be ready to manufacture the 'Little Boy' uranium 235 bomb. One can therefore imagine their consternation when a science fiction magazine published not only the basic principles of how to make such a bomb but also included a frightening amount of authentic detail! Instead of an atomic bomb requiring tons of uranium – the accepted belief of the scientific community outside of Los Alamos at that time – Cartmill's bomb required 'only a few pounds of U-235'. It is not surprising that the authorities were alarmed: this was the key fact that the Allies wished to keep secret from the Nazis. After the war, interrogation of Heisenberg and the other German nuclear physicists showed that they had indeed been unaware of this discovery. The smallness of the amount of U-235 required made all the difference between the feasibility and impossibility of actually making a uranium bomb. The unexpected discovery of this fact initiated

John Wood Campbell Jr. was indisputably the greatest editor of science fiction and nurtured a whole generation of SF writers including Isaac Asimov and Robert Heinlein. In 1938 he became editor of Astounding and, by his insistence on a much higher standard of writing and his help and support for writers like Asimov, created the modern SF genre. Under the pseudonym of Don A. Stuart he also wrote science fiction himself. Isaac Asimov rates Campbell's story 'Who goes there?' published in *Astounding* in 1938, as 'one of the very best science fiction stories ever written'.

Fig. 13.1 The cover of the March 1944 issue of *Astounding* that contained Cleve Cartmill's story 'Deadline'. In the story, Cartmill reveals the greatest secret of the Second World War, namely, that 'only a few pounds of U235' was required to make an atomic bomb. After the defeat of Germany, Heisenberg and other German nuclear physicists were detained at Farm Hall, near Godmanchester in England. It was there that they heard the news of the atomic bomb dropped on Hiroshima. The conversations between the German scientists were secretly recorded and make clear that Heisenberg initially believed that even a bomb containing as much as 30 kilograms of U235 'wouldn't go off, as the mean free path is too big'. After a week's intensive work, Heisenberg had realized the errors in his thinking and gave a lecture to his colleagues showing how a uranium bomb could, after all, be built. It was just as well that Heisenberg, unlike Werner Von Braun, was not a wartime subscriber to *Astounding!*

the chain of events that led to the establishment of the Manhattan project. It was in 1940, in Birmingham, England, that two German refugees, Otto Frisch and Rudolf Peierls, first calculated the critical mass required for a pure U-235 bomb. Frisch and Peierls were amazed at how small a mass was needed:

> We estimated the critical size to be about a pound, whereas speculations concerned with natural uranium had tended to come out with tons.

Frisch and Peierls were in awe of their own results. They calculated that a pound or so of U-235 would release the equivalent energy of thousands of tons of ordinary explosive. The explosion from one bomb would be large enough to destroy 'the centre of a big city' – the kiloton bomb had arrived. Frisch and Peierls speculated that an isotope separation plant with around 100 000 separation stages could produce 'a pound of reasonably pure uranium-235 in a modest time, measured in weeks'. Although the cost of such a plant seemed daunting, both the refugees were haunted by the fear that Hitler's physicists would get there first. As they said in their report to the UK government, 'even if this plant costs as much as a battleship, it would be worth having'. They were acutely aware of the need for secrecy so Peierls typed their report himself, rather than asking a secretary. Their report became known as the 'Frisch–Peierls Memorandum'. It led directly to Churchill setting up a group called the MAUD Committee to investigate the feasibility of building such an atomic bomb. There is a wonderful Kafka-esque twist to this tale – Frisch and Peierls were at first officially forbidden to read their own report on the grounds that they were enemy aliens!

To return to fiction, as a story 'Deadline' is unconvincing and implausible. The action takes place on another planet with a war between the '*Seilla*' and the '*Sixa*' mirroring the real war being played out on Earth between the *Allies* and the *Axis* powers of Hitler and Mussolini. The story begins with a spy being dropped deep in enemy territory. His mission is to destroy a secret atomic bomb developed by the increasingly desperate Sixa powers. Suddenly, in the pages of this unconvincing tale, one of the biggest military secrets of the Second World War is casually revealed. Cartmill describes how the Sixa had separated several pounds of U-235 using 'new atomic isotope separation methods'. Worse still, his hero explains how a bomb containing only a few pounds of U-235 could release 'as much as a hundred million pounds of TNT' and, 'set off on an island, lay waste the whole island'. The story also contained alarmingly plausible details about how such a uranium bomb might be constructed. The fictional bomb was made of two hemispheres of uranium separated by a neutron-absorbing layer made of cadmium. The two halves were then brought together to form a critical mass for a chain reaction by a small explosive trigger that destroyed the separating layer.

It is unlikely that Cartmill knew anything specific about either the Manhattan project or about the desperate fear that the Germans would get there first – but he clearly knew enough nuclear physics to make some good guesses. There is another curious coincidence. The characters in the story debate whether the explosion of such a bomb could accidentally ignite the atmosphere and destroy the whole planet. In fact, Edward Teller, who was later to lead the development of the hydrogen bomb, had proposed just such a doomsday possibility in 1942. One of the Manhattan Project leaders, Arthur Compton, wrote of the time that Robert Oppenheimer briefed him about this possibility:

> Was there really any chance that an atomic bomb would trigger the explosion of the nitrogen in the atmosphere or the hydrogen in the ocean? This would be the ultimate catastrophe. Better accept the slavery of the Nazis than to run a chance of drawing the final curtain on mankind!

The leader of the theory group, Hans Bethe, was sufficiently concerned that he performed a detailed check of Teller's calculations. Fortunately, Bethe found that Teller had made some unjustified assumptions and that such a catastrophic result was extremely unlikely! The official record of the bomb design programme dismisses the question with the statement: 'The impossibility of igniting the atmosphere was thus assured by science and common sense'. Despite such theoretical reassurance, there must have been some nervousness among the Manhattan physicists who assembled to watch the first nuclear test at the Trinity site in New Mexico. If Cleve Cartmill had heard nothing from any of the physicists involved in the Manhattan project – and surely he would not have published his story if he had – then it is remarkable how accurately he had envisaged the goals of the ultra-secret project. Perhaps just as surprising is the fact that US military intelligence monitored the pages of a science fiction magazine. It is perhaps more probable that some of their officers read it for interest in their spare time and noticed the similarity to the Manhattan project. Nevertheless, John Campbell was relieved that the officers who visited his officers did not notice the map on the wall showing the distribution of subscribers to *Astounding*. Had they done so they would have seen a suspicious looking cluster of pins labelled Post Office Box 1663, Santa Fe, New Mexico – the mail address of Los Alamos during the war. Even worse, one subscriber was the German rocket scientist Werner von Braun, who managed to import his copy of *Astounding* to Germany throughout the war. This is a case where good science fiction can clearly be bad for national security!

Atomic bombs are a powerful symbol of our understanding of the physics of the atom and the nucleus. Nuclear power, although not without its own environmental problems, is another, more positive, symbol. Asimov's famous *Foundation* trilogy was published in *Astounding* between May 1942 and January 1950. The series deals with the decline of the apparently invincible 'Galactic Empire' and of one man's attempt to shorten the length of the inevitable period of barbarism that would descend on the galaxy. The off-stage hero, Hari Seldon, has invented a new science called 'psychohistory', which allows him to predict and shape events in the decline of the Empire. A colony of 'encyclopedists' – the 'Foundation' – has been established on an obscure planet on the remote fringes of the galaxy. The Foundation has the seemingly harmless mission of producing the definitive 'Encyclopedia Galactica' to preserve scientific knowledge during the fall of the Empire predicted by Seldon. In fact, with Imperial rule breaking down in the periphery of the galaxy, the Foundation is forced to survive on its wits. Continued scientific innovation is no longer regarded as an irrelevant academic luxury but as essential for the Foundation's survival. In the star systems of the surrounding war-lords, and even in the heart of the Galactic Empire, the onset of scientific illiteracy is signalled by the loss of control of atomic power. Knowledge of atomic power is used by Asimov as a measure of the health of a civilization. In a modern context, two contemporary SF writers, Larry Niven and Jerry Pournelle, have used this same measure as a symbol of hope for our own civilization. *Lucifer's Hammer* is a modern 'end of the world' SF novel in which civilization is destroyed by a collision between the Earth and a comet. As an example of the state of the world, the action is set in California, which is fought over by bands of modern-day war-lords. In Asimov's trilogy, the Foundation, with its respect for science and mastery of nuclear power, serves as a symbol of hope for humankind in the days of the dying Galactic Empire. In Niven and Pournelle's story, hope for the future is symbolized by a group of scientists restoring power to the San Joaquin Nuclear Power Plant.

It is easy to point to the obvious problems and dangers of nuclear power. The nuclear accidents at Three Mile Island and at Chernobyl have raised awareness of these dangers in dramatic fashion. The 1979 film *The China Syndrome* was released just after the Three Mile Island incident. The basic premise of the film is that if the cooling system of the nuclear reactor fails, an uncontrollable nuclear reaction – a 'meltdown' leading to catastrophic nuclear contamination – could take place in the reactor core. The China syndrome of the title refers to the fanciful idea that temperatures in the reactor core would be high enough for the core to melt its way right through the Earth. Negative images of nuclear power, uranium mining and radioactivity are also present in films like *Silkwood* and *Thunderheart*. It is certainly true that the clean-up costs of the Hanford site in Washington State – where most of the plutonium was produced for the Manhattan Project – are estimated to be many billions of dollars. But there may be some hope on the horizon for the beleaguered nuclear industry. First, it is now obvious that coal- and gas-fired power stations have their own environmental problems and are major contributors to global warming. Second, physicists may be able to devise ways of making nuclear waste less dangerous. Research groups based at Los Alamos and at CERN are investigating 'nuclear transmutation'. The idea is that firing neutrons at dangerous radioactive elements can convert them into much more harmless nuclear waste. For example, reactors around the world produce technetium-99 as a by-product. This has a radioactive half-life of 200 000 years. Adding a neutron to the nucleus converts this to technetium-100 with a half-life of just 16 seconds, decaying into harmless and stable ruthenium-100. Supporters

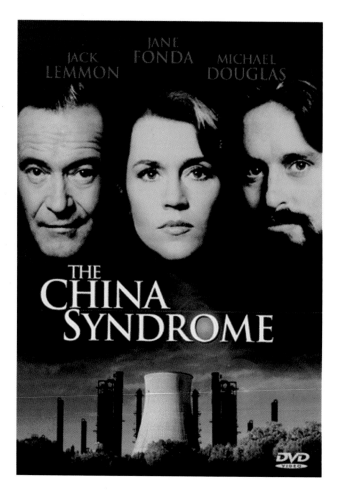

Fig. 13.2 In *The China Syndrome* movie, nuclear power is portrayed as a dangerous technology. The title refers to the fanciful idea that in a meltdown the reactor core could burn its way through the Earth. In Asimov's *Foundation* trilogy, nuclear power has a more uplifting image as a symbol of civilization.

of this approach claim that the volume of dangerous waste could be reduced by a factor of a 100. Furthermore, instead of having to find nuclear waste repositories that will be secure for hundreds of thousands of years, the lifetime required could be reduced to 'mere' hundreds of years. There is still a long way to go before physicists can claim to have solved the nuclear waste problem in an environmentally acceptable way. Nevertheless, with global warming becoming increasingly visible, nuclear power may yet play a significant role for energy generation in the twenty-first century.

Jonbar Points, multiple universes and Schrödinger's cat

Alternate worlds have long been a favourite device of science fiction writers. One version of this type of SF takes the form of a simple 'what if' question. What would have happened if Hitler and the Axis powers had

The novels of Philip K. Dick (1928–1982) are now enjoying a cult revival. Perhaps Dick is best known for his 1968 novel *Do Androids Dream of Electric Sheep?*, which was made into the memorable movie *Blade Runner* by director Ridley Scott. His short story 'We can remember it for you wholesale' was the basis for the movie *Total Recall*. His 1962 novel *The Man in the High Castle* is one of the best 'alternate worlds' SF novels and pictures a USA dominated by the Germans and the Japanese after their victory in the Second World War.

won the Second World War? This is the premise of *The Man in the High Castle*, one of Philip K. Dick's most famous novels. In the book, the USA has been partitioned between the Germans in the East and the Japanese in the West. The plot revolves around a mysterious underground novel called *The Grasshopper Lies Heavy*, which offers an alternative vision of world history in which Hitler and the Japanese were defeated. Which is the real reality? Can both worlds exist? In its own strange way, the book explores the Many World ideas of Everett and Wheeler – although it seems unlikely Dick would have known of this theory.

The 'classic' alternate world story focuses on the moment the choice of futures was made, sometimes called a 'Jonbar Point.' The term arises from a 1938 story by Jack Williamson published in *Astounding* under the title 'The legion of time'. The hero, a Harvard man called Lanning, is visited by two ladies from two different futures – one is from the desirable city of Jonbar and the other from the decadent city of Gyronchi. The two futures are mutually exclusive and only one will be realized – depending on how their potentialities are affected in Lanning's time. Lanning traces the moment of choice back to a boy in a meadow. The boy, John Barr, will either pick up a magnet from the grass and become a great scientist – leading of course to Jonbar – or will pick up a pebble and become a migratory worker. Lanning, battling valiantly with the enemy throughout, eventually manages to cast a magnet at the boy's feet and see 'the very light of science' dawn in the boy's eyes. This is the origin of the Jonbar Point. Such stories put forward the optimistic view that individual human actions can change history. The future is in our hands. The movie *Sliding Doors* is a recent update on this theme: in one future the heroine catches the subway train and in the other she does not.

Another type of alternate world novel assumes that all possible choices can be made at a Jonbar Point and that each choice generates a new 'parallel' universe. There are many variants of such multiple universe stories but most usually depend on some mechanism for crossing between the different universes. Clifford D. Simak's *Ring Around the Sun* and Keith

Fig. 13.3 Poster from the 'alternate universe' movie *Sliding Doors*. The 'Jonbar' Point for the heroine, played by Gwyneth Paltrow, is missing or catching a train.

Fig. 13.4 The UK TV series *Red Dwarf* is an ironic antidote to the earnestness of US series such as *Star Trek*. The episode 'Dimension jump' has the familiar 'loser' version of hologram Arnold Rimmer meeting his 'winner' duplicate Ace Rimmer. The 'Jonbar' Point could be traced back to the moment in childhood when the Ace version of Rimmer was held back a year at school. This picture shows Arnold with Ace, his quantum clone.

Laumer's *Worlds of the Imperium* both imagine a whole series of parallel but different 'Earths'. Much of Michael Moorcock's work takes place in what he calls the 'multiverse', an infinite series of alternate universes. Multiple universes also provide the basis for comedy in the British science fiction TV series *Red Dwarf*. Here, all the usual SF trappings are used merely as an excuse to generate new comic situations – an antidote to the seriousness of *Star Trek*. An example is the episode 'Dimension Jump'. Here, the 'no-hoper' version of Arnold Rimmer meets his alternate version 'Ace' Rimmer. The two Rimmers trace their 'Jonbar Point' back to the moment when one of them had to repeat a year in school. By contrast, the *Next Generation* series of *Star Trek* takes the topic more seriously. In an episode called 'Parallels', Worf's ship traverses a 'quantum fissure in space-time' and the barriers between different quantum realities 'break down'. As Worf jumps from one branch of the wavefunction to another, he experiences different quantum realities. Data sums up the many world interpretation of quantum mechanics with the statement: 'All things which can occur, do occur'.

    Is there any scientific basis for such multiple universes? In chapter 7 we introduced the 'measurement problem' in quantum mechanics. One proposed resolution of this problem is the many world interpretation of quantum mechanics due to Hugh Everett. Instead of an electron wavefunction collapsing to one particular point on measurement, Everett supposed that all possible results of the measurement happen, but each in a different parallel universe. Since measurements on quantum systems take place all the time, this leads to an incredibly large and continually increasing number of different universes. Although such an 'explanation' of the measurement problem has some attractions, the theory seems to have little or no predictive content, since one cannot explore or interact with these extra universes. David Deutsch, a physicist from Oxford, takes a more positive view of the multiverse theory. Deutsch is justifiably regarded as one of the pioneers of quantum computing. He was first to show how quantum computers can use 'quantum parallelism' – many possible computational paths being followed simultaneously – to achieve results more quickly than can be obtained with a conventional computer. Deutsch believes that the existence of quantum interference and quantum parallelism is only understandable from the many worlds quantum viewpoint. Other physicists disagree!

As we have seen, Erwin Schrödinger illustrated his problems with measurements, observers and quantum theory by reference to an experiment on his unfortunate cat. A few science fiction writers have boldly ventured to explore such quantum measurement problems. One of the first SF writers to take the problem as the basis for a story was the astronomer Fred Hoyle. Hoyle was one of a small number of successful research scientists who also wrote successful science fiction. Although his research on the problem of formation of the different elements in stars nearly gained him a Nobel Prize, Hoyle is probably best known to the general public as one of the creators of the 'Steady State' model of the universe. This theory accounted for the observed expansion of the universe by proposing that matter is continually being created. For some years, the Steady State theory was a serious rival to the alternative Big Bang theory. Indeed the name 'Big Bang' was coined by Hoyle as a derisive name for this theory in a radio interview. Nowadays, scientists believe that new evidence accumulated in the past 20 years or so points overwhelmingly towards the Big Bang theory. Hoyle's first successful science fiction novel was called *The Black Cloud* and was published in 1957. A sentient cloud of gas had arrived in the solar system and obscured the Sun, accidentally threatening all life on Earth. All of Fred Hoyle's deep distrust of the 'establishment' – both in science and in politics – are apparent in his portrayal of the hero scientist's struggles against red tape and bureaucracy. More relevant to our discussion is a short story published in a collection called *Element 79*. The story in question is titled 'A jury of five' and is interesting in that it addresses the measurement problem of quantum mechanics in a novel way. The plot centres on a car crash in which the police find only one body and are not sure which of the drivers is dead. The other has apparently taken a blow to the head and wandered off into the countryside in a confused state. The story is told from the ghostly point of view of the two drivers. They are able to hear and see what the police and their family and friends are doing but are unable to interact with any of them. The final scene is in the morgue. Five people connected with the two men are present when the sheet is lifted from the face of the body on the slab. One of the disembodied drivers, Adams, is a reticent and shy professor of philosophy in Oxford; the other is a hard-driving businessman who was cheating on his wife. Adams has eventually realised what must be happening:

> On their way into the city, Adams remarked, 'I believe I've got it straight at last. One of us is going to be found under that sheet,

Fig. 13.5 The cover of Fred Hoyle's *Element 79* collection of SF short stories. The title story is about a meteorite of pure gold, the element of the title, devastating much of Scotland but reviving the UK economy by giving the Government control of the world gold market. In his story 'A jury of five' Hoyle explores the role of the observer in quantum mechanics with a human version of Schrödinger's cat paradox.

dead. The other is going to be found wandering around the countryside, alive.'
'I don't bloody well understand.'
'I don't think it is decided yet, whether it's you or me.'
'How d'you mean?'
'It's going to depend on what they want.'

The jury of five go about their business and vote and Adams winds up dead. Hoyle explains as follows:

The decision rested on Adams, on his split-second reaction to Hadley's car blundering in front of him. Now, Adams' split-second reaction depended on electronic neurological activity in his brain, which in the last analysis turned on a single quantum event, on whether the event took place or not. Until the winding sheet was whipped away from the body in the morgue, the wave function representing the event was still in what physicists call a 'mixed state.' Let it be added, for the sake of the smart physicist, that a clue to the solution of the deepest problem of theoretical physics – the condensation of the Schrödinger wave function – is to be found in the manner in which our jury of five arrived at their decision.

Hoyle has constructed an intriguing re-working of the paradox of Schrödinger's cat. On this occasion, the corpse is not observed by the single friend of Wigner but by a majority vote by five distinct consciousnesses. Both the famous mathematician John Von Neumann and the physicist Eugene Wigner felt logically forced to take the position that the reduction of the wave packet takes place in the observer's consciousness. John Wheeler went further when he speculated:

> May the universe in some strange sense be 'brought into being' by the participation of those who participate? 'Participation' is the incontrovertible new concept given by quantum mechanics. It strikes down the 'observer' of classical theory, the man who stands safely behind the thick glass wall and watches what goes on without taking part.

Hoyle's story certainly highlights the problem of measurement. It also provides an interesting twist for those who believe that the role of a conscious observer is the way out of the difficulties of quantum mechanics.

Nanotechnology and quantum computers

What new quantum technology can SF writers of the present turn to for inspiration? Few writers have taken up the challenge of nanotechnology. Two exceptions are Greg Bear with his novel *Queen of Angels* and Neal Stephenson with *The Diamond Age*. Greg Bear's *Queen of Angels* is set in Los Angeles at the turn of the 'binary millennium' 2048. The intricate storyline inter-weaves the investigation of a murder with news of the nano-robotic exploration of planets in the Alpha Centauri system. The possibilities of extra-terrestrial intelligence and self-aware computers serve as sub-plots to the main action, which is concerned with mind manipulation, exploration and punishment. The nano-surgeon character describes the techniques he has developed to explore what he calls the 'Country of the Mind' – in this case the disturbed mind of a famous poet turned multiple killer:

> The advent of nano therapy – the use of tiny surgical prochines to alter neuronal pathways and perform literal brain restructuring – gives us the opportunity to fully explore the Country of the Mind. I could not find any method of knowing the state of individual neurons in the hypothalamic complex without invasive methods

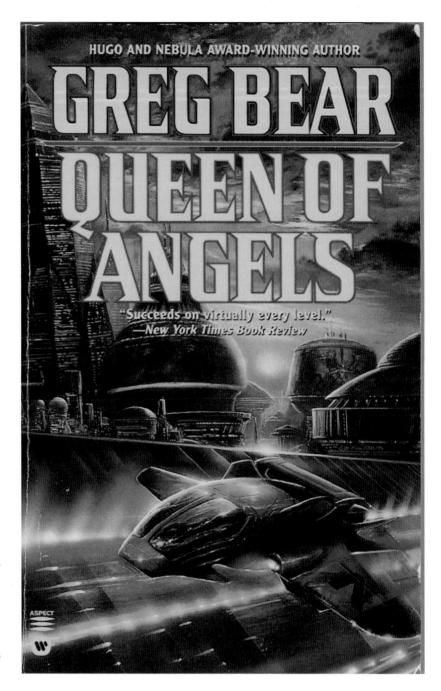

Fig. 13.6 The cover of Greg Bear's *Queen of Angels* first published in 1990. The novel is a murder story set against a backdrop of nano-technological miracles. As the author says in a footnote to the novel, 'The nanotechnology described here is highly speculative' and he refers to K. Eric Drexler's visionary book *The Engines of Creation.*

such as probes ending in a microelectrode, or radioactively tagged binding agents – none of which would work for the hours necessary to explore the Country. But tiny prochines capable of sitting within an axon or neuron, or sitting nearby and measuring the neuron's state, sending a tagged signal through microscopic 'living' wires to sensitive external receivers ... I had my solution. Designing and building them was less of a problem than I expected; the first prochines were nano therapy status-reporting units, tiny sensors which monitored the activity of surgical prochines and which did virtually everything I required.

Similar nanotechnological miracles are deployed in the unmanned 'AXIS' space probe to the Alpha Centauri system:

The AXIS 'mind' consists of a machine system and a biological system. During the years when AXIS accelerated on a furious torch of matter–antimatter plasma, the unmanned interstellar probe was controlled by a primitive, rugged and radiation proof inorganic computer.... Some six months before the beginning of AXIS's deceleration phase, AXIS allowed itself the luxury of powering up a small fusion generator, very little larger than a human thumb. This produced sufficient heat to allow nano-machine activity, and the creation of AXIS's huge, yet very thin and light superconducting wings.... AXIS waited for the proximity of Alpha Centauri B to begin to grow its biological thinker system.

The action switches from scene to scene with a pervasive backdrop of nano-technology – nano-wood, nano-cube books, nano-food and nano-perfume. The police also deploy an impressive array of technological gadgets – nano-molecular body armour, forensic robotic dustmice, nano-watchers embedded in the paint and flechette darts designed to change shape and burrow into a wound. When it comes to concealed weapons, nanotechnology gives these a new dimension:

She patiently watched the nano at work. The metal tubing of the bootrack had crumpled under the gray coating. The resulting pool of paste and deconstructed objects was contracting into a round complexity. Nano was forming an object within that convexity like an embryo with an egg.... The convexity grew lumpy now. She could make out the basic shape. To one side, excess raw material was being pushed into lumps of raw slag. Nano withdrew from the slag. Handle, loader, firing chamber, barrel and flightguide. To one side of the convexity a second lump not slag was forming. Spare clip.

A similar view of the future is to be found in Neal Stephenson's novel *The Diamond Age*, subtitled *A Young Lady's Illustrated Primer*. In this tale, nano-technology is used for art and recreation, for feeding and clothing the masses, for nano-warefare between clouds of 'smart' fog, and for the intelligent and interactive 'primer' of the title. The primer is an illicit subversive miracle of technology that teaches the reader everything from mythology and science to martial arts and techniques for survival in an often hostile environment:

A leaf of paper was about a hundred thousand nanometers thick; a third of a million atoms could fit into this span. Smart paper

Fig. 13.7 The cover of *The
Diamond Age* by Neal
Stephenson. The alternate
title is *A Young Lady's
Illustrated Primer*, referring to
the intelligent book
crammed with
nanotechnological miracles
that found its way to Nell,
the book's heroine of sorts.

consisted of a network of infinitesimal computers sandwiched
between mediatrons. A mediatron was a thing that could change its
color from place to place; two of them accounted for about
two-thirds of the paper's thickness, leaving an internal gap wide
enough to contain structures a hundred thousand atoms wide. Light
and air could easily penetrate to this point, so that works were
contained within vacuoles – airless buckminsterfullerene shells
overlaid with a reflective aluminium layer so that they would not
explode *en masse* whenever the page was exposed to sunlight. The
interiors of the buckyballs, then, constituted something close to a
eutactic environment. Here resided the rod logic that made the
paper smart. Each of these spherical computers was linked to its four
neighbors, north–east–south–west, by a bundle of flexible pushrods
running down a flexible, evacuated buckytube, so that the page as a

whole constituted a parallel computer made up of about a billion separate processors.

The primer designed by the nano-engineer John Hackworth was many times smarter than such smart paper. It was produced by a 'matter compiler' that plucks atoms from a conveyor one at a time, according to the instructions of a program, to assemble the desired structure. A world populated with nano-devices requires some adjustments to our present ways of thinking:

> Aerostat meant anything that hung in the air. This was an easy trick to pull off nowadays. Computers were infinitesimal. Power supplies were much more potent. It was almost difficult not to build things that were lighter than air. Really simple things like packaging materials – the constituents of litter, basically – tended to float around as if they weighed nothing, and aircraft pilots, cruising along ten kilometres above sea level, had become accustomed to the sight of empty, discarded grocery bags zooming past their windshields (and getting sucked into their engines).

Such technology also has a sinister side. Harv, the street-wise brother of Nell, the heroine, explains to her why the sky had suddenly turned a leaden colour:

> 'Mites,' he said 'or so they say down at the Flea Circus anyway.' He picked up one of the black things taken from the mask and flicked it with a fingertip. A cineritious cloud swirled out of it, like a drop of ink in a glass of water, and hung swirling in the air, neither rising or falling. Sparkles of light flashed in the midst of it like fairy dust. 'See, there's mites around all the time. They use the sparkles to talk to each other, 'Harv explained. 'They're in the air, in food and water, everywhere. And there's rules that these mites are supposed to follow, and those rules are called protocols. And there's a protocol from way back that says they're supposed to be good for your lungs. They're supposed to break down into safe pieces if you breathe one inside of you ... But there are people who break these rules sometimes. Who don't follow the protocols. And I guess if there's too many mites in the air all breaking down inside your lungs, millions – well maybe those safe pieces aren't so safe if there's millions. But anyways, the guys at the Flea Circus say that sometimes the mites go to war with each other.... This dust – we call it toner – is actually the dead bodies of all those mites.

So how realistic are these scenarios? As we saw in chapter 8, nano-engineering is slowly becoming a reality but clearly has a long way to go to reach this level of sophistication. At present, we are only in the early phases of acquiring sufficient control over atomic matter to be able to assemble bespoke artificial atoms and other nano-systems. Much more is needed before we can construct nano-systems that are able to run a program or exert some degree of control over their own environment. In Eric Drexler's vision of

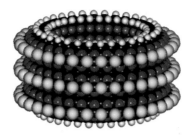

Fig. 13.8 A macromolecular sleeve bearing designed by K. Eric Drexler and Ralph C. Merkle, two of the most energetic advocates for molecular nanotechnology. Drexler's vision of molecular nanotechnology, *The Engines of Creation*, was published in 1986 but it was not until 1991 that Drexler was awarded his Ph.D. from MIT. The Electrical Engineering and Computer Science Department had refused to accept Drexler as an Interdisciplinary Ph.D. student so Marvin Minsky enrolled him in MIT's famous Media Lab. Although Drexler admits that his vision of the future is some decades away from realization, his colleague Ralph Merkle has now left the Xerox PARC research laboratory to work for a nanotechnology start-up company.

molecular nano-technology, the key advance required to realize his dream is the ability to construct a 'self-assembling system' or 'assembler' – a nano-system that can reliably reproduce copies of itself or run a program that tells it how to build other specific nano-systems. Some first steps have been taken in this direction but we are still a long way from realizing Drexler's vision. When – or if – nano-technology reaches this level, Drexler sees a utopian world without famine, since we can build food to order from any raw materials, without energy shortages, since we can build tiny efficient photo-cells for next to nothing, and without disease, since tiny robotic nano-surgeons can be injected into the bloodstream to eliminate diseased cells or viruses. It sounds too good to be true and, of course, it may indeed turn out to be a hopeless dream. But it is a future worth trying for, and, as Feynman said in his 1959 talk, nanotechnology is 'an invitation to enter a new field of physics'. Forty years later, apart from the efforts of still relatively small numbers of pioneering scientists, the original invitation of Feynman and the utopian vision of Drexler are still exciting challenges for the next generation of scientists and engineers.

What is left for science fiction to explore? To conclude this chapter we turn to Michael Crichton's novel *Timeline*. In *Jurassic Park* and *The Lost World*, Michael Crichton took up the challenge of genetic engineering: in

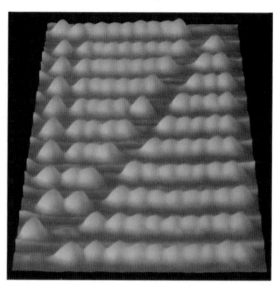

Fig. 13.9 A molecular abacus created by Jim Gimzeweski and his group at IBM Zurich. The 'beads' are buckyballs – the newly discovered stable form of carbon. The abacus can be manipulated at room temperature using an STM. Although this is an impressive achievement there is still a long way to go before we are able to create true nanoscale computers and other devices.

Timeline he turns to quantum teleportation and quantum computing. The novel blends futuristic quantum technology with time travel and medieval history to form a compelling tale. The setting of the novel is New Mexico, close to the Los Alamos National Laboratory, in our present reality one of the leading centres of research into quantum cryptography and quantum computing. A high-tech start-up company called the International Technology Corporation or ITC is mysteriously sponsoring historical research into the Hundred Years War between England and France. The action switches between the Dordogne excavation site in France and the ITC headquarters in New Mexico. The professor leading the excavation is called back to ITC to see the company president. Back in France, after some days without news from him, his students unearth an apparently medieval request for help from their professor! The students are flown to New Mexico to form a rescue party – to be sent back into the past to rescue him. The ITC executives have to explain their new technology to the sceptical students:

> Ordinary computers make calculations using two electron states, which are designed one and zero. That's how all computers work, by pushing round ones and zeros. But twenty years ago Richard Feynman suggested it might be possible to make an extremely powerful computer using all thirty-two quantum states of an electron. Many laboratories are now trying to build these quantum computers. Their advantage is unimaginably great power – so great that you can indeed describe and compress a three-dimensional living object into an electron stream. Exactly like a fax. You can then transmit the electron stream through a quantum foam wormhole and reconstruct it in another universe. And that's what we do. It's not quantum teleportation. It's not particle entanglement. It's direct transmission to another universe.

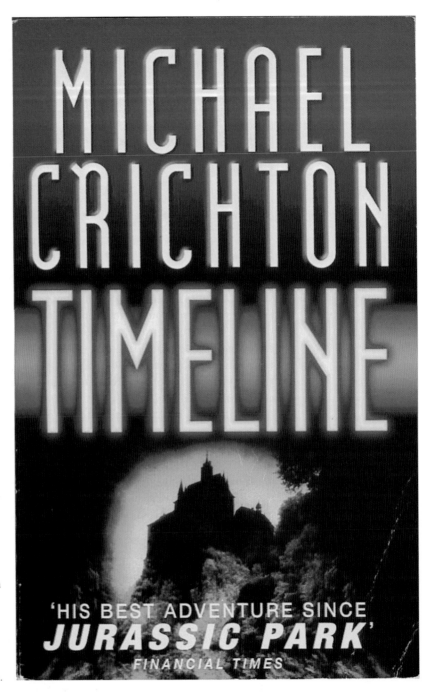

Fig. 13.10 The cover of Michael Crichton's *Timeline* – a novel that combines quantum teleportation with time travel. Medieval archaeology students are transported back in time to rescue their professor from a battlefield of the Hundred Years War in France. As reference for the science in Crichton's book he cites *The Fabric of Reality: The Science of Parallel Universes and Its Implications* by David Deutsch.

As we have seen, quantum computing is beginning to become a reality and small-scale quantum teleportation experiments have actually been done. What we have here is intelligent extrapolation beyond our present knowledge, mixing interesting quantum technologies – quantum computing,

Fig. 13.11 An artist's impression of a future 'New York Quantum Transit Authority'. Michael Crichton envisages a similar quantum teleportation system being used for time travel in his novel *Timeline*.

teleportation, entanglement, the quantum multiverse and wormholes – in a plausible sounding way. A false note is the reference to 'thirty-two quantum states of the electron' but perhaps this is meant to be a little puzzling! One of the hallmarks of good science fiction writing is the ability to look more than just one step ahead. Here, Crichton not only recognizes the vast computational task that would be involved in storing and compressing the informational equivalent of a person but also he has a solution:

'...you'd need massive parallel processing,' Gordon said, nodding ...
'You hook several computers together and divide the job up among

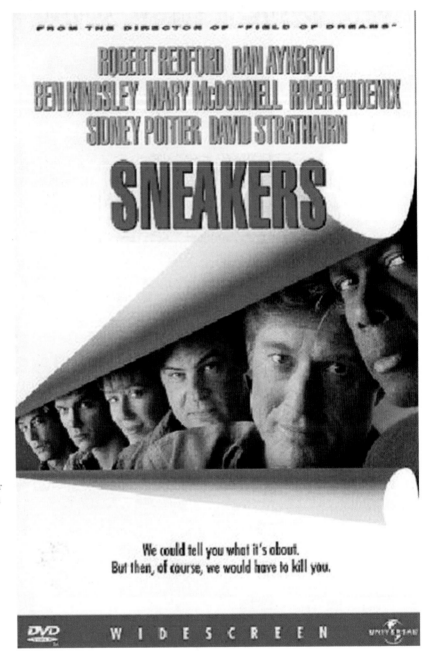

Fig. 13.12 Len Adleman is a Computer Science Professor in Los Angeles and is the 'A' in the RSA system for data encryption. The RSA system is based on the fact that factorization of very large numbers is computationally very difficult. If a new fast algorithm could be found for factorization – or a quantum computer constructed – much RSA encrypted data would be vulnerable to attack. Adleman was asked for advice on cryptography for the movie *Sneakers* starring Robert Redford. Much of the action is centred on the theft and retrieval of a 'quantum computer-like' box.

them, so it gets done faster. A big parallel-processing computer would have sixteen thousand processors hooked together. For a really big one, thirty-two thousand processors. We have **thirty-two billion** processors hooked together.' . . . Gordon sat and smiled. He was looking at Stern waiting. 'The only possible way to do that much processing,' Stern said, 'would be to use the quantum characteristics of individual electrons. But then you'd be talking about a quantum computer. And no one's ever made one.' Gordon just smiled. 'Have they?' Stern said.

Crichton also puts his finger on one of the potential problems for quantum computing. Quantum information is stored as delicate differences in quantum superpositions and, as for ordinary classical computers, the stored information is subject to errors. In ordinary computer memory, stray cosmic ray particles pass through the system and occasionally randomly flip a zero to a one or a one to a zero. For this reason, computer engineers have developed error detection and correction schemes that make it possible to find and correct for such errors. For quantum computers there are many more possibilities for errors to creep in, although it has been shown that it is possible – in principle – to correct for such errors. In transmitting something as complex as a human being through time and back again, it is obviously important for the transmitted human that the information does not change. In the development of their quantum transportation system, ITC have had problems with just such 'transcription errors' and Wellsey, the cat, is a living warning:

> 'Wellsey's split,' Kramer said to Stern. 'He was one of the first test animals that we sent back. Before we knew that you had to use water shields in a transit. And he's very badly split.'
> 'Split?'
> Kramer turned to Gordon. 'Haven't you told him anything?' 'Of course I told him,' Gordon said. He said to Stern, 'Split means he had very severe transcription errors.'

The use of the phrase 'transcription errors' to describe life-threatening deformations seems reminiscent of the military describing civilian casualties as 'collateral damage'. All in all, these speculations on quantum computing make an exciting backdrop to *Timeline*. Few authors can combine real, frontier scientific research results and place them in the context of an intelligent and imaginative SF drama as well as Michael Crichton. These skills are on display again in *Prey*, a book that explores the potentially dangerous downside of nanotechnology when combined with self-adaptive intelligent agent systems.

A final word

From our brief survey we have seen that although quantum physics pervades much of current SF, it is rarely centre stage. Nonetheless, quantum jumps and wave particle duality are now part of our everyday language – and the puzzling nature of quantum phenomena has even become the subject of some novels and plays. In the play *Hapgood*, the author, Tom Stoppard, plays on the fact that double agents exhibit a type of wave particle duality

Fig. 13.13 For the frontispiece to his play *Hapgood*, Tom Stoppard uses a quotation from Feynman: 'We choose to examine a phenomenon which is impossible, absolutely impossible, to explain in any classical way, and which has in it the heart of quantum mechanics. In reality it contains the only mystery...' 'Any other situation in quantum mechanics, it turns out, can be explained by saying, "You remember the case of the experiment with the two holes? It's the same thing."' In the play, Kerner, a physicist double – or triple agent – says to Blair, the spycatcher, 'You get what you interrogate for.' This is very reminiscent of our discussion of the problems of quantum measurement.

like light. The physicist tells the spycatcher: 'A double agent is like a trick of the light. You get what you interrogate for.' The same search for the 'true' version of events appears in Michael Frayn's recent play *Copenhagen*. Here, the action focuses on the events that occurred in an interview between Heisenberg, the famous quantum physicist who stayed behind to work in Nazi Germany, and Niels Bohr, his Danish mentor in quantum theory, that took place in Nazi-occupied Copenhagen. But in these plays quantum mechanics is really only used as a metaphor for the complexities of interactions between people. In the novels by Bear, Stephenson and Crichton, our entire future has come to be dominated by quantum technology. This century will see major advances in both bioinformatics and genomics. Equally, if only a part of the dreams of nanotechnologists come true, it is likely that our

Fig. 13.14 The programme cover for the premier of Michael Frayn's play Copenhagen at the Cottesloe Theatre, London. The play is centred on a real event: Heisenberg did go to Copenhagen in 1941 to meet with Bohr and they did discuss the possibility of atomic weapons. In their published recollections of the meeting they disagree about the content and even the location of their conversation. Although Heisenberg was neither a Nazi nor a Nazi sympathizer, he undoubtedly believed that it was important for Germany to win the war. According to Stefan Rozental, on that same visit Heisenberg had said that 'the occupation of Denmark, Norway, Belgium and Holland was a sad thing but as regards the countries in East Europe it was a good development because these countries were not able to govern themselves'. Reportedly, Christian Moller replied 'So far we have only learned that it is Germany which cannot govern itself'. After the war, another visit by Heisenberg to Bohr was arranged during which, like in the play, 'the two men spent an evening trying to reconstruct without success their discussion in 1941'. This is a case in which Heisenberg's uncertainty principle certainly applied!

engineers of the future will be true 'quantum engineers'. It is not for nothing that the physicist Paul Davies has said that:

> The nineteenth century was known as the machine age, the twentieth century will go down in history as the information age. I believe that the twenty-first century will be the quantum age.

Epilogue

A poet once said 'The whole universe is in a glass of wine'. We will probably never know in what sense he meant that, for poets do not write to be understood. But it is true that if we look at a glass closely enough we see the entire universe. There are the things of physics: the twisting liquid which evaporates depending on the wind and weather, the reflections in the glass, and our imagination adds the atoms. The glass is a distillation of the Earth's rocks, and in its composition we see the secret of the universe's age, and the evolution of the stars. What strange array of chemicals are there in the wine? How did they come to be? There are the ferments, the enzymes, the substrates, and the products. There in wine is found the great generalization: all life is fermentation. Nobody can discover the chemistry of wine without discovering, as did Louis Pasteur, the cause of much disease. How vivid is the claret, pressing its existence into the consciousness that watches it! If our small minds, for some convenience, divide this glass of wine, this universe, into parts – physics, biology, geology, astronomy, psychology, and so on – remember that Nature does not know it! So let us put it all back together, not forgetting ultimately what it is for. Let it give us one more final pleasure: drink it and forget it all!

Richard Feynman

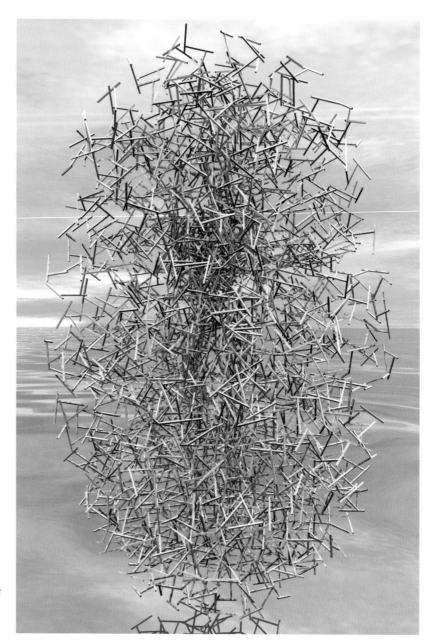

Artist Antony Gormley worked with engineers Neil Thomas and Gary Eliot to create this 33 m high sculpture 'Quantum Cloud'. It is formed from 4000 identical 1.6 m long sections of galvanized steel and is evocative of a quantum mechanical probability distribution. It stands on four cast iron caissons in the River Thames close to the Millenium Dome and the Isle of Dogs.

Appendix I
The size of things

Powers of ten

The distance scales that we encounter in the quantum world are much smaller than anything we meet in everyday life. Similarly, the distances involved in any discussions of stars and galaxies are enormously greater than distances between distant places on Earth. A convenient way of writing distances that are either very small or very large is to use a notation based on powers of ten.

Large numbers can be written in terms of a power of ten as follows:

$$\text{ten} = 10 = 10^1$$
$$\text{one hundred} = 100 = 10 \times 10 = 10^2$$
$$\text{one thousand} = 1000 = 10 \times 10 \times 10 = 10^3$$

To see how this works, consider how we would write a very large number like the velocity of light in this notation. The velocity of light is usually denoted by the symbol c and is about 300 million metres per second:

$$c = 300\,000\,000\,\text{m/s} = 3 \times 10^8\,\text{m/s}$$

The advantage of using the powers of ten notation is evident.

Small numbers can be expressed in a very similar way:

$$\text{one-tenth} = 1/10 = 10^{-1}$$
$$\text{one-hundredth} = 1/100 = 1/10 \times 1/10 = 10^{-2}$$
$$\text{one-thousandth} = 1/1000 = 1/10 \times 1/10 \times 1/10 = 10^{-3}$$

The scale of quantum objects is typified by Planck's constant. This is usually denoted by the symbol h and has a magnitude of about 4.2 thousand trillionths of an 'electron volt second':

$$h = 4.2/1\,000\,000\,000\,000\,000\,\text{eV s}$$

In powers of ten notation this is much less cumbersome:

$$h = 4.2 \times 10^{-15} \, \text{eV s}$$

We shall now use this notation to give you some idea of the scale of the various objects that appear in this book.

Mass scales

A familiar unit of mass is the kilogram or 'kg'. This is roughly two pounds in the old non-metric units. Atoms and nuclei are very much lighter than this and a kilogram is a very large and inconvenient unit for such masses. The basic building blocks of atoms and nuclei are protons, neutrons and electrons and these have approximately the masses shown below:

$$\text{proton and neutron mass} = 1.7 \times 10^{-27} \, \text{kg}$$
$$\text{electron mass} = 9.1 \times 10^{-31} \, \text{kg}$$

These very tiny masses contrast with the enormous masses of the planets and stars. Again, a kilogram is not a very convenient unit for such masses as can be seen from the following examples:

$$\text{mass of Earth} = 6 \times 10^{24} \, \text{kg}$$
$$\text{mass of Jupiter} = 2 \times 10^{27} \, \text{kg}$$
$$\text{mass of the Sun} = 2 \times 10^{30} \, \text{kg}$$

Distance scales

In everyday life, the normal unit of length is the metre. This is about the height of a small child and is just over three feet in non-metric units. In Table A1.1 we contrast the scales of things encountered in this book.

Speed scales

An ordinary walking pace is about one metre per second or about 2.5 miles per hour. Table A1.2 compares the speeds of various objects up to the speed of light, according to Einstein a speed that cannot be exceeded.

Electromagnetic spectrum

Light is a type of electromagnetic radiation corresponding to a wave of fluctuating electric and magnetic fields. The wavelengths of visible light form only a very small part of the whole spectrum of electromagnetic radiation. We list in Table A1.3 the typical wavelength ranges of the different

Table A1.1 *Typical distance scales*

Large scales		Small scales	
10^4 m = 10 km	Neutron star, black hole	10^{-4} m	Smallest object visible to naked eye
10^7 m = 10^4 km	White dwarf, Earth	10^{-6} m	Smallest object visible under light microscope
10^8 m = 10^5 km	Jupiter	10^{-8} m	Large molecules
10^9 m = 10^6 km	Sun, normal stars	10^{-10} m	Atoms
10^{11} m = 10^8 km	Red giant, Earth–Sun distance	10^{-14} m	Nucleus
		10^{-15} m	Proton, neutron
10^{16} m = 1 light year (ly)	Distance travelled by light in one year, distance to nearest stars		
10^{21} m = 10^5 ly	Size of our Galaxy		
10^{21} m = 10^7 ly	Clusters of galaxies		
10^{26} m = 10^{10} ly	Most distant galaxies, quasars		

m = metres.

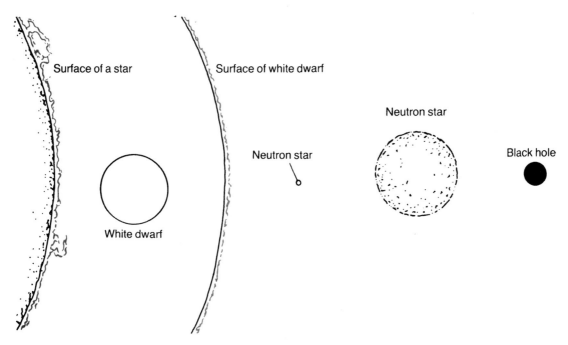

Fig. A1.1 Relative sizes of stars.

Table A1.2 *Speed scales*

3 m/s	Reasonable jogging pace
200 m/s	Jet aircraft
300 m/s	Speed of sound in air
10 000 m/s	Escape velocity for rocket
30 000 m/s	Speed of Earth round Sun
2×10^7 m/s	Typical speed of electron in television tube
3×10^8 m/s	Speed of light in vacuum

m/s = metres per second.

Table A1.3 *Electromagnetic spectrum*

Type of radiation	Typical wavelength range	Typical photon energy
Gamma	10^{-11}–10^{-14} m	10^6 eV = 1 MeV
X-ray	10^{-8}–10^{-11} m	10^3 eV = 1 keV
Ultraviolet	4×10^{-7}–10^{-8} m	10 eV
Visible	8×10^{-7} (red)–4×10^{-7} m (violet)	1 eV
Infrared	10^{-4}–4×10^{-7} m	10^{-1} eV
Microwave	1–10^{-4} m	10^{-4} eV
Radio	10^{-3} m and longer	10^{-8} eV

m = metres; eV = electron volts.

Table A1.4 *Time intervals*

Long times		Short times	
10^3 s	Time for light from Sun to reach Earth, neutron lifetime	10^{-6} s	Lifetime of a muon
3.1×10^7 s	One year	10^{-9} s	Typical lifetime of atomic excited state
10^9 s	Human lifetime	10^{-10} s	Time for light to travel one foot (30 cm)
10^{11} s	Origin of city civilizations in the Near East, lifetime of carbon 14	10^{-15} s	period of oscillation of visible light
10^{14} s	First hominids	10^{-18} s	Time for light to cross atom
10^{17} s	Origin of the solar system, lifetime of uranium 238	10^{-24} s	Time for light to cross nucleus
10^{18} s	Origin of the universe		

s = seconds.

types of electromagnetic waves. We have also listed the corresponding photon energy associated with each type of radiation, which reminds us that light interacts in a quantum manner.

Time intervals

A convenient unit for time intervals in normal life is a second – roughly one heartbeat. In Table A1.4 we contrast long and short time intervals.

Appendix 2
Solving the Schrödinger equation

This appendix is intended for readers with some knowledge of calculus – mainly how to differentiate and integrate sines and cosines. We consider the problem of finding the quantum probability amplitudes for a particle 'in a box'. This involves solving the Schrödinger equation for a particle of some energy E, confined to a certain region by an electrical potential V:

$$E\,\psi(x) = -\frac{\hbar^2}{2m}\frac{d^2\psi(x)}{dx^2} + V(x)\psi(x)$$

We consider the case where the electron can only move in one space dimension, rather than in three dimensions as in real life. In spite of the apparent artificiality of this situation, this example illustrates many of the features that occur in more realistic problems. A straightforward three-dimensional generalization of this example can be used to illustrate many features of the quantum physics not only of electrons in a metal, but also of neutrons and protons in the nucleus.

The one-dimensional box potential is shown in Fig. 4.8. Outside the box the potential is assumed to be infinitely high so that there is no probability of finding the particle in these regions. The particle is constrained to be within the box, i.e. the quantum amplitude can only be non-zero between $x = 0$ and $x = L$, the two ends of the box. Inside this region, the particle can move freely: in other words, the quantum amplitude for the particle in the box is determined by the Schrödinger equation with no potential term:

$$E\,\psi(x) = -\frac{\hbar^2}{2m}\frac{d^2\psi(x)}{dx^2}$$

To see what form of solution is allowed, we first 'tidy up' the equation by introducing the new variable k:

$$k^2 = \frac{2mE}{\hbar^2}$$

In terms of k, the Schrödinger equation for the particle in the box now reads:

$$\frac{d^2\psi(x)}{dx^2} = -k^2\psi(x)$$

which shows the mathematical structure more clearly. Differentiating the wavefunction twice has to give the same function back again, multiplied by $-k^2$. This equation is well known in classical physics. It is the equation for simple harmonic motion and the solutions are either sines or cosines or a mixture of both of them. Thus, the most general form of the quantum wavefunction in this case is a sum of sine and cosine terms:

$$\psi(x) = A\sin(kx) + B\cos(kx)$$

where A and B are arbitrary constants. This is the general form of the solution; we want the solution for our particular problem. We therefore require the quantum amplitude to vanish for all values of x less than zero and greater than L. The solution we want must satisfy these 'boundary conditions'. Imposing the boundary condition at $x = 0$ requires:

$$\psi(x = 0) = A\sin(0) + B\cos(0) = 0$$

This means that the constant B must be zero, since the sine of zero vanishes but the cosine of zero does not. Now we know that $B = 0$, the boundary condition at $x = L$ gives us the condition:

$$\psi(x = L) = A\sin(kL) = 0$$

In order to have a non-zero wavefunction at all, we need A to be non-zero. This boundary condition therefore implies that $\sin(kL)$ must vanish in order for us to have an acceptable solution to our problem. This will happen when kL is an integral number times π radians (π radians $= 180°$). Thus, the allowed wavefunctions for our problem are those for which k satisfies the condition:

$$k = n\pi/L \ \text{ for } n = 1, 2, 3\ldots$$

Since k is related to the energy E by the equation:

$$E = \hbar^2 k^2 / 2m$$

we see that only certain values of the energy are allowed. Substituting the condition for k, we find that the energy is 'quantized' and its magnitude determined by the quantum number n according to the formula:

$$E = n^2 \frac{\pi^2 \hbar^2}{2mL^2} \ \text{ for } n = 1, 2, 3\ldots$$

These are the energy levels for the box potential shown in chapter 3 along with their associated sine wavefunctions. All we have done in solving the Schrödinger equation is to determine which wavelengths can fit in a box of length L. This is the origin of energy quantization.

The constant A may be determined as follows. According to Max Born, the probability of finding the particle at any given position within the box is given by the square of the wavefunction at that point. Since the sum of the probabilities of finding the particle at all possible positions within the box must add up to unity, we must have a 'normalization' condition of the form:

$$\int_0^L \psi(x)^2 dx = 1$$

Inserting the wavefunction:

$$\psi(x) = A \sin(kx)$$

and using the trigonometric formula:

$$\cos(2kx) = 1 - 2\sin^2(kx)$$

with $k = n\pi/L$, we can perform the integration to obtain the result:

$$A^2 L/2 = 1$$

Thus, the constant A is determined to be:

$$A = \sqrt{\frac{2}{L}}$$

and the quantum wavefunctions for a particle confined to a box of length L have the form:

$$\psi_n(x) = \sqrt{\frac{2}{L}} \sin(n\pi x/L) \text{ for } n = 1, 2, 3\ldots$$

For a particle in a three-dimensional box, the wavefunctions will just have extra sine factors for the y and z directions. This 'infinite square well' example is a very special case: in general, the mathematics involved in the solution of the Schrödinger equation for more realistic three-dimensional potentials is much more complicated. Nonetheless, this example is helpful in that it contains much of the essential physics and demonstrates the principles involved in solving the Schrödinger equation.

Glossary

alpha particle   A helium nucleus which consists of two protons and two neutrons. In alpha radioactive decay an unstable nucleus emits a rapidly moving alpha particle.

amplitude   The maximum displacement of a wave motion above its average value. The amplitude determines the amount of energy carried by the wave. High-amplitude waves carry more energy than low-amplitude waves as is immediately apparent comparing a stormy sea with calm water. (See *probability* for probability amplitude.)

angular momentum   A measure of the amount of rotational motion in a system. Angular momentum is conserved for isolated systems. In quantum theory angular momentum is quantized.

antiparticle   A particle with the same mass as the particle concerned but with the opposite 'charge-like' properties. When a particle and its antiparticle meet they can annihilate to energy. Thus, for example, there is an antineutron despite the neutron having zero electric charge, because neutrons have a property corresponding to the conservation of the total number of baryons in the universe. Antineutrons have the opposite value of this 'baryonic charge'.

atom   A nucleus with a bound system of electrons. Normal atoms are neutral and form the smallest identifiable amount of a chemical element.

baryon   A strongly interacting fermion such as a neutron or proton. Baryons are made up of three quarks.

beta decay   A radioactive process in which a neutron (or proton) is changed to a proton (or neutron) emitting an electron (or antielectron). The emitted electron is called a beta particle. Beta radioactive decay proceeds by the weak interaction and is always accompanied by antineutrino (neutrino) emission.

Big Bang model   Proposes that the universe began at some definite time (about 15 billion years ago). The early universe is both extremely hot and dense, but is continually cooled by a universal cosmic expansion. This expansion is now observed as the recession of the galaxies.

bit (and byte) The smallest unit of data – a single binary digit that can be either 0 or 1. The word 'bit' is a contraction of *binary digit*. A byte consists of eight bits of data.

black hole An object where gravity dominates all other forces and produces a collapse to a singularity where the known laws of physics break down. Once light or anything else enters a critical region about the singularity, at a distance called the Schwarzschild radius, it can never escape. This property gives rise to the name 'black' hole.

Bose–Einstein condensate A dilute gas of atoms so cold that all the atoms behave as a single entity in a collective quantum fashion.

boson Particles with the property that any number of them can occupy the same quantum state. The force carrying particles such as photons and gluons are bosons. Composite objects with an even number of 'matter-like' fermions such as ^{4}He (two protons, two neutrons and two electrons) are also bosons.

brown dwarf A 'failed star' – an object more massive than any planet but insufficiently massive to ignite hydrogen fusion in its core.

charmed quark A quark type, which like the 'up' quark has a charge equal to 2/3 that of proton, but with an additional property called 'charm', which is conserved in strong interactions but may be destroyed by the weak force.

classical theories Physics before the quantum theory and relativity, usually used to designate the key areas of Newtonian mechanics, electromagnetic theory and thermodynamics.

colour charge Colour charge is a property possessed by quarks and gluons. The colour charge gives rise to the force between quarks (and gluons) in a theory called quantum chromodynamics (QCD). In classical electromagnetic theory the ordinary electric charge plays a similar role to the colour charge in QCD.

conservation When the total amount of some quantity always remains the same it is said to be conserved. For example, the total energy of an isolated system is conserved.

Copenhagen interpretation The conventional interpretation of quantum mechanics in which the measurement process in a 'classical apparatus' collapses the quantum wavefunction to one observed outcome.

cosmic rays High-energy particles originating from extra-terrestrial sources. Some of the relatively low-energy cosmic rays undoubtedly come from the Sun, but the origin of the very-high-energy cosmic rays remains a matter of some controversy.

dark matter | Estimates of matter in the galaxy lead astronomers to believe there is matter in the form of dark matter that does not show up on photographs. The nature of this dark matter is a key problem for particle physics and cosmology.

decoherence | The idea that a quantum system can never be totally isolated from its environment. Advocates of this proposed solution to the quantum measurement problem argue that this coupling causes the delicate phase relations of quantum amplitudes to decay very rapidly.

deuterium | An isotope of hydrogen with a nucleus containing one proton and one neutron.

down quark | The lightest quark with charge equal to $-1/3$ that of the proton.

Doppler effect | The change in the wavelength of waves arriving at a receiver when the source and receiver are in relative motion. The Doppler shift causes an increase in the wavelength when the source and receiver are moving apart and conversely a decrease in wavelength when they approach.

Einstein–Podolsky–Rosen paradox (EPR) | A famous 'thought experiment' that highlights the paradox of a quantum mechanical 'faster than light' connection between rapidly separating quantum objects. The EPR proposal was transformed by John Bell into an experimentally testable inequality that sheds light on the basic nature of quantum mechanics.

electromagnetic radiation | See appendix 1.

electromagnetism | One of the fundamental forces of Nature that occurs when charged particles interact. The electrical attraction between electrons and protons that holds atoms together is an example of the electromagnetic force.

electron | A negatively charged elementary particle with no strong interaction (with the nucleus). Electrons are constituents of all atoms, surrounding the central nucleus and giving the atom its size, strength and chemical properties. Electrons are very light compared with the nucleus.

electron-volt (eV) | A unit of energy comparable with the binding energy of electrons in atoms. An electron-volt is exactly the energy needed to remove an electron from a one volt potential well.

electroweak theory　The theory unifying electromagnetism and the weak nuclear force. The relationship between the two forces is hidden at normal energies. At high energies, such as the energies prevailing in the early universe, the two forces function as a single electroweak force.

elementary particle　A particle without any internal structure and considered to be one of the fundamental building blocks of matter. At the present time quarks and leptons (electrons, neutrinos, etc.) are thought to be elementary.

energy　The capability of doing things. The total amount of energy is conserved though it may be transformed into different forms such as electrical potential or kinetic energy.

exclusion principle (Pauli)　No two fermions can have the same quantum numbers. Applied to electrons in atoms this principle explains the periodic table of the elements.

fermion　The elementary 'matter-like' particles or any composite object containing an odd number of elementary fermions. Thus, electrons and quarks are fermions, and so are protons and neutrons because they contain an odd number (3) of quarks. All elementary fermions have spin 1/2.

Feynman diagram　A pictorial representation of a contribution to the quantum amplitude for a process built up in terms of simple pieces. Feynman's contribution was to give definite rules for calculating the amplitude from the simple components in the diagram.

field　Any quantity that extends smoothly over an extended region of space. A field should be contrasted with a particle, which we think of as having a well-defined location, rather than being spread out over space.

fission　The division of a large nucleus into two roughly equal moderate sized nuclei with possibly some small fragments in addition. Nuclear fission can be induced in certain nuclei by firing neutrons into them and this forms the basis of nuclear reactors.

flavour　A property which distinguishes the various types of quark – up, down, strange, charmed, bottom and top. This property is also applied to leptons, where it distinguishes the various types (electron, muon, etc.).

force　Anything which causes a change in the motion of a body. There are four fundamental forces: gravity, weak, electromagnetic and strong.

frequency (f) of a wave　The number of wave crests passing a particular point in one second. In quantum theory the energy of a quantum particle is proportional to the frequency of the quantum wave.

fusion	Any nuclear reaction in which two or more nuclei come together to form a nucleus with more neutrons plus protons than each of the ingredients. Fusion reactions of hydrogen to make helium are the energy source of the Sun.
galaxy	A large collection of more than 10 million stars in a region of space well separated from other such collections. Our Sun is one of about 200 billion stars making up our Milky Way Galaxy.
gamma ray	A photon of very high energy (see table A1.3 of appendix 1).
gauge theory	Electromagnetism is the simplest gauge theory. In the gauge formulation the electromagnetic field arises from the demand that the phase difference of a charged particle quantum wave between different points of space-time is unmeasurable. A similar approach can be given for theories of the weak interaction and also the colour force between quarks (QCD).
gluon	The quantum particle associated with the 'colour' force between quarks.
Grand Unified Theory (GUT)	A theory that suggests a link between the electroweak and strong forces. The idea is similar to that of electroweak unification and the relationship between the two forces remains hidden until extremely high energies, well beyond any conceivable accelerator experiments.
ground state	The ground state of a quantum system (e.g. atom) is its condition (described by a wavefunction) when it is in its lowest energy level. The quantum condition when the system is in any other energy level is called an excited state.
hadron	Any strongly interacting particle. Hadrons are divided into mesons and baryons. Thus, a pion is a meson and a proton is a baryon and both are hadrons.
half-life	The time required for half the atoms in a sample of a specific radioactive element to disintegrate.
Hall effect	Production of a voltage across a current-carrying conductor or semiconductor by a magnetic field at right angles to the current. At very low temperatures the Hall voltage becomes quantized – this is the 'Quantum Hall effect'.
Higgs particle or boson	A hypothetical particle predicted by the Glashow, Salam and Wienberg electroweak theory that is needed to give mass to the W and Z particles.
hydrogen	The lightest chemical element. The nucleus of ordinary hydrogen consists of a single proton.

ion	An atom in which the number of electrons is not equal to the number of protons. An ordinary atom is neutral because the positive charge of the protons in the nucleus is exactly matched by an equal number of oppositely charged electrons surrounding the nucleus.
interference	A characteristic property of waves whereby the total wave height of two overlapping waves is the sum of the individual heights. For example, if two identical waves overlap with the crests of one coinciding with the troughs of the other then the two waves will cancel and this is an example of destructive interference. In other situations interference can increase the wave motion.
insulator	Material that is a poor conductor of electricity.
isotope	Nuclei with the same number of protons but different numbers of neutrons are the isotopes of a given element.
kaon (K-meson)	A type of strange meson.
kinetic energy	Energy due to the motion of an object.
lambda (Λ)	A type of strange baryon.
laser	A device for producing coherent light by stimulated emission. The individual 'photon waves' making up coherent light all vibrate and move in step, so we have effectively a single electromagnetic wave.
lepton	A fermion such as an electron or neutrino which is not influenced by the strong interaction.
light year	The distance travelled by light in one year. In more familiar units it is roughly 10 million million kilometres. Stars in our region of the Galaxy are typically separated by a few light years.
magnetic moment	A magnet can be viewed as a pair of magnetic north and south poles separated by a small distance. The magnetic moment is a quantity which describes how the magnet will be influenced by a magnetic field.
magnetic monopole	A hypothetical particle consisting of a single isolated magnetic pole.
Many Worlds interpretation	The idea that each measurement of a quantum system causes the universe to split into multiple copies corresponding to all possible outcomes of the experiment, but in different non-interacting worlds.
measurement problem	The problem of understanding how a quantum wavefunction corresponding to many possible final outcomes 'collapses' to a single

definite state. The collapse is not governed by the Schrödinger equation and many possible solutions have been proposed to explain this problem (see *Copenhagen interpretation*, *decoherence* and *Many Worlds interpretation*).

meson
Any strong interacting boson. All mesons are unstable and consist of a bound state of a quark and antiquark.

metal
Material which is both a good conductor of heat and electricity and is also shiny (a good reflector of light). In metals, large numbers of electrons are free to move about the whole volume of metal.

molecule
A bound system of two or more atoms.

momentum
A measure of the quantity of motion defined as the product of mass and velocity. In the absence of external forces the total momentum of the system under consideration is conserved.

Moore's Law
The prediction by Gordon Moore, co-founder of Intel, that the number of transistors on a chip and its performance doubles every 18 months. This 'law' has been true for the past 30 years but, unlike a real law of physics, will ultimately fail when the smallest feature size on a chip approaches atomic dimensions.

muon (μ)
A lepton similar to the electron but over 200 times as massive. A muon is unstable and decays to an electron and pair of neutrinos.

nanotechnology
Technology employing devices smaller than about 100 nanometers. Devices on this scale are much smaller than the macroscopic objects we are familiar with and are governed by a combination of classical and quantum physics.

neutral current
A weak interaction that leaves the charge of the interacting particles unaltered.

neutrino
Electrically neutral lepton. Neutrinos interact only by the weak interaction (and gravity) and so (at low energies) are enormously penetrating. Neutrinos come in three varieties called the electron neutrino, the muon neutrino and the tau neutrino.

neutron
Electrically neutral baryon with almost the same mass as the proton. Neutrons together with protons are the constituents of the nucleus.

neutron star
A star composed predominantly of neutrons and with a diameter of about 16 km (10 miles) and a mass comparable with the Sun. Neutron stars are believed to be formed as a consequence of supernova explosions.

nucleus	Dense core of an atom made of neutrons and protons held together by the strong force.
nuclear reaction	A collision of nuclei which results in the redistribution of the protons and neutrons so that different nuclei emerge from the collision.
omega minus (Ω^-)	The lowest mass baryon made of three strange quarks.
particle	A small object with a well-defined location at each instant of time like a bullet. A quantum object sometimes behaves like this but on other occasions it behaves more like a wave. In this book we also use the word to describe any quantum object with identifiable properties such as a photon or electron.
periodic table	Tabulation of the different kinds of atoms (elements) in terms of increasing number of protons in the nucleus. Atoms with similar chemical and physical properties occur regularly and are placed in columns (see Fig. 6.1).
phase	A wave motion passing a particular point will undergo an up–down motion. At any given time and place the state of the wave's motion is given by the phase. When one wavelength has passed, the motion will have undergone a complete cycle and this suggests we measure phase by an angle, with $360°$ corresponding to one cycle. Two waves are in phase when the positions of their crests coincide, otherwise the phase difference expresses the difference in their state of motion.
photon	The quantum particle associated with light waves or, more generally, electromagnetic radiation.
Planck's constant (h)	The fundamental constant of quantum mechanics.
plasma	A mixture of ions and electrons. In the interior of stars ordinary ions cannot exist and the plasma is made of nuclei and electrons.
pion (pi-meson or π)	The lightest meson. It comes in three varieties distinguished by an electric charge $+1$, 0 and -1 times that of the proton.
polarization	Polarized light is a beam of light for which the electric field oscillates in a single direction or rotates about the direction of motion. At the quantum level, an individual photon can exist in one of two polarization states corresponding to the two spin states of a zero mass, spin one, quantum particle.
positron	Positively charged antiparticle of the electron.

potential energy	Energy a system has by virtue of its position or state. For example, the height of an object above the Earth (or other massive object) determines its gravitational potential energy.

probability	A number expressing the likelihood of an event taking place. Quantum theory involves intrinsic uncertainty and so probabilities are essential to its description. All possible quantum outcomes have a number associated with them called the probability amplitude. The square of this number gives the probability of the event.

proton	A particle with opposite charge to the electron but about 1836 times more massive. Protons experience the strong force and are joined together with neutrons in the nuclei of atoms. The nucleus of ordinary hydrogen consists of one proton.

proton–proton cycle	A sequence of nuclear reactions whose overall effect is to convert hydrogen into helium. This sequence is responsible for the generation of energy in the Sun.

pulsar	Rapidly varying periodic source of radio waves believed to be due to a rotating neutron star.

QCD (quantum chromodynamics)	The quantum theory of the interaction of quarks and gluons. The quarks and gluons carry a property similar to the electric charge called 'colour' (although it has nothing to do with the familiar concept of colour).

QED (quantum electrodynamics)	The quantum theory of electromagnetic interactions.

quantized	A physical quantity which can have only certain discrete values in a particular system is described as quantized. Thus, a hydrogen atom has certain discrete energy levels and so we say that its energy is quantized.

quantum dot	A type of quantum 'nanocircuit' in which electrons are confined to a small region. The confinement space can be made so small that the energy quantization is measurable and the result is a sort of 'artificial atom'.

quantum number	A whole or half integer number or set of numbers which specify the state of a quantum system. For example, the quantized energy levels of a hydrogen atom are denoted by a sequence of positive integers starting with $n = 1$ for the ground state.

quark	The elementary particles which are believed to form the basic building blocks of hadronic matter such as protons and pions. Quarks carry a fractional electric charge.

quasar (quasistellar object)
A star-like object with a large red shift. If the red shift is interpreted as due to the Hubble expansion of the universe then quasars radiate about 100 times the energy of a conventional galaxy from a central region not much larger than the solar system! Large black holes are involved in most explanations of quasars.

qubit
The basic unit of quantum information. In a two state quantum system, a *quantum bit* or qubit can correspond to a classical bit like 1 or 0, or a quantum mechanical superposition of both.

radioactivity
Spontaneous disintegration of certain nuclei with the emission of alpha, beta or gamma radiation.

rectifier
An electronic component which allows the flow of current in one direction but prevents its passage in the reverse direction.

red shift
The displacement of the wavelength of received light to the red end of the spectrum, compared with its emission wavelength. The most common cause of red shift is the Doppler effect due to the source of light moving away from the receiver.

refraction
The change in the direction of light or other electromagnetic radiation as it passes from one transparent medium to another.

Relativity
Einstein's fundamental theories of space and time. Special relativity is concerned with non-accelerating systems. General Relativity or GR applies to accelerating systems and turns out to be a theory of gravity.

Schrödinger's cat
A thought experiment in which a cat is confined to a box for which there is a 50:50 chance of a quantum process triggering its death. The measurement problem of quantum mechanics is highlighted by the apparent necessity for the cat to exist in a quantum superposition of both dead and alive states until the box is opened and its state observed.

Schrödinger's equation
The basic equation of quantum mechanics which describes the behaviour of a particle in a potential.

semiconductor
A material whose ability to conduct electric current is intermediate between metals and insulators and also increases with increasing temperature.

spin
Fundamental property of a particle corresponding to an intrinsic angular momentum.

spontaneous emission
Photons emitted from an isolated atom (or other quantum system) in a transition from an excited state to a lower energy state.

Standard Model	The accepted theory of particle physics – quantum chromodynamics for the quark strong interactions and the electroweak theory for the electromagnetic and weak forces.
strangeness	Property of hadrons conserved in the strong interaction but destroyed by the weak interaction. It is associated with the strange quark.
string theory	A set of theories based on the idea that the most basic entities are not point particles but one-dimensional objects like strings. Such theories are attractive since they offer the prospect of a consistent quantum theory of gravity and the unification of all four forces of Nature.
strong interaction (force)	The force that holds the nucleus together and is responsible for the interaction of hadrons. The strong force is a remnant of the inter-quark 'colour' force.
superconductor	A material (usually a metal or alloy) whose electrical resistance vanishes below a certain critical transition temperature.
superfluid	A fluid that flows without friction and has a high thermal conductivity below a certain critical temperature.
supernova	A catastrophic stellar explosion in which the power output becomes comparable with that of a whole galaxy for about a month and thereafter gradually declines. The explosion may leave behind a neutron star or black hole.
supersymmetry	A new type of symmetry that requires that the equations of the fundamental theory should be unchanged when fermions are replaced by bosons and vice versa. This hypothesis implies that there should be many as-yet-undiscovered supersymmetric partners of familiar particles or 'sparticles'. Thus the fermionic partner of a photon is called a photino and the bosonic partner of a quark is called a squark!
temperature	A measure of hotness. If two bodies in contact have different temperatures then heat will flow from the hotter body to the colder one. The temperature of a body is a measure of the average kinetic energy of its atoms (or other constituents).
transistor	A semiconductor device in which the flow of current between two contacts can be controlled by a potential applied to a third contact.
tunnelling	The ability of quantum objects to pass through regions which classically are energetically forbidden.

uncertainty principle   If an experiment specifies the position of a particle to a certain accuracy
(Heisenberg)   (Δx) then it automatically results in the momentum becoming
unpredictable to an accuracy greater than Δp, such that the product
$\Delta x \Delta p$ is greater than a certain minimum value of the order of Planck's
constant. A similar principle applies to the uncertainties in energy and
time measurements.

up quark   The lightest quark, with charge equal to 2/3 that of the proton.

virtual particle   A particle violating the conservation of energy and existing only for a
brief instant of time so the energy–time uncertainty principle is satisfied.

W particle   Massive charged particle that together with the neutral Z particle is
associated with the weak interaction, just as the photon is associated with
the electromagnetic interaction.

wave motion   Any kind of oscillating or moving disturbance which at any given instant
is spread over a region of space. The simplest kind of wave corresponds to
a periodic up–down disturbance. The distance between adjacent crests is
called the wavelength.

weak force   One of the fundamental interactions. The weak force is responsible for
beta decay and any interaction involving neutrinos.

white dwarf   A dense compact remnant star with a typical mass comparable with that
of the Sun but of a size comparable with that of the Earth. The star is
held up by the Pauli principle applied to electrons and is still hot but is
gradually cooling.

X-rays   Relatively energetic form of electromagnetic radiation or photons (see
appendix 1).

Z particle   Neutral massive boson associated with the weak interaction.

zero-point motion   Vibrational motion of atoms at absolute zero temperature due to
Heisenberg's uncertainty principle.

Zeeman effect   Splitting of single spectral lines into two or more components when the
atoms are subject to a magnetic field.

Quotations and sources

Chapter 4

p. 47   'Atoms are completely impossible . . .', Richard Feynman, *The Feynman Lectures on Physics*, Vol. 3, Ch. 2, p. 6 (Addison-Wesley 1966)

p. 48   'It was quite the most . . .', Ernest Rutherford, quoted in *The Evolution of the Nuclear* Atom by G. K. T. Conn and H. D. Turner (American Elsevier 1965)

p. 52   'Everything became clear', Niels Bohr, quoted in *What Little I Remember* by Otto Frisch, Ch. 2 *Atoms* (Cambridge University Press 1979)

p. 70   'Here, right now . . .', Hans Dehmelt in *Atomic Physics*, 1984

p. 71   'The well-defined identity . . .', Hans Dehmelt in *Science*, **247**, 1990

Chapter 5

p. 73   'It is possible . . .', Richard Feynman, *The Feynman Lectures on Physics*, Vol. 3, Ch. 8, p. 12 (Addison-Wesley 1966)

p. 81   'Measuring at night . . .', Gerd Binnig and Heinrich Rohrer, *Reviews of Modern Physics*, July 1987

p. 81   'a change in the distance . . .', Geerd Binnig and Heinrich Rohrer, quoted in *Nano!* by Ed Regis, Ch. 11 (Bantam Press 1995)

p. 81   'our microscope . . .', ibid.

p. 81   'I could not stop looking . . .', Geerd Binnig and Heinrich Rohrer, *Reviews of Modern Physics,* July 1987

p. 88   'If alpha particles . . .', Ernest Rutherford, *Philosophical Magazine* **37**, 581 (1919); quoted in *From X-Rays to Quarks* by Emilio Segre (W. H. Freeman and Co., 1980), p. 110

p. 90   'Cockcroft and Walton . . .', Ernest Lawrence, *Lawrence and Oppenheimer* by Nuell Pharr Davis, Cape 1969 p. 45

p. 90   'That's what physicists . . .', James Brady, ibid., p. 45

p. 103   'that every miner . . .', quoted in *The Physics of the Atom* by M. R. Wehr, J. A. Richards and T. W. Adair (Pearson Education 1984)

Chapter 6

p. 107   'It is the fact . . .', Richard Feynman, *The Feynman Lectures on Physics*, Vol. 3, Ch. 2, p. 7 (Addison-Wesley 1966)

p. 109   'How can one avoid . . .', Wolfgang Pauli, quoted in *From X-Rays to Quarks* by Emilio Segre, Ch. 7 (W. H. Freeman and Co. 1980)

p. 109   'I have already sent . . .', Paul Ehrenfest, quoted in *Quantum Profiles* by Jeremy Bernstein (Princeton University Press 1991)

p. 124   'The general aim of the program was to obtain as complete an understanding as possible of semiconductor phenomena, not in empirical terms, but on the basis of atomic theory.' John Bardeen, *Nobel Lectures, Physics 1942–1962*, Elsevier Publishing Company, Amsterdam, 1967, p. 319

p. 126   'With the advent of the transistor . . .', G. W. A. Dummer, *Electronic Components in Great Britain* 1952

Chapter 7

p. 131   '. . . there are certain situations . . .', Richard Feynman, *The Feynman Lectures on Physics*, Vol. 3, Ch. 21, p. 1 (Addison-Wesley 1966)

p. 134   'A splendid light . . .', Albert Einstein in a letter to M. Besso, 6 September 1916; *Albert Einstein–Michele Besso Correspondence 1903–1955*, edited by P. Speziali (Hermann 1972) p. 82

p. 134   'What I most admired . . .', Albert Einstein in a letter to V. Besso, 21 March 1955; *Albert Einstein–Michele Besso Correspondence 1903–1955*, edited by P. Speziali (Hermann 1972) p. 37

p. 140   '. . . if he thought . . .', S. N. Bose in a letter to A. Einstein, 4 June 1924; quoted by Abraham Pais in his book *Subtle Is the Lord* (Oxford University Press 1982), p. 423

p. 140   '. . . an important advance', ibid., p. 423

p. 141   'from a certain . . .', Albert Einstein in a letter to P. Ehrenfest, 29 November 1924

p. 141   'The theory is pretty . . .', Albert Einstein in a letter to P. Ehrenfest, 29 November 1924

p. 142   'If the beaker . . .', Kurt Mendelssohn, quoted in *Quantum Physics of Atoms, Molecules, Solids, Nuclei, and Particles* by R. Eisberg and R. Resnick (John Wiley & Sons, Inc. 1974) p. 439

Chapter 8

p. 157   'We have always . . .', Richard Feynman, *Simulating Physics with Computers* (*International Journal of Theoretical Physics*, **21**, 1982); reprinted in *Feynman and Computation*, edited by Tony Hey (Perseus Books 1999)

p. 159   'He [God] does not play dice', Albert Einstein, in a letter to Max Born, 1926, in *The Born–Einstein Letters* edited by I. Born (Walker 1971)

p. 159   '. . . to prescribe to God . . .', Niels Bohr quoted by W. Heisenberg in *Physics and Beyond*, (Harper and Row 1971) p. 81

p. 160   'There is one lucky break . . .', Richard Feynman, *The Feynman Lectures on Physics*, Vol. 1, Ch. 37, p. 1 (Addison-Wesley 1966)

p. 164   'Heisenberg split', John Bell, 'Quantum mechanics for cosmologists' in *Quantum Gravity 2*, edited by C. Isham, R. Penrose and D. Sciama (Oxford, Clarendon Press 1981); reprinted in *Speakable and Unspeakable in Quantum Mechanics* (Cambridge University Press 1986) p. 123

p. 164   'shifty boundary', John Bell in the Preface to *Speakable and Unspeakable in Quantum Mechanics*, p. viii

p. 164   'wavy quantum states', John Bell ibid., p. viii

p. 164   'rotten', John Bell, in conversation, quoted in *Quantum Profiles* by Jeremy Bernstein (Princeton University Press 1991)

p. 164   'observations not only ...', Pascual Jordan; quoted by Max Jammer in *The Philosophy of Quantum Mechanics*, (Wiley 1974) p. 161

p. 165   'The entire formalism ...', Niels Bohr; from *The Philosophy of Quantum Mechanics* by Max Jammer

p. 165   'There is no quantum ...', quoted by Aage Petersen, *Bulletin of the Atomic Scientists*, (September 1963) p. 8

p. 165   'In the experiments ...', Werner Heisenberg; quoted in *The Quantum World* by J. C. Polkinghorne (Longman 1984)

p. 165   'Does the Moon only exist ...', Albert Einstein, quoted by Abraham Pais in *Review of Modern Physics*, **51**, 907 (1979); see also *Subtle Is the Lord* by Abraham Pais p. 5

p. 166   'While we have ...', Albert Einstein, Boris Podolsky and Nathan Rosen, 'Can quantum-mechanical description of physical reality be considered complete?', *Physical Review* **47**, 777, 1935

p. 166   'spooky, action-at-a-distance', Albert Einstein; quoted in 'Spooky actions at a distance', in *The Great Ideas Today* (Encyclopedia Brittanica Inc. 1988); reprinted in *Boojums All the Way Through* by N. David Mermin (Cambridge University Press 1990); *Born–Einstein Letters*, p. 158

p. 168   'Einstein and Bohr ...', John Bell in conversations with Tony Hey at CERN in 1974

p. 173   'If all this ...', Erwin Schrödinger; quoted by Werner Heisenberg in *Physics and Beyond* (Harper and Row 1971) p. 73

p. 173   'A cat is penned up ...', Erwin Schrödinger, *Naturwiss*, **23** (1935); English translation by J. D. Trimmer, **124**, 323 (1980)

p. 174   'The traditional description ...', Richard Feynman, *Feynman Lectures on Gravitation* edited by Brian Hatfield (Addison Wesley 1995) p. 14

p. 175   'every quantum transition ...', Bryce DeWitt; quoted in *The Ghost in the Atom*, edited by P. C. W. Davies and J. R. Brown (Cambridge University Press 1986) p. 36

p. 176   'These are very wild ...', Richard Feynman in *Feynman Lectures on Gravitation*, edited by Brian Hatfield (Addison Wesley 1995) p. 15

p. 176   'does not associate ...', John Bell 'Quantum mechanics for cosmologists', p. 135

p. 176   'if such a theory ...', John Bell, ibid., p. 136

p. 176   'hinges on observing ...' interview with David Deutsch in *The Ghost in the Atom*, edited by Davies and Brown, p. 98

p. 178   'So long as ...', John Bell in 'On wave packet reduction in the Coleman–Hepp model', *Helvetica Physica Acta*, **48**, 93 (1975); reprinted in *Speakable and Unspeakable in Quantum Mechanics*, p. 48

p. 179   'exactly when and where ...'. John Bell, ibid., p. 51

Chapter 9

p. 181   'What I want to talk about ...', Richard Feynman, *There's Plenty of Room at the Bottom*, reprinted in *Feynman and Computation*, edited by Tony Hey (Perseus 1999), p. 63

p. 181   'I am not inventing ...', Richard Feynman, ibid.

p. 181   'to the first guy who makes...', Richard Feynman, ibid.

p. 181   'to the first who can take ...', Richard Feynman, ibid.

p. 183   '... it would be ...', Richard Feynman, ibid.

p. 185   'When this ...', Robert Noyce, quoted in *The Genesis of the Integrated Circuit,* M. Wolff (IEEE Spectrum, August 1976); see also www.intel.com/intel/museum

p. 187   'Integrated circuits ...', Gordon Moore, 'Cramming more components onto integrated circuits', *Electronics Magazine*, 1965; see also www.intel.com/intel/museum

p. 187   'In 1968, I was invited ...', Carver Mead, in 'Feynman as a Colleague' published in *Feynman and Computation*, edited by Tony Hey (Perseus 1999)

p. 188   'A very small addendum ...', Arthur Rock, 'Intel Processor Hall of Fame', www.intel.com/intel/museum

p. 189   '... life gets very interesting', Gordon Moore, *Scientific American Interview,* October 1997

p. 190   'Interconnect has been ...', Semiconductor Industry Association Roadmap 1999

p. 193   '... all of the information ...', Richard Feynman, 'There's Plenty of Room at the Bottom', p. 66

p. 197   'I'm not happy ...', Richard Feynman, in *Simulating Physics with Computers*, reprinted in *Feynman and Computation*, edited by Tony Hey (Perseus 1999), p. 151

p. 198   'wonderful, intense ...', Ed Fredkin, in conversation

p. 198   'It was very hard ...', Ed Fredkin, 'Feynman, Barton and the Reversible Schroedinger Equation', published in *Feynman and Computation*, edited by Tony Hey, p. 139

p. 198   'The trouble with you ...', Richard Feynman quoted by Ed Fredkin, ibid.

p. 198   'Can you do it ...', Richard Feynman, in 'Simulating Physics with Computers'. ibid.

p. 200   'I would not call ...', Erwin Schrödinger, *Proceedings of the Cambridge Philosophical Society*, **31**, 555 (1935)

p. 201   'I am tempted ...', Charles Bennett, 'Quantum Information Theory', published in *Feynman and Computation*, edited by Tony Hey, p. 179

p. 204   'Stand by ...', advertisement in *Scientific American*, February 1996.

p. 204   'In any organisation ...', Charles Bennett, in conversation.

p. 206   'We should always ...', Richard Feynman, in *Feynman Lectures on Gravitation*, edited by Brian Hatfielfd (Addison-Wesley 1995)

Chapter 10

p. 207   'One of the most impressive ...', Richard Feynman, *The Feynman Lectures on Physics*, Vol. 1, Ch. 3, p. 7 (Addison-Wesley 1966)

Chapter 11

p. 227   'It was as though ...', Richard Feynman, *Theory of Positrons*, (*Physical Review*, **76**, 1949)

Chapter 12

p. 245   'Now we are in a position ...', Richard Feynman, from a BBC *Horizon* programme produced by Christopher Sykes, edited transcript reprinted in *The Listener*, 26 November 1981.

p. 253   Cecil Powell
'When they were recovered....' Quoted in *From X-Rays to Quarks* by Emilio Segre, W. H. Freeman & Company (1980) p. 250

Chapter 13

p. 285   'You read too many novels!' Richard Feynman in conversation with Tony Hey at Caltech, 1972

p. 286   'It contains ...', Captain S. P. Meek, *Submicroscopic*, reprinted in *Before the Golden Age* edited by Isaac Asimov (Doubleday & Co., 1974) p. 66

p. 286   'The work of Bohr and Langmuir ...', ibid, p. 68

p. 287   'It is probable ...', Frederick Soddy, *Atomic Transmutation* (New World 1953) p. 95.

p. 289   'The dozen years ...', James Gunn, *Illustrated History of Science Fiction*, (Prentice Hall, 1975)

p. 289   'During the Golden Age ...', Isaac Asimov, *Before the Golden Age* edited by Isaac Asimov (Doubleday & Co., 1974) p. xv.

p. 291   'We estimated ...', Otto Frisch and Rudolf Peierls, quoted in *Bird of Passage* by Rudolf Peierls (Princeton University Press 1985)

p. 292   'Was there really ...', Arthur Compton, *Atomic Quest*, p. 127 (Oxford University Press 1956).

p. 290   '... wouldn't go off ...', Werner Heisenberg, *The Farm Hall Transcripts*, quoted by Thomas Powers in his book *Heisenberg's War*, Ch. 36 p. 445 (Jonathan Cape 1993).

p. 297   'On their way ...', Fred Hoyle, 'Jury of Five', in *Element 79* by Fred Hoyle (New American Library 1967) p. 136.

p. 298   'The decision rested on ...', Fred Hoyle, ibid, p. 140.

p. 299   'May the universe ...', John Wheeler, quoted by Gary Zukav in *The Dancing Wu Li Masters* (Mass Market Paperback, 1994)

p. 299   'The advent of nano-therapy ...', Greg Bear, *Queen of Angels* (Warner Books 1991) p. 198

p. 301   'The AXIS "mind" consists ...', Greg Bear, ibid, p. 19

p. 301   'She patiently watched ...', Greg Bear, ibid, p. 258

p. 301   'A leaf of paper ...', Neal Stephenson, *The Diamond Age* (Bantam 1995) p. 64

p. 303   'Areostat meant anything ...', Neal Stephenson, ibid, p. 56

p. 303   '"Mites," he said ...', Neal Stephenson, ibid, p. 60

p. 305   'Ordinary computers ...', Michael Crichton, *Timeline* (Arrow Books 2000) p. 138

p. 307   '"... you'd need ..."', Michael Crichton, ibid, p. 137

p. 309   '"Wellsey's split ..."', Michael Crichton, ibid, p. 445

p. 310   'The occupation of Denmark ...', Werner Heisenberg, quoted in a letter from Stefan Rozental to M. Gowing, 1984

p. 310   'So far we have ...', Christian Moller, ibid.

p. 310   'the two men spent ...', Abraham Pais, *Niels Bohr's Times,* Ch. 21 (Oxford University Press 1991).

p. 311   'We choose to examine ...', Richard Feynman, *The Feynman Lectures on Physics, Volume III,* Ch. 1, p. 1–1 (Addison Wesley 1965)

p. 311   'Any other situation ...', Richard Feynman, *The Character of Physical Law,* Ch. 6, p. 130 (MIT Press 1965)

p. 311   'You get what you interrogate for', Tom Stoppard, *Hapgood,* p. 10 (Faber and Faber 1988)

p. 311   'The nineteenth century was known as the machine age, the twentieth century will go down in history as the information. I believe the twenty-first century will be the quantum age.' Paul Davies's Foreword to *Quantum Technology* by Gerard Milburn, Allen & Unwin 1996, p. viii

Epilogue

p. 313   'A poet once said ...', Richard Feynman, *The Feynman Lectures on Physics Volume I*, Ch. 3, p. 10 (Addison Wesley 1965)

Suggestions for further reading

Quantum mechanics

R. P. Feynman (1965). *The Character of Physical Law* (MIT Press)
This book comprises seven lectures presented at Cornell University in 1964. Even 20 years on, the book still sparkles with Feynman's unique style and wit.

R. P. Feynman (1965). *The Feynman Lectures on Physics* (Addison-Wesley)
All three volumes are notable for their insight and novel presentation of all aspects of physics. Volume 3 contains Feynman's unusual approach to quantum mechanics – most students find it rather difficult and prefer a more conventional approach.

R. P. Feynman (1985). *QED* (Princeton University Press)
An entertaining yet serious attempt at a 'popular' exposition of quantum electrodynamics or QED. As ever, Feynman tries to achieve maximum clarity and simplicity without compromise by distortion of the truth.

R. P. Feynman (1996). *The Feynman Lectures on Computation*, edited by Tony Hey and Robin Allen (Addison Wesley)
A 'Feynmanesque' overview of computer science from a physicist's point of view – Feynman spent the last 5 years of his life giving these lectures.

A. P. French and E. F. Taylor (1978). *An Introduction to Quantum Physics* (Norton, USA; Nelson, UK)
A traditional quantum mechanics book, but more wordy than most, and quite accessible for the majority of its chapters.

J. C. Polkinghorne (1984). *The Quantum World* (Longman)
A clear and careful introduction to the conceptual problems of quantum mechanics. The famous paradoxes of Schrödinger's cat, Wigner's friend and Einstein, Podolsky and Rosen are all discussed in detail.

G. Gamow (1965). *Mr Tompkins in Paperback* (Cambridge University Press)
The famous physicist George Gamow's entertaining account of Mr Tompkins' imagined explorations of relativity and quantum mechanics.

G. Gamow and R. Stannard (1999). *The New World of Mr Tompkins* (Cambridge University Press)
An updated account of George Gamow's classic text.

David Lindley (1996) *Where Does the Weirdness Go?* (Basic Books)
A readable and clear account of the paradoxes and problems raised by
quantum mechanics.

Gerard Milburn (1996) *Quantum Technology* (Allen and Unwin)
(published in the USA as *Schroedinger's Machines* (Freeman))
An up to date account of the new quantum technologies including ion
traps, quantum nanocircuits, quantum cryptography and quantum
computing.

Hans Christian von Baeyer (1992) *Taming the Atom* (Random House)
An excellent and readable history of atomism from the earliest beginnings
up to the new atom manipultion experiments.

Historical background

O. Frisch (1979). *What Little I Remember* (Cambridge University Press)
E. Segré (1980). *From X-rays to Quarks* (Freeman)
These two fascinating autobiographical accounts of the early days of
quantum mechanics are well worth reading.

R. P. Feynman (1985). *Surely You're Joking, Mr Feynman!* (Norton)
Feynman's delightful and entertaining collection of anecdotes containing
many of the legendary 'Feynman stories' and much else besides.

A. Pais (1982). *Subtle is the Lord – The Science and the Life of Albert Einstein*
(Oxford University Press)
Probably the definitive book on Einstein's contributions to the
foundations of quantum mechanics and his development of the general
theory of relativity.

P. Goodchild (1980). *J. Oppenheimer – Shatterer of Worlds* (BBC Publications)
The book from the BBC television series about a fascinating piece of
contemporary history.

Richard Rhodes (1986). *The Making of the Atomic Bomb* (Simon and Schuster),
The definitive, Pulitzer Prize-winning account of the development of the
atomic bomb and the Manhattan project.
S. Augarten (1984). *Bit by Bit – An Illustrated History of Computers* (Tickner
and Fields)
An absorbing account of the history of computing from the early pioneers
like John von Neumann and Alan Turing, to modern heroes like Jobs and
Wozniak, the creators of the personal computer.

Companion to this volume

A. Hey and P. Walters (1997). *Einstein's Mirror* (Cambridge University Press)

Photo-credits

We would like to express our thanks to the following individuals and organizations for allowing us to use their material in this book:

Chapter 1

Newton	Trustees of the British Museum
Young	Trustees of the British Museum
Thompson	Cavendish Laboratory, University of Cambridge
Fig. 1.1	Education Development Center. Inc., Newton MA
Fig. 1.2	NASA
Fig. 1.3	unable to trace the copyright holders
Fig. 1.4	Oxford University Press
Fig. 1.6	The Open University Press, from Discovering Physics (S271)
Fig. 1.8	Oxford University Press
Fig. 1.10	Prof. Dr. C. Jönsson
Fig. 1.12	The New York Academy of Sciences, from Dr Hannes Lichte, (the experiment was conducted using a Mollenstedt-type electron biprism interferometer);

Chapter 2

Heisenberg	Cavendish Laboratory, University of Cambridge
Planck	Photo Deutsches Museum Munchen
Feynman	Prof. R. Feynman
Fig. 2.2	unable to trace the copyright holders
Fig. 2.4	Oxford University Press
Fig. 2.5	National Optical Astronomy Observatories
Fig. 2.6	Albert Rose, from monograph Vision: Human and Electronic
Fig. 2.7	Patrick Seitzer, NOAO
Fig. 2.10	Prof. Tony Hey
Fig. 2.11	Richard F. Voss

Chapter 3

de Broglie	National Portrait Gallery
Schroedinger	Pfaundler, Innsbruck
Plaque	Roger Stalley
Gell-Mann	Prof. M. Gell-Mann
Zweig	Prof. G. Zweig

Fig. 3.1	Oxford University Press
Fig. 3.5	Science Photo Library, from T. Brain
Fig. 3.6	Ministry of Agriculture, Fisheries and Food (SEM Unit, Slough Laboratories)
Fig. 3.7	Schoken Books Inc, from Scharf, D. Magnifications, 1977
Fig. 3.8	SLAC
Fig. 3.9	SLAC

Chapter 4

Rutherford	Cavendish Laboratory, University of Cambridge
Coat of Arms	Niels Bohr Institute
Dehmelt	University of Washington/Davis Freeman
Paul	unable to trace the copyright holders
Fig. 4.4	US Atomic Energy Commission
Fig. 4.5	NASA
Fig. 4.6	The Open University Press, from Discovering Physics (S271)
Fig. 4.7	Education Development Center. Inc., Newton MA
Fig. 4.9	Dr C. M. Hutchins
Fig. 4.10	Prof. Thomas D. Rossing
Fig. 4.12	Prof. Don Eigler, IBM Almaden Research Center
Fig. 4.13	Prof. Don Eigler, IBM Almaden Research Center
Fig. 4.16	Orion nebula, Anglo Australian Observatory, Photograph by David Malin
Fig. 4.19	NIST

Chapter 5

Binnig and Rohrer	IBM Research
Quare, Binnig and Gerber	Prof. Jim Gimzewski
Gamow	Maurice M. Shapiro, Colorado Associated Press
Cockcroft, Rutherford and Walton	Ullstein Bilderdienst
Lawrence and Livingstone	Lawrence Berkeley Laboratory, University of California
Meitner and Hahn	Ullstein Bilderdienst
Fig. 5.3	unable to trace the copyright holders
Fig. 5.4	unable to trace the copyright holders
Fig. 5.5	Education Development Center. Inc., Newton MA
Fig. 5.7	Prof. Jim Gimzewski
Fig. 5.8	Naval Research Laboratory
Fig. 5.9	Don Eigler, IBM Almaden Research Center

Fig. 5.10	U.S. Naval Research Laboratory
Fig. 5.11	Zyvex
Fig. 5.12	Patrici Molinas-Mata
Fig. 5.13	Don Eigler , IBM Almaden Research Center
Fig. 5.14	The Regents of the University of California
Fig. 5.15	Alex Rimberg, JC Nabity Lithography Systems
Fig. 5.17	Proceedings of The Royal Society
Fig. 5.18	Cavendish Laboratory, University of Cambridge
Fig. 5.19	Rutherford Appleton Laboratory
Fig. 5.24	Chicago Historical Society
Fig. 5.25	The Daily Mail
Fig. 5.26	Hiroshima-Nagasaki Publishing Committee (Eiichi Matsumoto)
Fig. 5.27	Physicians for Social Responsibility
Fig. 5.28	US Atomic Energy Commission
Fig. 5.29	The Slide Centre
Fig. 5.30	Proceedings of The Royal Society
Fig. 5.31	British Broadcasting Corporation

Chapter 6

Mendeleev	Ann Ronan Picture Library
Pauli and Wife	Associated Press
Fermi	Argonne National Laboratory
Stern, Postcard	unable to trace the copyright holders
Shockley, Brattain and Bardeen	Associated Press
Oppenheimer and von Neumann	Institute for Advanced Study, Princeton NJ
Fig. 6.1	unable to trace the copyright holders
Fig. 6.2	unable to trace the copyright holders
Fig. 6.9	Copyright Earth Satellite Corporation under the GEOPIC trademark
Fig. 6.11	Schoken Books Inc, from Scharf, D. Magnifications
Fig. 6.12	E. Leitz Inc
Fig. 6.17	A T & T Bell Laboratories
Fig. 6.18	Texas Instruments
Fig. 6.19	Fairchild
Fig. 6.20	Intel
Fig. 6.21	Roger Pearce, Southampton
Fig. 6.22	Times Mirror Magazines Inc, reprinted from Popular Science with permission © 1946
Fig. 6.23	IBM Corporation

Chapter 7

Townes	Associated Press
Maiman	National Portrait Gallery

Gabor	National Portrait Gallery
Bose	Indian Academy of Sciences
Einstein	BBC Hutton Picture Library
Osheroff	Prof. Douglas D. Osheroff
Richardson	Doug Hicks, Cornell University
Lee	Doug Hicks, Cornell University
Kleppner	Prof. Daniel Kleppner, MIT
Townes	Prof. Charles H. Townes
Chu	Prof. Steve Chu, Stanford
Phillips	Prof. William Phillips, NIST
Cohen-Tannoudji	Prof. Claude N. Cohen-Tannoudji, Département de Physique – Ecole Normale Supérieure
Ketterle	Prof. Wolfgang Ketterle
Wieman	Prof. Carl Wieman
Cornell	Prof. Eric Cornell
	Prof. John Bardeen
Bednorz	IBM Zurich Research Laboratory
Muller	IBM Zurich Research Laboratory
Josephson	Cavendish Laboratory, University of Cambridge
Klitzing	Prof. Klaus von Klitzing
Fig. 7.1	NASA
Fig. 7.4	Fiat Auto
Fig. 7.8	Dr. Patrick Walters
Fig. 7.9	John Wiley & Sons Inc, Smith Principles of Holography
Fig. 7.12	Mendelssohn, K. Cryophysics. Interscience
Fig. 7.13	Prof J. F. Allen, American Institute of Physics
Fig. 7.15	Mark Helfer, NIST
Fig. 7.16	Illustration by Michael R. Matthews. Figure courtesy of the JILA BEC group
F7.17	Prof. Wolfgang Ketterle, MIT
F7.18(a)	MIT News Office
F7.18(b)	Prof. Wolfgang Ketterle, MIT
F7.18(c)	Prof. Wolfgang Ketterle, MIT
Fig. 7.19	Prof J. F. Allen, American Institute of Physics
Fig. 7.21	Molecular Universe
Fig. 7.23	Prof. T. H. Geballe

Chapter 8

Born	Niels Bohr Library
Bell	CERN
Solvay	Solvay S. A.
Bohm	Mark Edwards, Still Pictures

Chapter 9

| Kilby | Texas Instruments |
| Moore | Intel Corporation |

Turing	Turing Archive
Scientists	Andre Berthiaume
Zeilinger	FIRST LOOK Productions
Fig. 9.1	Melanie Jackson Agency, Caltech
Fig. 9.2	Tom Newman, Stanford
Fig. 9.3(a)	Prof. Don Eigler, IBM Almaden Research Center
Fig. 9.3(b)	Prof. Don Eigler, IBM Almaden Research Center
Fig. 9.4	IBM
Fig. 9.5	Prof. Wilson Ho, Cornell, University of California, Irvine
Fig. 9.15	David J. Wineland, NIST
Fig. 9.16	NEC

Chapter 10

Eddington	National Portrait Gallery
Bethe	Cornell University Department of Physics
Hoyle	Prof. Sir Fred Hoyle
Chandrasekar	Niels Bohr Library
Bell	Royal Observatory Edinburgh
Fig. 10.1	NASA (montage by Stephen Meszaros)
Fig. 10.3	Brookhaven National Laboratory
Fig. 10.5	Kalmbach Publishing Co., Milwaukee WI, from Astronomy Magazine with permission (artwork by John Clarke)
Fig. 10.6	Mount Wilson and Las Campanas Observatories, Carnegie Institution of Washington
Fig. 10.7	Lick Observatory
Fig. 10.8	Lick Observatory
Fig. 10.9	Lick Observatory
Fig. 10.9	Mount Wilson and Las Campanas Observatories, Carnegie Institution of Washington
Fig. 10.10	Anglo Australian Observatory, Photograph by David Malin
Fig. 10.11	Image courtesy Paul Scowen, Jeff Hester (Arizona State University) and the Mt. Palomar Observatories
Fig. 10.12	Mullard Radio Observatory, Cavendish Laboratory, University of Cambridge
Fig. 10.13	Lick Observatory
Fig. 10.16	Rob Hynes, University of Southampton

Chapter 11

Einstein	BBC Hulton Picture Library
Dirac and Heisenberg	Niels Bohr Archive, Copenhagen
Glaser and Anderson	Prof. Emilio Segre
Casimir	Prof. Hendrik Casimir

Gamow and	Dr Robert Herman and Dr R. A. Alpher
'YLEM'	Dr Robert Herman and Dr R. A. Alpher
Fraunhofer	Photo Deutsches Museum Munchen
Hawking	Mason's News Service
Fig. 11.3	Prof. C. D. Anderson
Fig. 11.4	CERN
Fig. 11.5	Lawrence Berkeley Laboratory, University of California
Fig. 11.11	National Optical Astronomy Observatory
Fig. 11.12	Mount Wilson and Las Campanas Observatories, Carnegie Institution of Washington
Fig. 11.13	National Optical Astronomy Observatory
Fig. 11.15	Macdonald & Co. (Publishers) Ltd.

Chapter 12

Maxwell	Cavendish Laboratory, University of Cambridge
Weyl	Hilbert, Constance Reid, Springer-Verlag
Yang	Prof. C. N. Yang
Yukawa	Associated Press
Glashow and Weinberg	Harvard University Department of Physics
Salam	Prof. Abdus Salam
Ting	Prof. Samuel Ting
Goldhaber, Perl and Richter	SLAC
t'Hooft	Prof. Gerard t'Hooft
Fig. 12.3	The Institute of Physics, from Rep. Prog. Phys. 13, 350 (1950)
Fig. 12.4	CERN
Fig. 12.5	Prof. C. D. Rochester, FRS and Sir Clifford Butler, FRS
Fig. 12.6	CERN
Fig. 12.7	Brookhaven National Laboratory
Fig. 12.8	Brookhaven National Laboratory
Fig. 12.9	CERN
Fig. 12.10	CERN
Fig. 12.11	Prof. Abdus Salam
Fig. 12.12	CERN
Fig. 12.13	CERN
Fig. 12.17	Roger Cashmore and the TASSO collaboration
Fig. 12.18	Roger Cashmore and the TASSO collaboration
Fig. 12.20	Karl Kuhn and J. S. Faughn, from Physics in Your World
Fig. 12.23	CERN
Fig. 12.24	CERN
Fig. 12.25	CERN
Fig. 12.26	CERN
Fig. 12.27	CERN

Fig. 12.28	CERN
Fig. 12.29	UK AEA, Photo by Eric Jenkins
Fig. 12.30	Super-Kamiokande project
Fig. 12.31	Anglo Australian Observatory, Photograph by David Malin

Chapter 13

Wells	TimePix
Szilard	unable to trace the copyright holders
Gernsback	Reproduced by permission of Gernsback Publications, Inc.
Wood Campbell Jr	unable to trace the copyright holders
Dick	unable to trace the copyright holders
Fig. 13.2	Columbia TriStar Home Entertainment
Fig. 13.3	Paramount Home Entertainment
Fig. 13.4	Grant Naylor Productions Ltd.
Fig. 13.5	unable to trace the copyright holders
Fig. 13.6	Time Warner Books
Fig. 13.7	unable to trace the copyright holders
Fig. 13.8	Prof. Merkle, Zyvex
Fig. 13.9	IBM Zurich Research Laboratory
Fig. 13.10	Timeline by Michel Crichton, published by Century. Reprinted by permission of The Random House Group Ltd.
Fig. 13.11	Phil Saunders
Fig. 13.12	Universal Studios Home Video
Fig. 13.13	Faber and Faber
Fig. 13.14	'Poster from the 1998 Royal National Theatre production of Copenhagen by Michael Frayn. Image of Werner Heisenberg, 1947, designed by Michael Mayhew'

Name index

Subject index

Page numbers in **bold** refer to entries in the Glossary.